MCBU
Molecular and Cell Biology Updates

Series Editors:

Prof. Dr. Angelo Azzi
Institut für Biochemie
und Molekularbiologie
Bühlstr. 28
CH-3012 Bern
Switzerland

Prof. Dr. Lester Packer
Dept. of Molecular
and Cell Biology
251 Life Science Addition
Membrane Bioenergetics Group
Berkeley, CA 94720
USA

Oxidative Stress, Cell Activation and Viral Infection

Edited by   C. Pasquier
                    R.Y. Olivier
                    C. Auclair
                    L. Packer

Birkhäuser Verlag
Basel · Boston · Berlin

Editors

Dr. C. Pasquier
INSERM U. 294
16, rue Henri Huchard
F - 75018 Paris
France

Dr. R.Y. Olivier
Unité d'Oncologie Virale
Dept. SIDA & Rétrovirus
25-28, rue du Dr. Roux
F - 75015 Paris
France

Dr. C. Auclair
Institut Gustave Roussy
INSERM U.140
39, rue Camille Desmoulins
F - 94805 Villejuif Cedex
France

Prof. Dr. L. Packer
Dept. of Molecular and Cell Biology
251 Life Science Addition
Membrane Bioenergetics Group
Berkeley, CA 94720
USA

A CIP catalogue record for this book is available from the Library of Congress, Washington D.C., USA

Deutsche Bibliothek Cataloging-in-Publication Data

Oxidative Stress, cell activation and viral infection / ed. by
C. Pasquier ...– Basel ; Boston ; Berlin : Birkhäuser, 1994
  (Molecular and cell biology updates)
  ISBN 978-3-0348-7426-7

NE: Pasquier, Catherine [Hrsg.]

© 1994 Birkhäuser Verlag, PO Box 133, CH - 4010 Basel, Switzerland
Softcover reprint of the hardcover 1st edition 1994

Camera-ready copy prepared by the authors
Printed on acid-free paper produced from chlorine-free pulp

ISBN 978-3-0348-7426-7   ISBN 978-3-0348-7424-3 (eBook)
DOI 10.1007/978-3-0348-7424-3

9 8 7 6 5 4 3 2 1

Table of Contents

Foreword

Since the appearance of photosynthesis on our planet, all living organisms have been facing a new abundant, extremely reactive element, oxygen. This element is used for the synthesis of highly energetic compounds, but can also generate molecules capable of damaging constituents of living structures, including proteins, nucleic acids and lipids.

It is therefore no surprise that all biological organisms have evolved multiple and sophisticated ways to reduce the detrimental effects of oxygen. For cells and tissues of highly organized animals, particularly cells of the immune system, superoxidation products play an important role, via oxidative stress, in activation, inflammation and viral and bacterial infections.

We must be grateful to Drs. C. Pasquier, C. Auclair, L. Packer and R. Olivier for having brought together many specialists in an international meeting held in Paris in March 1993, at the Ministry of Research.

This book comprises an impressive amount of recent knowledge, a real mine for the reader in this fast developing field of research.

Furthermore, we should not forget that this basic research may lead in the future to new therapeutic approaches to the most important pathologies of the latter part of this century, including AIDS.

Luc Montagnier
September 1993

Preface

Life in the presence of oxygen mainly depends on the tricky balance between the flux of oxy-radicals coming from various origins and the efficiency of the multiple and concerted cellular protection mechanisms. Oxy-radicals have been of immense interest since they were discovered to be involved in number of pathophysiological processes including inflammation, cancer, ageing, and neurodegenerative diseases. During the two past decades, investigations were mainly focused on the deleterious effects of oxy-radicals and on the constant battle between aerobic cells and molecular oxygen. A considerable amount of work has been devoted to this subject, and the mechanisms of oxy-radical-mediated damage at the tissular, cellular and molecular levels have been elucidated. Many investigations on the protection of cells against oxy-radicals injury have also been carried out. These include extensive studies on enzymatic and non-enzymatic antioxidant status in cells as well as the development of various exogenous antioxidants as therapeutic agents.

In the past few years the concept of biological oxidation has suddenly advanced from the hazards of oxygen utilization to oxidation as a natural process of metabolism regulation at the molecular level. Evidence is growing that redox mechanims play a major role in fundamental cellular events such as binding of transcription factors to DNA, protein binding to mRNA, hormone-receptor interactions, translational regulation processing of antigens, etc. Oxidizing species are on the way to being elevated to the status of major mediators of various cellular functions. The most striking example is the redox regulation of transcription factors such as NF-κ-B and AP-1 which activate gene transcription in response to peroxides. The involvement of the cellular nuclear factor κB in the activation of HIV gene transcription related to the altered antioxidant status currently observed in AIDS patients, has led to new insights into both the regulation of HIV replicative cycles and the design of therapeutic protocols.

These new orientations clearly require collaborative efforts at the interface of various scientific domains. The international conference on oxidative stress, cell activation and viral infection held in Paris on March 1993 was organized with this in mind. The philosophy of the conference was to promote contacts and exchanges between scientists working in the fields of oxy-radicals and those from other disciplines such as molecular and cellular biology, as well as virology. This book presents the most exciting developments in these different research areas.

Christian AUCLAIR, Catherine PASQUIER
Paris, July 1993

DNA Glycosylases Involved in the Repair of Oxidized Bases in Escherichia coli

S. Boiteux

URA147 CNRS, U140 INSERM, Groupe "Réparation des lésions radio et chimioinduites" Institut Gustave-Roussy, 94800-Villejuif, France

Summary

Oxidative damages to DNA are considered as one of the most important causes of spontaneous mutations and may play a role in aging and related diseases such as cancer in human. In DNA, sugars and bases are damaged by free radicals yielding modified bases, abasic sites and chain breakage. To counteract the lethal and mutagenic effects of oxidative lesions in DNA, cells have developed defence strategies including DNA repair systems. In Escherichia coli, the repair of oxidized DNA bases is mostly mediated by three DNA glycosylases. Endonuclease III excises oxidized pyrimidines and the Fpg protein excises oxidized purines. The MutY protein that cooperates with the Fpg protein to repair 8-oxoguanine residues in DNA.

Introduction

Oxygen derived species such as the superoxide radical (0_2^-), hydrogen peroxide (H_2O_2), and singlet oxygen ($^1O_2$) are formed in cells during aerobic metabolism (Halliwell and Gutteridge, 1989). The toxicity of O_2^- or H_2O_2 is thought to result from their metal-ion-catalysed conversion into the highly reactive hydroxyl radical (OH˙) that causes base and sugar damage in DNA (Dizdaroglu, 1991). Generation of free radicals and reactive oxygen species has biological consequences including lethality, mutagenesis or carcinogenesis (Breimer, 1990; Piette,1991). Fraga *et al.*,(1990) have calculated that approximately 90,000 oxidative DNA damage events occur per cell per day in the rat. This estimation is based on the measurement of 8-oxoguanine residues excreted in urine (Fraga *et al.*,1990). The mutagenic potential of the major oxidation product, 8-oxoguanine, was demonstrated in vitro and in vivo (Shibutani *et al.*, 1991; Michaels and Miller, 1992a).

The abundance and the mutagenic potency of DNA lesions induced by free radicals and reactive oxygen species suggest the need for DNA repair mechanisms. In Escherichia coli, the repair of oxidized base in DNA is mostly mediated by the base excision repair pathway. The first step in this repair pathway is catalysed by DNA glycosylases (Myles and Sancar, 1989; Boiteux 1993). The nucleotide excision repair pathway mediated by the UvrABC complex may also play a role in the repair of oxidized bases when the DNA glycosylases are inactive or saturated (Lin and Sancar, 1989; Kow *et al.*, 1990; Czeczot *et al.*, 1991). This review summarizes the properties and the biological functions of the three DNA glycosylases of Escherichia coli that are involved in the repair of oxidized DNA bases.

Results and Discussion

Base excision repair of oxidative damage to DNA

Base excision repair is a multi-step mechanism that is initiated by the action of a DNA-glycosylase which cleaves the (N-1'C) chemical bond to release the damaged base from DNA and thereby produce an abasic site (Fig.1). In Escherichia coli, the excision of oxidized bases is catalysed by two DNA glycosylases, the endonuclease III and the Fpg protein (Boiteux, 1993). The resulting abasic site is incised by a specific endonuclease and the 5'-deoxyribosephosphate residue is enzymaticaly excised (Sadigursky and Franklin, 1992; Graves *et al.*, 1992). The one nucleotide gap formed is filled by a DNA polymerase and the continuity of the DNA strand is restored by DNA ligase action (Fig.1).

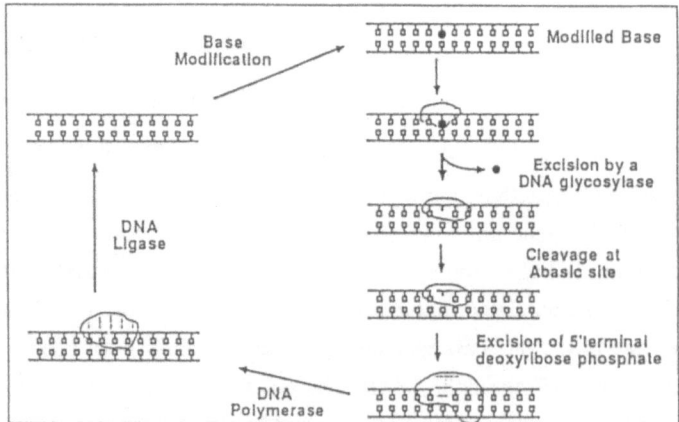

Figure 1. Base excision repair pathway in Escherichia coli : The excision of the oxidized base in DNA (w) is catalysed either by endonuclease III or the Fpg protein according to the chemical nature of the DNA base damage. Cleavage at the abasic site is catalysed by the exonuclease III or the endonuclease IV which are the major activities that incise 5'-side of the abasic site. The 5'-terminal deoxyribosephosphate is excised either by the Exonuclease I or the Fpg protein. Then, DNA repair is completed by DNA polymerase and DNA ligase action.

Endonuclease III

The endonuclease III is a monomeric protein of 23 kDa, containing 211 amino acid residues (Breimer and Lindahl, 1984; Asahara *et al.*,1989). This protein has a $(4Fe-4S)^{+2}$ iron-sulfur cluster that is bound entirely within the carboxy-terminal loop with a ligation pattern (Cys-X_6-Cys-X_2-Cys-X_5-Cys-X_8-COOH) (Kuo *et al.*,1992). The iron-sulfur cluster of endonuclease III is required to have a functional protein, since the iron-free apoprotein has no enzyme activity nor DNA binding activity (Fu *et al.*,1992). Endonuclease III has a DNA glycosylase activity that excises a large number of oxidized pyrimidines generated by exposure to ionizing radiation and oxidizing agents (Breimer and Lindahl, 1985; Boiteux, 1993). Endonuclease III also releases some ultraviolet-light-induced DNA base damage including cytosine and uracil-hydrates (Boorstein *et al.*,1989). All damaged DNA bases excised by endonuclease III are derived from pyrimidine bases and show ring-saturation, ring-contraction or ring-fragmentation. These modifications result in the formation of non-planar structures which may be recognized by this repair enzyme. Endonuclease III is endowed with both DNA glycosylase activity and abasic site nicking activity (Kow and Wallace, 1987; Bailly and Verly, 1987).

In spite of the ability of endonuclease III to remove several pyrimidine lesions that are potentialy lethal, the bacterial mutant lacking endonuclease III (nth) is not unusually sensitive to ionizing radiation, oxidizing agents or ultraviolet-radiation (Cunningham and Weiss, 1985). However, the survival of OsO_4 treated RF-I phage ËX174 DNA was reduced when transfected into the nth mutant (Kow *et al.*,1990). The introduction of an uvrA mutation into the nth background resulted in a further decrease in survival over nth alone (Kow *et al.*,1990). Although the nth mutant is not unusualy sensitive to oxidizing agents it has a spontaneous mutator phenotype. A 4-fold to 22-fold increased reversion of the argE3 mutant was observed (Cunningham and Weiss, 1985). The chemical nature of the DNA lesion responsible for the mutator phenotype has not been yet determined.

The Fpg protein

The Fpg protein was initially identified as a DNA glycosylase that excises the imidazole ring-opened form of the major methylation product N7-methylguanine (2,6-diamino-4-hydroxy-5-N-methylformamidopyrimidine or Fapy) (Chetsanga and Lindahl, 1979; Boiteux *et al.*,1984). The Fpg protein is a monomer of 30.2 kDa containing 269 amino acids that has an isoelectric point of 8.5 (Boiteux *et al.*,1987; Boiteux *et al.*,1990). Atomic absorption spectroscopy analysis shows that there is one zinc/Fpg monomer (Boiteux *et al.*, 1990). The amino acid sequence of the Fpg protein reveals a zinc-finger consensus sequence at the carboxy terminal end with a (Cys-X_2-Cys-X_{16}-Cys-X_2-Cys-X_2-COOH) motif (O'Connor *et al.*,1993; Boiteux, 1993). The gas chromatography/mass spectrometry (GC/MS) method has been used to analyse the repair of various oxidative DNA base damage by the Fpg protein (Boiteux et al.,1992). The Fpg protein releases purine lesions including imidazole ring-opened guanine, imidazole ring-opened adenine and 8-oxoguanine (Boiteux *et al.*,1992). In contrast, none of the nine pyrimidine-derived lesions identified by GC/MS are excised by the Fpg protein (Boiteux *et al.*,1992). Although the Fpg protein is a small monomer, it is endowed with three enzymatic activities, a DNA glycosylase activity, an activity that nicks DNA at abasic site and a 5'-terminal deoxyribosephosphate excising activity. The Fpg protein catalyses the nicking of both the phophodiester bond 3'-side and 5'-side of abasic site in the DNA so that the base-free deoxyribose is released yielding a gap limited by 3'-phosphate and 5'-phosphate (O'Connor and Laval, 1989). The third enzyme activity associated with the Fpg protein catalyses the excision of a 5'-deoxyribosephosphate from DNA and may play a role in a late steps of base excision repair (Graves *et al.*,1992).

A mutant defective in Fpg protein was isolated and localized at 81.7 min on the Escherichia coli linkage map (Boiteux and Huisman, 1989). The bacterial mutant is not unusually sensitive to oxidizing agents and ionizing radiation (Boiteux and Huisman, 1989). However, the survival of MB-light treated pBR322 plasmid DNA was greatly reduced when transformed into the fpg uvrA double mutant (Czeczot *et al.*,1991). In contrast, the survival of MB-light treated DNA was not significantly reduced when tranformed into fpg or uvrA single mutant compared with the wild-type host (Czeczot et al.,1991). These results suggest that MB-light induces lethal DNA base modifications repaired by the Fpg protein or the UvrABC excision complex. The chemical nature of the lethal lesion(s) has not been established, but, imidazole ring-opened guanine is a candidate. The major lesion in MB-light-treated DNA, 8-oxoguanine, is highly mutagenic but has little or no lethal effects (Michaels and Miller, 1992a). The role of the Fpg protein in the prevention of the mutagenic effects of oxidative damage was assessed by the spontaneous mutator phenotype displayed by fpg(mutM) mutant that accumulates GC-->TA transversions (Michaels *et al.*,1992c). The mutator phenotype of the fpg mutant is most likely due to the formation of 8-oxoguanine residues in DNA. The incorporation of dAMP opposite 8-oxoguanine in course of DNA replication implies that this oxidized base specifically produce GC-->TA transversion events (Shibutani *et al.*,1991).

The MutY protein

The MutY protein does not excise oxidized bases in DNA as endonuclease III and the Fpg protein. The MutY(MicA) protein has a DNA glycosylase activity that excises adenine in Guanine/Adenine, Cytosine/Adenine, 8-oxoguanine/Adenine and 8-oxoadenine/Adenine mispairs (Radicella *et al.*,1988; Au *et al.*,1988; Michaels *et al.*,1992b). The MutY protein is a 39 kDa protein which contains 350 amino acids (Michaels *et al.*,1990; Tsai-Wu *et al.*,1991). Like endonuclease III, MutY protein has a $(4Fe-4S)^{2+}$ iron-sulfur cluster (Tsai-Wu et al.,1992). The

MutY protein possesses both a DNA glycosylase activity and an activity that cleaves DNA at the 3'-side of an abasic site (Tsai-Wu *et al.*,1992). The mutY locus was localized at 64 min on the genetic map of Escherichia coli (Nghiem *et al.*,1988; Radicella *et al.*,1988). A bacterial mutant defective in MutY protein has a spontaneous mutator phenotype and accumulates GC-->TA transversions (Nghiem *et al.*,1988). The role of the MutY protein in the repair of 8-oxoguanine is suggested by several lines of evidence. The MutY protein excises adenine in the 8-oxoguanine/Adenine mispair (Michaels *et al.*,1992b). The double mutant fpg mutY shows an extreme and specific GC-->TA mutator phenotype (Michael *et al.*,1992c). The way the MutY and Fpg proteins cooperate to repair 8-oxoguanine residues in DNA is described in Fig.2. The role of the MutY protein is to prevent the mutation fixation after replication of 8-oxoguanine residues and to provide another possibility of repair by the Fpg protein (Fig.2). The endogeneous reactive species that attacks guanine residues in DNA is not yet identified.

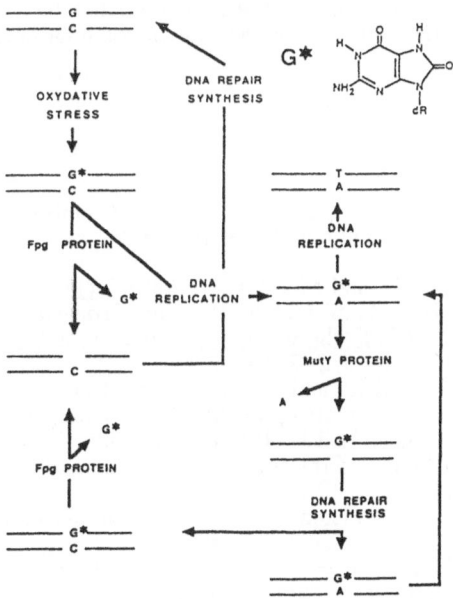

Figure 2. Fpg and MutY proteins cooperate for the repair of 8-Oxoguanine : Free radicals and reactive oxygen species may attack guanine residues yielding 8-oxoguanine (G*). 8-Oxoguanine may be excised by the Fpg protein. If the Fpg protein is saturated, 8-oxoguanine is replicated yielding G*/Adenine (G*/A) or G*/cytosine (G*/C) mismatches. 8-Oxoguanine paired with adenine (G*/A) is a substrate for the MutY protein that excises the adenine residue. In contrast, 8-oxoguanine paired with adenine (G*/A) is a poor substrate for the Fpg protein (Tchou et al.,1991). The excision of the adenine by the MutY protein prevents mutation fixation (GC-->TA) and may allow the formation of a (G*/Cytosine) mismatch that is repaired by the Fpg protein.

Conclusions

The biological function of the Endonuclease III, the Fpg protein and the MutY protein of Escherichia coli is to prevent the lethal and mutagenic effects of oxidative damage to DNA. Together endonuclease III and the Fpg protein may excise most of the lesions induced by ionizing radiation or oxidizing agents in DNA. These results show how cells cope with a variety of oxidative DNA damage with a small number of repair enzymes that have a broad substrate specificity. The repair of 8-oxoguanine in DNA involves two DNA glycosylases, the Fpg and the MutY proteins, which cooperate to remove this mutagenic lesion. The spontaneous mutation rate of mutants lacking these repair proteins provides evidence that free radicals generated in normal growth condition are a significant threat to the cell.

Acknowledgements

This work was supported by the Centre National de la Recherche Scientifique, Institut National de la Santé et de la Recherche Médicale and the Association pour la Recherche sur le Cancer.

References

Asahara,H., Wistort,P.M., Bank,J.F., Bakerian,R.H. and Cunningham,R.P. (1989) Purification and characterization of E.coli endonuclease III from the cloned nth gene. *Biochemistry* 28: 4444-4449.

Au,K.G., Cabrera,M., Miller,J.H. and Modrich,P. (1988) Escherichia coli mutY gene product is required for specific A-G-->C.G mismatch correction. *Proc.Natl.Acad.Sci.US* 85: 9163-9166.

Bailly,V. and Verly,W.G. (1987) Escherichia coli endonuclease III is not an endonuclease but a ß-elimination catalyst. *Biochem. J.* 242: 565-572.

Boiteux,S., Belleney,J., Roques,B.P. and Laval,J. (1984) Two rotameric forms of open ring 7-methylguanine are present in alkylated polynucleotides. *Nucleic Acids Res.* 12: 5429-5439.

Boiteux,S., O'Connor,T.R., and Laval,J. (1987) Formamido- pyrimidine DNA glycosylase of E.coli: cloning and sequencing of the fpg structural gene and overproduction of the protein. *EMBO J.* 6: 3177-3183.

Boiteux,S. and Huisman,O. (1989) Isolation of a formamidopyrimidine-DNA glycosylase (fpg) mutant of E.coli K12. *Mol. Gen. Genet.* 215: 300-305.

Boiteux,S., O'Connor,T.R., Lederer,F., Gouyette,A. and Laval,J. (1990) Homogeneous Fpg protein: a DNA glycosylase which excises imidazole ring-opened purines and nicks DNA at apurinic/ apyrimidinic sites. *J. Biol. Chem.* 265: 3916-3922.

Boiteux,S., Gajewski,E., Laval,J. and Dizdaroglu,M. (1992) Substrate specificity of the Escherichia coli Fpg protein: Excision of purine lesions in DNA produced by ionizing radiation or photosensitization. *Biochemistry* 31:106-110.

Boiteux,S. (1993) Properties and biological functions of the NTH and FPG proteins of E.coli: two DNA glycosylases that repair oxidative damage in DNA. J. Photochem. Photobiol. B: in press.

Boorstein,R.J., Hilbert,T.P., Cadet,J., Cunningham,R.P. and Teebor,G.W. (1989) UV-induced pyrimidine hydrates in DNA are repaired by bacterial and mammalian DNA glycosylase activities. *Biochemistry* 28: 6164-6170.

Breimer,L.H. and Lindahl,T. (1984) DNA glycosylases activities for thymine residues damaged by ring saturation, fragmentation or ring contraction are functions of endonuclease III in Escherichia coli. *J. Biol. Chem.* 259: 5543-5548.

Breimer,L.H. and Lindahl,T. (1985) Thymine lesions produced by ionizing radiation in double stranded DNA, *Biochemistry* 24: 4018-4022.

Breimer,L.H. (1990) Molecular Mechanisms of Oxygen Radical Carcinogenesis and Mutagenesis: The Role of DNA Base Damage. *Mol. Carcinogenesis* 3: 188-197.

Chetsanga,C.J. and Lindahl,T. (1979) Release of 7-methylguanine residues whose imidazole rings have been opened from damaged DNA by a DNA glycosylase from Escherichia coli. *Nucleic Acids Res.* 6: 3673-3683.

Cunningham,R.P. and Weiss,B. (1985) Endonuclease III (nth) mutants of Escherichia coli. *Proc. Natl. Acad. Sci. US* 82: 474-478.

Czeczot,H., Tudek,B., Lambert,B., Laval,J. and Boiteux,S. (1991) Escherichia coli Fpg protein and UvrABC endonuclease repair DNA damage induced by methylene blue plus visible light in vivo and in vitro. *J. Bacteriol.* 173: 3419-3424.

Dizdaroglu,M. (1991) Chemical determination of free radical-induced damage to DNA. *Free Radical Biology and Medicine* 10: 225-242.

Fraga,C.G., Shigenaga,M.K., Park,J.W., Degan,P. and Ames,B.N. (1990) Oxidative damage to DNA during aging: 8-hydroxy-2'-deoxyguanosine in rat organ DNA and urine. *Proc. Natl. Acad. Sci. US* 87: 4533-4537.

Fu,W., O'Handley,S., Cunningham,R.P. and Johnson,M.K. (1992) The role of the iron-sulfur cluster of Escherichia coli Endonuclease III. *J. Biol. Chem.*, 267: 16135-16137.

Graves,R.J., Felzenswalb,I., Laval,J. and O'Connor,T.R. (1992) Excision of 5'-terminal deoxyribose phosphate from damaged DNA is catalysed by the Fpg protein of Escherichia coli. *J. Biol. Chem.* 267: 14429-14435.

Halliwell,B. and Gutteridge,J.M.C. (1989) *Free Radicals in Biology and Medicine*, Clarendon Press, Oxford.

Kow,Y.W. and Wallace,S.S. (1987) Mechanism of action of E.coli endonuclease III. *Biochemistry* 28: 8200-8206.

Kow,Y.W., Wallace,S.S. and van Houten,B. (1990) UvrABC nuclease complex repairs thymine glycol an oxidative DNA base damage. *Mutation Res.* 235: 147-156.

Kuo,C.F., McRee,D.E., Fisher,C.L., O'Handley,S.F., Cunningham,R.P. and Tainer,J.A. (1992) Atomic structure of the DNA repair (4Fe-4S) enzyme endonuclease III. *Science* 258: 434-440.

Lin,J.J. and Sancar,A. (1989) A new mechanism for repairing oxidative damage to DNA: (A)BC excinuclease removes AP sites and Thymine glycols from DNA. *Biochemistry* 28: 7979-7984.

Michaels,M.L., Pham,L., Nghiem,Y., Cruz,C. and Miller,J.H. (1990) MutY an adenine glycosylase active on G-A mispairs, has homology to endonuclease III. *Nucleic Acids Res.* 18: 3841-3845.

Michaels,M.L. and Miller,J.H. (1992a) The GO system protects organisms from the mutagenic effect of the spontaneous lesion 8-hydroxyguanine. *J. Bacteriol.* 174: 6321-6325.

Michaels,M.L., Tchou,J., Grollman,A.P. and Miller,J.H. (1992b) A repair system for 8-Oxo-7,8-dihydroguanine, *Biochemistry* 31 (1992) 10964-10968.

Michaels, Cruz,C., Grollman,A.P. and Miller,J.H. (1992c) Evidence that MutY and MutM combine to prevent mutations by an oxidative damaged form of guanine. *Proc. Natl. Acad. Sci. US* 89: 7022-7025.

Myles,M. and Sancar,A. (1989) DNA repair, *Chem. Res. Toxicol.* 2: 197-226.

Nghiem,Y., Cabrera,M., Cupples,C.G. and Miller,J.H. (1988) The mutY gene: a mutator locus in Escherichia coli that generates GC-->TA transversions. *Proc. Natl. Acad. Sci. US* 85: 2709-2713.

8

O'Connor,T.R. and Laval,J. (1989) Physical association of the formamidopyrimidine DNA glycosylase of Escherichia coli and an activity nicking DNA at apurinic/ apyrimidinic sites. *Proc. Natl. Acad. Sci. US* 86: 5222-5226.

O'Connor,T.R., Graves,R.J., de Murcia,G., Castaing,B. and Laval,J. (1993) Fpg protein of Escherichia coli a zinc finger protein whose cysteine residues have a structural and/or functional role. *J. Biol. Chem.*: in press.

Piette,J. (1991) Biological Consequences associated with DNA Oxydation mediated by Singlet Oxygen. *J. Photochem. Photobiol.B* 11: 241-260.

Radicella,J.P., Clark,E.A. and Fox,M.S. (1988) Some mismatch repair activities in Escherichia coli. *Proc. Natl. Sci. US.* 85: 9674-9678.

Sadigursky,M. and Franklin,W.A. (1992) DNA deoxyphosphodiesterase of E.coli is associated with exonuclease I. *Nucleic Acids Res.* 20: 4699-4703.

Shibutani,S., Takeshita,M. and Grollman,A.P. (1991) Insertion of specific bases during DNA synthesis past the oxidation-damaged base 8-oxodG. *Nature* 349: 431-434.

Tchou,J., Kasai,H., Shibutani,S., Chung,M.H., Laval,J., Grollman,A.P. and Nishimura,S. (1991) 8-Oxoguanine (8-hydroxy-guanine) DNA glycosylase and its substrate specificity. *Proc. Natl. Acad. Sci. US* 88: 4690-4694.

Tsai-Wu,J.J., Radicella,J.P. and Lu,A.L. (1991) Nucleotide sequence of the Escherichia coli micA gene required for A/G-specific mismatch repair: Identity of MicA and MutY. *J. Bacteriol.* 173: 1902-1910.

Tsai-Wu,J.J., Liu,H.S. and Lu,A.L. (1992) Escherichia coli MutY protein has both N-glycosylase and apurinic/apyrimidinic endonuclease activities on A.C and A.G mispairs. *Proc. Natl. Acad. Sci. US* 89: 8779-8783.

Oxidative Stress, Cell Activation and Viral Infection
C. Pasquier et al. (eds)
© 1994 Birkhäuser Verlag Basel/Switzerland

Potential Role of Protein Oxidation and Proteasome in Antigen Processing

K. J. A. Davies

Department of Biochemistry & Molecular Biology, The Albany Medical College, Albany, New York 12208 U.S.A.

Summary

The hypothesis is presented that in antigen presenting cells, normal background rates of protein oxidation may supply endogenous protein substrates for the 19S proteasome proteolytic complex. Certain antigen presenting cells, such as macrophages, may actually use an oxidative burst to "mark" endogenous cytoplasmic proteins for partial degradation by the 19S proteasome. Proteasome seems to be able to generate peptides of approximately nine amino acids in length, which can combine with major histocompatibility complex (MHC) class I molecules and β_2-microglobulin. Such MHC class I antigen complexes next appear to migrate through the endoplasmic reticulum and the Golgi apparatus, assisted by specific ABC (ATP-binding cassette) superfamily transporter proteins, to the cell surface. Thus antigen presenting cells may use oxidative marking (either from background protein oxidation, or through an oxidative burst) of precursor self-antigens to direct cell surface class I presentation for surveillance by $CD8^+$ T cells.

The 670 kDa Multicatalytic Proteinase Complex, Proteasome

Cells exposed to oxidative stress exhibit rapid rates of protein degradation and can be shown to contain oxidatively modified proteins (Davies, 1985; Davies, 1986; Fucci *et al.*, 1983; Levine *et al.*, 1981; Stadtman, 1986). Extracts of cells rapidly degrade both native and foreign proteins that have been oxidatively modified *in vitro* (Davies, 1985; Davies, 1986; Fucci *et al.*, 1983; Levine *et al.*, 1981; Stadtman, 1986; Davies and Goldberg, 1987a; Davies and Goldberg, 1987b; Davies, 1987; Davies *et al.*, 1987a; Davies and Delsignore, 1987; Davies *et al.*, 1987b; Davies and Lin, 1988a; Davies and Lin, 1988b; Davies, 1988; Marcillat *et al.*, 1988; Murakami *et al.*, 1990; Pacifici and Davies, 1987; Pacifici *et al.*, 1989; Pacifici and Davies, 1990; Rivett, 1985; Rivett, 1989; Salo *et al.*, 1988; Salo *et al.*, 1990; Taylor and Davies, 1987; Wolff and Dean, 1986; Wolff *et al.*, 1896). Proteases have been isolated from both bacteria and eucaryotic cells that selectively recognize and degrade the oxidatively modified forms of endogenous and foreign protein (Davies and Lin, 1988b; Murakami *et al.*, 1990; Pacifici *et al.*, 1989; Rivett, 1985; Salo *et al.*, 1990).

From mammalian cells (as well as many other eucrayotes) a high molecular-mass multicatalytic proteinase complex has been isolated on the basis of its ability to selectively degrade oxidatively modified proteins (Murakami *et al.*, 1990; Pacifici *et al.*, 1989; Rivett, 1985; Salo *et al.*, 1990). This laboratory originally called the high molecular-mass proteinase, macroxyproteinase or

M.O.P (Pacifici *et al.*, 1989; Salo *et al.*, 1990) and many other names have been proposed by various groups. The term proteasome (Arrigo *et al.*, 1988) is now receiving quite widespread acceptance and will be used throughout the remainder of this paper.

Proteasome is a large, approximately 670 kDa, multisubunit proteolytic complex. Some 15 or more subunits (each in multiple copy) make up the complex, and each subunit ranges in size between 21 and 35 kDa (Arrigo *et al.*, 1988; Pacifici *et al.*, 1989; Rivett, 1985; Rivett, 1989; Salo *et al.*, 1990). The 19S proteasome exhibits no ATP stimulation (in fact, ATP is usually slightly inhibitory) and does not preferentially degrade ubiquitinated substrates (Arrigo *et al.*, 1988; Pacifici *et al.*, 1989; Rivett, 1985; Rivett, 1989; Salo *et al.*, 1990). Nevertheless it is now clear that proteasome is the core enzyme complex of the 26S, ATP-stimulated, ubiquitin conjugate degrading enzyme or UCDEN (Arrigo *et al.*, 1988). Proteasome is primarily found in the cell cytoplasm, although some is also present in the nucleus (Rivett, 1989). The term multicatalytic proteinase complex is often applied to proteasome which exhibits inhibitor profiles reminiscent of serine proteases, sulfhydryl proteases, and metallo-peptidases (Pacifici *et al.*, 1989; Rivett, 1985; Rivett, 1989; Salo *et al.*, 1990; Arrigo *et al.*, 1988). It is, however, clear that proteasome is an entirely non-classical or unique proteolytic complex. Even the structure of proteasome, a hollow cylinder with active enzyme sites on the interior (Rivett, 1989; Arrigo *et al.*, 1988), has left most biochemists entirely puzzled about its origin and biological functions. To further confuse the story it is now clear that proteasome is identical to a structure previously called prosome which has been reported to play a role in RNA processing (Rivett, 1989; Arrigo *et al.*, 1988).

Protein Oxidation and Antigen Processing by Proteasome

With all this in mind it should come as no surprise that another, altogether new, function for proteasome has recently been proposed; the processing of class I antigens. In the following pages I will attempt to review some of the recent literature on the processing of class I molecules, as it relates to a possible involvement of proteasome. Readers should be aware that antigen presenting cells such as macrophages, dendritic cells, and B lymphocytes, contain high concentrations of proteasome. In this paper I wish to present the novel hypothesis that normal rates of protein oxidation within cells may "mark" proteins for proteolysis by intracellular proteasome. It should be noted that, in an oxygen environment, protein oxidation is a normal component of protein turnover.Some antigen presenting cells, such as macrophages, may actually use their oxidative burst to mark endogenous proteins for processing by proteasome. Short peptides (8-10 amino acids) generated by proteasome (from oxidatively modified endogenous proteins) may then be combined with class I major histocompatibility complex (MHC) molecules (plus β_2 microglobulin) and transported to the macrophage cell surface for presentation as effective class I antigens. While the work on protein oxidation and degradation by proteasome and the work on a potential role for proteasome in class I antigen processing, presented in these pages are each thought to be internally valid, the concept for a linkage between these two areas of research remains largely untested. Thus readers are asked to view this chapter solely as an hypothesis which, hopefully, will stimulate actual experimental testing.

The Processing of Class I Antigens by Proteasome

The processing and presentation of class I antigens has recently been postulated to involve proteasome (Robertson, 1991). Class I major histocompatibility complex (MHC) glycoproteins bear the responsibility for processing and presenting antigenic peptides derived from endogenously synthesized proteins. Such proteins include self-antigens and antigens encoded by intracellular parasites such as viruses (Parnham, 1990). In contrast, MHC class II molecules are involved in the presentation of exogenous antigens (Hackett, 1991). Both class I and class II molecules, combined with appropriate antigenic peptides, must be transported to the cell surface where they are "presented" for recognition by T cells. Antigen presenting cells include macrophages, dendritic cells, and B lymphocytes.

For some time it has been known that two proteins encoded in the MHC class II region of the human and mouse genome are actually vital components of the MHC class I pathway. Named LMP-1 and LMP-2 (for low molecular-weight proteins) by Monaco and McDevitt (1982) these gene products turned out to be two polymorphic subunits of a large cytoplasmic complex containing 16 distinct subunits. Monaco and McDevitt originally speculated that LMP-1 and LMP-2 might play an integral role in antigen processing (Monaco and McDevitt, 1986).

Brown, Driscoll, and Monaco (1991) noted that many properties of the LMP complex bear striking similarity to those exhibited by proteasome. Using antibodies directed against either LMP or proteasome complexes Brown *et al.* (1991) demonstrated that two-dimentional electrophoresis patterns of immunoprecipitates from murine macrophages were strikingly similar. Some subunits were, however, specific to either the LMP immunoprecipitate or the proteasome precipitate indictating that (despite several probable common subunits) the two complexes are not entirely identical. Brown *et al.* (1991) further demonstrated that anti-proteasome antiserum was able to preclear LMP antigens from macrophage extracts, although

MHC class I antigens were not precleared and anti-LMP antiserum did not preclear proteasome from the extracts.

In related work Martinez and Monaco (1991) demonstrated that the *LMP-2* gene shows high sequence identity with genes encoding rat and human proteasome subunits. In the same issue of Nature in which Brown *et al.* (1991) reported the LMP-2/proteasome similarities, Glynne *et al.* (1991) reported that a gene (*RING10*), sharing considerable sequence identity with a consensus sequence of several proteasome subunits, maps between two putative transporter genes in the MHC class II region. These two putative transporter genes appear to encode membrane proteins of the ABC (ATP-binding cassette) superfamily. The ABC genes may encode the two halves of a heterodimeric protein transporter, necessary for transporting MHC class I molecules with bound peptide antigen through the endoplasmic reticulum and Golgi apparatus on their way to the cell membrane (Deverson *et al.*, 1990; Trowsdale *et al.*, 1990; Spies *et al.*, 1990; Monaco *et al.*, 1990; Spies and DeMars, 1991). A second proteasome gene was subsequently shown by Ortiz-Navarrete *et al.* (1991) to reside in the MHC region (between the *K* locus and the MHC class II region), and to be inducible by interferon-_. Next Kelly *et al.* (1991) demonstrated that a third proteasome-related gene, which they called *RING12*, resides in the class II region of the MHC, immediately centromeric of the *RING4* locus. Interestingly, Kelly *et al.* (1991) reported that *RING12*, like *RING10* (Glynne *et al.*, 1991), the proteasome gene localized by Ortiz-Navarrete *et al.* (1991), the LMP genes, and the MHC class I molecules, are all induced by treatment with interferon-_ (Kelly *et al.*, 1991).

Ortiz-Navarrete *et al.* (1991) also demonstrated that a T2 (B cell-derived) cell line, with impaired supply of MHC class I peptides and reduced cell surface expression of MHC class I molecules, lacked the *RING12* protein. Importantly, the T2 line carries a large homozygous deletion encompassing the MHC class II region. These same authors also showed that interferon-_ can increase both proteasome synthesis and MHC class I cell surface expression in mouse CMT (H-2[b]) lung carcinoma cells (Ortiz-Navarrete *et al.*, 1991). In these cells a depressed association of class I heavy chains with β_2-microglobulin has been postulated to result from an inadequate supply of suitable peptides (presumably due to the lack of key proteasome subunits?).

The expression of MHC class I proteins is unextricably linked with autoimmunity. Not suprisingly, defects in the expression of class I molecules are associated with autoimmune diseases. Building on the work described in the preceding paragraphs, Faustman *et al.* (1991) used the nonobese diabetic (NOD) mouse model of autoimmune diabetes to further probe possible proteasome/antigen processing relationships. Faustman *et al.* (1991) found that expression of MHC class I proteins is inversely correlated with the severity of diabetes in the NOD mouse. Importantly the NOD mouse was characterized as carrying a mutation in the MHC class I peptide transporter. Other mice, exhibiting a defect in β_2-microglobulin expression, also developed late onset autoimmune diabetes. Humans with type I diabetes also tended to have low expression of MHC class I molecules (Faustman *et al.*, 1991). The clear conclusion drawn by Faustman *et al.* (1991) was that any mutations or deletions in the class II region of MHC can produce serious defects in the ability to generate MHC class I molecules, process class I antigen peptides, or transport class I/peptide/β_2-microglobulin complexes.

Despite the apparently strong association between proteasome subunits, MHC class II region encoded gene products, and elements necessary for the processing of class I antigens, two reports have recently appeared which question the proposed role of proteasome in antigen processing (Arnold *et al.*, 1992; Momburg *et al.*, 1992). Momburg *et al.* (1992) used the lymphoblastoid human B cell-derived mutant T2 cell line, in which the two ABC transporter polypeptide genes (*TAP1* and *TAP2*) and two proteasome subunit genes (*LMP2* and *LMP7*) have been deleted from the MHC class II region. The T2 cells exhibited decreased expression and cell surface presentation of class I molecules. When transfected with rat ABC transporter cDNA's for rat *TAP1ᵃ* and *TAP2ᵃ* (*mtp1ᵃ* and *mtp2ᵃ*) the T2 cells regained class I antigen processing capabilities, despite the fact that two key proteasome subunits were still missing (Momburg *et al.*, 1992). Arnold *et al.* (1992) used essentially the same experimental approach as did Momburg *et al.* (1992) and obtained essentially identical results.

It appears possible (perhaps even probable) that the results of Arnold *et al.* (1992) and Momburg *et al.* (1992), although technically strong, do not actually rule out an important role for proteasome in antigen processing. This is because at least one alternate pathway for the cleavage of precursor class I antigens has been proposed (Henderson *et al.*, 1992). Precursor class I antigens appear to be cleaved to peptides of nine amino acids in length by proteasome before combining with class I molecules and β_2-microglobulin (Hunt *et al.*, 1992). In proteasome deficient mutants, however, Henderson *et al.* (1992) have reported that cleavage can occur in the endoplasmic reticulum by an unknown proteolytic enyzme or enzymes (perhaps signal peptidases?). This putative alternate pathway appears to generate peptides of 10-12 amino acids in length which apparently still bind to class I molecules and β_2-microglobulin, stabilize the complex (as do the nonomeric peptides generated by proteasome) and permit appropriate class I antigen presentation on the cell surface.

In a system containing at least two proteolytic pathways capable of generating appropriate class I peptides, the information that can be gleaned from gene knock-out experiments such as those performed by Arnold *et al.* (1992) and Momburg *et al.* (1992) must be carefully analyzed and interpreted. The experiments of Arnold *et al.* (1992) and Momburg *et al.* (1992) certainly indicate that proteasome is not the sole or obligatory antigen cleaving proteinase in the class I pathway. They do not, however, indicate that the "alternate" pathway operates independently of proteasome in normal cells. The relative contributions of the cytoplasmic proteasome and putative endoplasmic reticulum proteases to type I antigen processing remain to be carefully evaluated. The experiments of Hunt *et al.* (1992) however, showing that most peptides bound to class I molecules are nine amino acids in length, may tend to implicate proteasome processing of antigens as the quantitatively more significant pathway.

Antisense Experiments Indicate a Role for Proteasome in Processing Class I Antigens

Although "alternate" pathways for antigen processing may, indeed, explain the apparently incongrous results of Arnold *et al.* (1992) and Momburg *et al.* (1992), it also appeared possible that the gene knock-out experiments performed by the authors may have affected affected the cells in more complex ways than were intended. It is important to note, for example, that the MHC class II region mutants studied by these authors involved extensive deletions of multiple genes in the MHC class II region. Some possible confirmation of this possibility may be gleaned from proteasome knock-out experiments in yeast, which prove fatal to the survival of the organism (data not shown). Thus, in at least one eucaryote, complete elimination of a proteasome subunit (and, therefore, elimination of the proteasome complex's functions) is a lethal manipulation.

We hypothesized that a more gentle experiment would be to use antisense oligonucleotides, for limited time periods, to decrease the effective (operational) proteasome concentrations within cells. For these studies (not previously published) we used Jurkat T cells incubated with or without sense or antisense oligonucleotides to the LMP-2 protein, the rat liver proteasome C2 subunit (RL C2), or the rat liver proteasome C5 subunit (RL C5). For the RL C2 subunit two different sense oligonucleotides and two different antisense oligonucleotides were synthesized: sense/antisense from amino acid position 16 to 25, and sense/antisense from amino acid position 130 to 139. All oligonucleotides used in this study were synthesized in the laboratory and are described in Table I.

Table I. Antisense and Sense Oligonucleotides Synthesized for this Study

Oligo-nucleotide	Type	Proteasome Subunit	Nucleotide Sequence
DV 48	Sense	LMP-2	5' ACA ACC ATC ATG GCA GTG GAG TTT GAC GGG GGT G 3'
DV 49	Antisense	LMP-2	5' GAC ACC CCC GTC AAA CTC CAC TGC CAT GAT GGT TGT 3'
DV 50	Sense	RL C2 (16-25)	5' CAG GGC AGG ATT CAT CAA ATC GAA TAT GCA ATG GAA G 3'
DV 51	Antisense	RL C2 (16-25)	5' AGC TTC CAT TGC ATA TTC GAT TTG ATG AAT CCT GCC CTG 3'
DV 54	Sense	RL C2 (130-139)	5' AGA CCG TAT GGT GTT AGT CGT CTC ATT GCT A 3'
DV 55	Antisense	RL C2 (130-139)	5' AGC AAT GAG CAG ACT AAC ACC ATA CGG TCT A 3'
DV 52	Sense	RL C5	5' CCT TAC GGT GTC TAC AAC ATC AAT GAG GGA CTT G 3'
DV 53	Antisense	RL C5	5' ATC AAG TCC CTC AAT GAT GTT GTA GAC ACC GTA AGG 3'

Sense and antisense oligonucleotides were synthesized on an Applied Biosystems Model 380A DNA Synthesizer, using a 61 to 64 step 1.0 μM cycle, and purified on an oligonucleotide purification cartridge (Applied Biosystems).

In order to test the possible importance of proteasome in MHC class I antigen processing and presentation we incubated Jurkat T cells without oligonucleotide, with sense oligonucleotides (a secondary control), or with antisense oligonucleotides to the LMP-2 protein, the rat liver proteasome C2 subunit (RL C2), or the rat liver proteasome C5 subunit (RL C5) as shown in Table II. The results of Table II show that cells incubated without oligonucleotides, or with sense oligonucleotides, all exhibited a single fluorescence peak of high intensity (High Affinity Peak in Table II). These data indicate that all Jurkat cells without oligonucleotide, or with sense oligonucleotide, expressed a single class or homogenous pool of MHC class I molecules (plus peptide antigen and β_2-microglobulin) on their surfaces. In contrast, three of the antisense oligonucleotides caused the appearence of a secondary fluorescence peak (Low Affinity Peak in Table II), of lower intensity, in 25% to 33% of the population. This secondary peak is taken to represent a pool of unusual class I antigens which may not be well recognized by CD8[+] T cells as self antigens. Interestingly, both of the rat liver proteasome C2 subunit (RL C2) antisense oligonucleotides, and the rat liver proteasome C5 subunit (RL C5) antisense oligonucleotide, caused the appearence of the low affinity fluorescence peak, with concommitant decrease in the population of cells exhibiting the high affinity peak, but the LMP-2 antisense oligonucleotide was without effect. Perhaps the extent of sequence identity between *LMP-2* and proteasome subunits is not sufficient to affect proteasome subunit transcription/translation. It is also possible that LMP-2 and proteasome subunits are merely similar structures (performing similar recognition and cleavage functions) but that the two are in fact different entities, despite their obvious high degree of sequence identity.

Table II. Effects of Proteasome Antisense Oligonucleotides on Antigen Presentation

Oligo-nucleotide	Type	Proteasome Subunit	High Affinity Peak (% of Cells)	Low Affinity Peak (% of Cells)
None	--	--	100	0
DV 48	Sense	LMP-2	100	0
DV 49	Antisense	LMP-2	100	0
DV 50	Sense	RL C2 (16-25)	100	0
DV 51	Antisense	RL C2 (16-25)	66	33
DV 54	Sense	RL C2 (130-139)	100	0
DV 55	Antisense	RL C2 (130-139)	66	33
DV 52	Sense	RL C5	100	0
DV 53	Antisense	RL C5	75	25

Jurkat T cells were cultured in RPMI medium containing 10% bovine calf serum, and maintained at 37°C in a 5% CO_2 atmosphere. Cells were incubated, in 0.1 ml volumes of 10% RPMI medium, without oligonucleotides, or with 50 µM sense or antisense oligonucleotide for 72 hours at 37°C. After incubation cells were washed and recovered by centrifugation at 1,000 rpm for 10 min (4°C). Cell pellets were resuspended in 25 µl of PBSA and 10 µl of monoclonal antibody against class I MHC antigen (Accurate Chemical and Scientific Corp., catalogue No. MAS 15326), and incubated for 30 min at 4°C. Next the Jurkat cells were washed by centrifugation (1,000 rpm for 10 min at 4°C) in 1.0 ml of PBSA. The cell pellets were next resuspended in 25 µl of PBSA and incubated with FITC-coupled rabbit anti-mouse-IgG (10 µl of 1:10 diluted FITC-coupled antibody was added) for 30 min at 4°C. A final PBSA centrifugal wash was performed, as above, and the cells were finally suspended in 0.5 ml of PBSA at 4°C. Class I molecule surface presentation was assessed by fluorescence activated cell sorter (FACS). The data presented are averages from several experiments.

Whatever may be the case for LMP-2, it is clear that incubating Jurkat T lymphocytes with proteasome antisense oligonucleotides adverseley affected class I antigen presentation. This is presumably because the 72 hour antisense incubation decreased the effective concentration of active proteasome. In this regard it should be noted that proteasome has a turnover time of approximately 7 to 10 days, so the three-day antisense experiments should not have been able to decrease proteasome concentrations by more than about 40%. Thus, the proposal that proteasome is integrally involved in the normal processing and cell-surface presentation of MHC class I antigens, appears entirely tenable.

Conclusions

In summary it would appear that endogenous antigens may undergo processing in the cell cytoplasm by proteasome, to generate peptides of nine amino acids in length. Nonomeric peptides next seem to travel (or be transported) to the endoplasmic reticulum where they associate with MHC class I molecules and β_2-microglobulin. Such complexes must traverse the Golgi apparatus and be inserted in the cell membrane. Transporter proteins, of the (ATP-binding cassette) ABC superfamily are required for this intracellular trafficking. Properly presented class I antigens on the cell surface are recognized by CD8[+] T cells as self antigens.

It appears an unlikely coincidence that many, if not most, antigen presenting cells exhibit an oxidative burst in response to many stimuli, and express a proteolytic complex (proteasome) that selectively recognizes and degrades oxidatively modified proteins. Whether there is any role for background rates of protein oxidation, or for the oxidative burst of macrophages, in "marking" endogenous antigens for processing by proteasome remains to be tested. Oxidative modification (whether from normal, oxygen-driven, intracellular protein oxidation, or from an enzymatic oxidative burst) clearly generates a universal recognition sequence for proteasome, in the form of exposed hydrophobic core residues (Davies, 1986: Davies *et al.*, 1987b: Davies and Lin, 1988b; Pacifici *et al.*, 1989). Such a universal recognition sequence would appear essential if proteasome is to be able to process all the endogenous class I antigens expressed by macrophages and other antigen presenting cells.

Acknowledgements

This research was supported by grant number ES-03598 from the National Institutes of Health/National Institute of Environmental Health Sciences, U.S.A. to K.J.A.D.

References

Arnold, D., Driscoll, J., Androlewicz, M., Hughes, E., Cresswell, P., and Spies, T. (1992) Proteosome subunits encoded in the MHC are not generally required for the processing of peptides bound by MHC class I molecules. *Nature* 360: 171-173.

Arrigo, A., Tanaka, K., Goldberg, A.L. and Welch, W.J. (1988) Identity of the 19S 'prosome' particle with the large multifunctional protease complex of mammalian cells (the proteasome). *Nature* 331: 192-194.

Brown, M.G., Driscoll, J., and Monaco, J.J. (1991) Structural and serological similarity of MHC-linked LMP and proteasome (Multicatalytic proteinase) complexes. *Nature* 353: 355-357.

Davies, K.J.A.. (1985) Free radicals and protein degradation in human red blood cells. In Cellular and molecular aspects of aging: the red cell as a model. (Eaton, Konzen, and White, eds.) Liss, NY, pp. 15-27

Davies, K.J.A. (1988) Protein oxidation, protein cross-linking, and proteolysis in the formation of lipofuscin. In Lipofuscin-1987: State of the art. (Nagy, ed.) Elsevier, Amsterdam, pp. 109-133.

Davies, K.J.A. (1986) Intracellular proteolytic systems may function as secondary antioxidant defenses: A hypothesis. *J. Free Radicals Biol. Med.* 2: 155-173.

Davies, K.J.A. (1987) Protein damage and degradation by oxygen radicals. I. General aspects. *J. Biol. Chem.* 262: 9895-9901.

Davies, K.J.A., and Delsignore, M.E. (1987) Protein damage and degradation by oxygen radicals. III. Secondary and tertiary structure. *J. Biol. Chem.*. 262: 9908-9913.

Davies, K.J.A., and Goldberg, A.L. (1987a) Oxygen radicals stimulate intracellular proteolysis and lipid peroxidation by independent mechanisms in erythrocytes. *J. Biol. Chem.* 262: 8220-8226.

Davies, K.J.A., and Goldberg, A.L. (1987b) Proteins damaged by oxygen radicals are rapidly degreaded in extracts of red blood cells. *J. Biol. Chem.* 262: 8227-8234.

Davies, K.J.A., Delsignore, M.E., and Lin, S.W.(1987a) Protein damage and degradation by oxygen radicals. II. Primary structure. *J. Biol. Chem.* 262: 9902-9907.

Davies, K.J.A., Lin, S.W., and Pacifici, R.E. (1987b) Protein damage and degradation by oxygen radicals. IV. Degradation of denatured protein. *J. Biol. Chem.* 262: 9914-9920.

Davies, K.J.A., and Lin, S.W. (1988a) Degradation of oxidatively denatured protein in *Escherichia coli. Free Radicals Biol. Med.* 5:215-223.

Davies, K.J.A., and Lin, S.W. (1988b) Oxidatively denatured proteins are degraded by an ATP-independent pathway in *Escherichia coli. Free Radicals Biol. Med.* 5:225-236.

Deverson, E.V., Gow, I.R., Coadwell, W.J., Monaco, J.J., Butcher, G.W., and Howard, J.C. (1990) MHC class II region encoding proteins related to the multidrug resistance family of transmembrane transporters. *Nature* 348: 738-741.

Faustman, D., Li, X, Lin, H.Y., Fu, Y., Eisenbarth, G., Avruch, J., and Guo, J. (1991) Linkage of faulty major histocompatibility complex class I to autoimmune diabetes. *Science* 25: 1756-1761.

Fucci, L., Oliver, C.M., Coon, M.J., and Stadtman, E.R. (1983) Inactivation of key metabolic enzymes by mixed-function oxidation reactions: Possible implication in protein turnover and aging. *Proc. Natl. Acad. Sci. USA* 80: 1521-1525.

Glynne, R., Powis, S.H., Beck, S., Kelly, A., Kerr, L.-A., and Trowsdale, J. (1991) A proteasome-related gene between the two ABC transporter loci in the class II region of the human MHC. *Nature* 353: 357-360.

Hackett, C.J. (1991) Later for the rendezvous. *Nature* 349: 655-656.

Henderson, R.A., Michel, H. Sakaguchi, K., Shabanowitz, J., Appella, E., Hunt, D.F., and Engelhard, V.H. (1992) HLA-A2.1-Associated peptides from a mutant cell line: A second pathway of antigen presentation. *Science* 255: 1264-1266.

Hunt, D.F., Henderson, R.A., Shabanoxwtz, J., Sakaguchi, K., Michel, H., Sevilir, N., Cox, A.L., Appella, E., and Engelhard, V.H. (1992) Characterization of peptides bound to the class I MHC molecule HLA-A2.1 by mass spectometry. *Science* 255: 1261-1263.

Kelly, A., Powis, S.H., Glynne, R., Radley, E., Beck, S., and Trowsdale, J. (1991) Second proteasome-related gene in the human MHC class II region. *Nature* 353: 667-668.

Levine, R.L., Oliver, C.N., Fulks, R.M., and Stadtman, E.R. (1981) Turnover of bacterial glutamine synthetase: Oxidative inactivation precedes proteolysis. *Proc. Natl. Acad. Sci. USA* 78: 2120-2124.

Marcillat, O., Zhang, Y., Lin, S.W., and Davies, K.J.A. (1988) Mitochondria contain a proteolytic system which can recognize and degrade oxidatively denatured proteins. *Biochem. J.* 254: 677-683.

Martinez, C.K., and Monaco, J.J. (1991) Homology of proteasome subunits to a major histocompatibility complex-linked LMP gene. *Nature* 353: 664-667.

Momburg, F., Ortiz-Navarrete, V., Neefjes, J., Goulmy, E., van de Wal, Y., Spits, H., Powis, S.J., Butcher, G.W., Howard, J.C., Walden, P., and Hämmerling, G.J. (1992) Proteasome subunits encoded by the major histocompatibility complex are not essential for antigen presentation. *Nature* 360: 174-177.

Monaco, J.J., and McDevitt, H.O. (1982) Identification of a 4th class of proteins linked to the murine major histocomparibility complex. *Proc. Natl. Acad. Sci. USA* 79: 3001-3005.

Monaco, J.J., and McDevitt, H.O. (1986) The LMP antigens; A stable MHC-controlled multisubunit protein complex. *Hum. Immun.* 15: 416-426..

Monaco, J.J., Cho, S., and Attaya, M. (1990) Transport protein genes in the murine MHC: possible implications for antigen processing. *Science* 250: 1723-1726.

Murakami, K., Jahngen, J.H., Lin, S.W., Davies, K.J.A., and Taylor, A. (1990) Lens proteasome shows enhanced rates of degradation of hydroxyl radical modified alpha-crystallin. *Free Radical Biol. Med.* 8: 217-222.

Ortiz-Navarrete, V., Seelig, A., Gernold, M., Frentzel, S., Kloetzel, P.M., and Hämmerling, G.J. (1991) Subunit of the '20S' proteasome (multicatalytic proteinase) encoded by the major histocompatibility complex. *Nature* 353: 662-664.

Pacifici, R.E., and Davies, K.J.A. (1987) The measurement of protein degradation in response to oxidative stress. In Oxygen radicals in biology and medicine: basic life sciences. Int. Congr. on Oxygen Radicals, LaJolla, Plenum, New York, pp. 531-535.

Pacifici, R.E., and Davies, K.J.A. (1990) Protein degradation as an index of oxidative stress. *Meth. Enzymol.* 186: 485-502.

Pacifici, R.E., Salo, D.C., and Davies, K.J.A. (1989) Macroxyproteinase (MOP): A 670 kDa proteinase complex that degrades oxidatively denatured proteins in red blood cells. *Free Radical Biol. Med.* 7: 521-536.

Parnham, P. (1990) Transporters of delight. *Nature* 348: 674-675.

Rivett, A.J. (1985) Purification of a liver alkaline protease which degrades oxidatively modified glutamine synthetase: Characterization as a high molecular weight cysteine protease. *J. Biol. Chem.* 260: 12600-12606.

Rivett, A.J. (1989) The multicatalytic proteinase of mammalian cells. *Arch Biochem. Biophys.* 268: 1-8.

Robertson, M. (1991) Proteasomes in the pathway. *Nature* 353: 300-301.

Salo, D.C., Lin, S.W., Pacifici, R.E., and Davies, K.J.A. (1988) Superoxide dismutase is preferentially degraded by a proteolytic system from red blood cells following oxidative modification by hydrogen peroxide . *Free Radical Biol. Med.* 5: 335-339.

Salo, D.C., Pacifici, R.E., Lin, S.W., Giulivi, C., and Davies, K.J.A. (1990) Superoxide dismutase undergoes proteolysis and fragmentation following oxidative modification and inactivation. *J. Biol. Chem.* 265: 11919-11927.

Spies, T., and DeMars, R. (1991) Restored expression of major histocompatibility class I molecules by gene transfer of a putative peptide transporter. *Nature* 351: 323-324.

Spies, T., Bresnahan, M., Bahram, S., Arnold, D., Blanck, G., Mellins, E., Pious, D., and DeMars, R. (1990) A gene in the human major histocompatibility complex class II region controlling the class I antigen presentation pathway. *Nature* 348: 744-747.

Stadtman, E.R. (1986) Oxidation of proteins by mixed-function oxidation system: Implication in protein turnover, ageing and neutrophil function. *TIBS* 11: 11-12.

Taylor, A., and Davies, K.J.A. (1987) Protein oxidation and loss of protease activity may lead to cataract formation in the aged lens. *Free Radical Biol. Med.* 3: 371-377.

Trowsdale, J., Hanso, I., Mockridge, I., Beck, S., Townsend, A., and Kelly, A. (1990) Sequences encoded in the class II region of the MHC related to the 'ABC' superfamily of transporters. *Nature* 348: 741-744.

Wolff, S.P. and Dean, R.T. (1986) Fragmentation of proteins by free radicals and its effect on their susceptibility to enzymatic hydrolysis. *Biochem. J.* 234: 399-403.

Wolff, S.P., Garner, A., and Dean, R.T. (1986) Free radicals, lipids, and protein degradation. *TIBS* 11: 27-31.

Oxidative Stress, Cell Activation and Viral Infection
C. Pasquier et al. (eds)
© 1994 Birkhäuser Verlag Basel/Switzerland

Involvement of hydrogen peroxide in the actions of TGFβ1

K. Nose, M. Ohba*, M. Shibanuma, and T. Kuroki*

*Department of Microbiology, Showa University School of Pharmaceutical Sciences,
Hatanodai 1-5-8, Shinagawa-ku, Tokyo 142, Japan
and *Deparment of Cancer Cell Research, Institute of Medical Sciences,
University of Tokyo, Shirokanedai 1-5-8, Minato-ku, Tokyo 108, Japan*

Summary

Several lines of evidence indicate that active oxygens could work as a competence factor, inducing the early response genes as well as DNA synthesis in quiescent cells. Competence factors, in fact, stimulate cells to produce active oxygens. On the other hand, hydrogen peroxide was found to cause reversible inhibition of DNA synthesis in mouse osteoblastic cells when added in late G1 phase, just like TGFb1. TGFb1 stimulated cells to produce hydrogen peroxide in cells at late G1 phase. These results suggested that hydrogen peroxide could be one of the second messengers of TGFb1-actions. To understand the common molecular mechanisms for the effects of hydrogen peroxide and TGFb1, we took two approaches. One was to isolate genes that are induced by both stimuli in late G1 phase, and the other was to detect cellular proteins whose phosphorylation was increased by these stimuli. Using differential hybridization method, one possible candidate gene (HIC-5) was isolated that encodes a novel Zn-finger protein with a molecular weight of approximately 55 kDa. Expression of HIC-5 gene was increased about 2-fold by hydrogen peroxide and TGFb1 with a peak of 4 to 6 hr after treatment. From analysis of phosphorylation of proteins, a unique protein with a molecular weight of 30 kDa was detected whose phosphorylation was increased both by hydrogen peroxide and TGFb1 only in late G1 phase. From several criteria, this was identified as a small molecular weight heat shock protein, HSP28. The increase in HSP28 phosphorylation induced by TGFb1 was suppressed by catalase, at least in part. These results indicate that hydrogen peroxide and TGFb1 share some common signal transduction pathways.

Introduction

Active oxygens are extremely reactive and modify various biochemical substances. This high reactivity is generally thought to have harmful effects on living organisms by damaging their membrane, enzymes and DNA (Imlay & Linn, 1988; Cerutti, 1985; Hyslop et al., 1988). On the other hand, accumulating evidence now suggests that active oxygens at low concentrations act as mediators of cellular responses and growth signals. Addition of hydrogen peroxide, like addition of insulin, stimulated glucose transport and lipid synthesis in isolated rat adipocytes (Mukhwejee et al., 1978; May et al., 1979). Pancreatic islet cells generate hydrogen peroxide under condition to elevate cytoplasmic free Ca^{2+} level and to activate protein kinase C (Yamada, 1988). Exogenous addition of active oxygens to resting fibroblastic or epidermal cells stimulated cells to induced proto-oncogenes, c-fos and c-myc, as well as DNA synthesis (Shibanuma et al., 1988; 1990; Crawford et al., 1988). Active oxygens are generated from cultured leukemia cells (Shibanuma et al., 1987), fibroblastic cells (Shibanuma et al., 1988), or keratinocytes (Szatrowski & Nathan, 1991) following stimulation of cells with mitogens.

While these findings suggested that active oxygens can work as one of second messengers of growth factor signalings in quiescent cells, we observed that hydrogen peroxide at low doses reversively inhibited DNA synthesis in Balb 3T3 or mouse osteoblastic MC3T3 cells at late G1 phase. The late G1 phase is critical for regulation of cell cycle progression, and is called a "check point" (Pardee, 1988). A well known factor that inhibits DNA synthesis in late G1 phase is TGFβ1, and it is a multifunctional polypeptide that regulates a wide variety of biological responses (for reviews, Sporn et al., 1987; Moses et al., 1990). Its most remarkable effect is inhibition of DNA synthesis in late G1 phase of cell cycle (Laiho et al., 1990; Pietenpol et al., 1990). These findings prompted us to examine production of active oxygens from cells treated with TGFβ1. In the present communication we describe evidence that hydrogen peroxide was produced into the culture medium following stimulation of MC3T3 cells with TGFβ1, and that hydrogen peroxide conveyed signals of TGFβ1 to induce inhibition of DNA synthesis, phosphorylation of HSP28 and expression of specific genes.

Materials and Methods

Cell culture and chemicals
Balb 3T3 clone A31-1-1 cells were grown in Eagle's MEM supplemented with 10% fetal bovine serum. Cells were made confluent by cultivation for 7 days following inoculation without medium renewal. Mouse osteoblastic cells (MC3T3-E1, Kodama et al. 1968) were grown in Dulbecco's modified MEM (DMEM) supplemented with 10% fetal bovine serum at 37°C, in 5% CO_2. Cells were made quescent by cultivating pre-confluent cells for 4 days in serum-free DMEM containing 0.1% bovine serum albumin, and stimulated to progress through cell cycle by addition of fresh serum.

For measurement of DNA synthesis, cells were inoculated into 35 mm plastic dishes, at a density of 2×10^4 cells/dish, and incubated for 3 days. Cells were treated with agents, labeled with [^{3}H]thymidine (0.1 μCi/ml) for 12 hr, and then rinsed with ice-cold phosphate-buffered saline, and treated with 5 % trichloroacetic acid in the cold. After disruption of the cells by sonication, the precipitate was filtered through glass fiber filters. The radioactivity incorporated into the acid-precipitable fraction was measured.

Measurement of hydrogen peroxide concentration in medium and cells.

Subconfluent cultures in 60 mm dishes, preincubated in serum-free medium(MEM containing 1 mg/ml bovine serum albumin), were treated with TGFβ1. Concentrations of hydrogen peroxide in the medium were determined by the method of Zaitsu and Ohkura (1978) with slight modifications. Briefly, an aliquot (2 ml) of the medium was incubated in a reaction mixture consisting of 5 mM 3-(p-hydroxyphenyl)-propionic acid and 1 unit of horseradish peroxidase in Tris-HCl (pH8.5) for 5 min, and then fluorescence intensity was measured at 404 nm with excitation at 320 nm. The amount of hydrogen peroxide was calculated from a calibration curve for the linear relationship between fluorescence intensity and hydrogen peroxide concentrations in the range of 0.2 - 10 μM.

Extraction of RNA and analysis

Total RNA was extracted by guanidium-hot phenol method (Maniatis et al., 1982), and separated in agarose gel containing formaldehyde. The RNA was then transferred to a nylon membrane and hybridized with [^{32}P]-labeled probes as described previously (Shibanuma et al., 1986) in 50 % foramide at 42°C for 24 hr. Mouse α-tubulin (Mα1, Lewis et al., 1985) was used as a reference to monitor the amounts of RNA in each lane of the gels. Filters were washed as described before and autoradiographed.

Differential screening of cDNA library

The cDNA libraries were constructed using 5 μg of poly(A)$^+$ RNA obtained from MC3T3 cells treated with TGFβ1 (2 ng/ml) for 4 hr. The library was prepared by a directional cloning stratery with poly(dT) linked to synthetic Xho I and GAGA sequences, using a kit from Stratagene (La Jolla, CA) following the manufacturer's manual, and double-strand cDNA was cloned into Eco RI/Xho I sites of λZAP vector. For screening, single-stranded cDNA was used as a probe after hybrid selection. Poly(A)$^+$RNAs were prepared from untreated cells and cells treated for 4 hr with 2 ng/ml TGFβ1, and used as templates for synthesis of cDNAs in the presence of [^{32}P]dCTP and reverse transcriptase from avian myeloblastosis virus under conditions described previously (Shibanuma et al., 1992a). The reaction was terminated by incubation at 65°C in 0.2N NaOH for 10 min. After extraction and ethanol precipitation, the labeled cDNAs were hybridized with 20-fold excess of poly(A)$^+$RNA from untreated cells, and single stranded cDNA was separated by use of hydroxyapatite. Replica filters were hybridized with the labeled differential cDNA probes, and clones that gave stranger signals with cDNA from TGFβ1-treated cells were isolated.

Among 6 clones that was induced by TGFβ1, one (HIC-5) was found to be induced by hydrogen peroxide, too. The DNA insert was cloned into pBluescript for sequencing.

Expression vectors for HIC-5 were constructed using cytomegarovirus promoter (pRc/CMV, Promega), and were introduced into cells by a conventional calcium phosphate precipitation method (Graham & van der Eb, 1973). The pRc/CMV plasmid contains a neomycin-resistant gene, and transfected cells were cultured in the presence of 800 μg/ml of geneticin to assess the growth of cells that incorporated the plasmids.

Two-dimensional gel electrophoresis
Quiescent cultures were labeled with [^{32}P]orthophosphate at a concentration of 0.3 mCi/ml for 6 hr. The labeled cells were then treated with hydrogen peroxide or TGFβ1 for the indicated times and extracts were prepared for the examination by 2-dimensional gel elctrophoresis following the procedures of O'Farrell (1975). Gels were dried and autoradiographed.

Results

Stimulation of DNA synthesis in quiescent Balb 3T3 cells by H_2O_2
Active oxygens were found to induce immediate early response genes (Shibanuma et al., 1988), and it was expected that they stimulated resting cells to progress through cell cycle. Competence factors for Balb 3T3 cells like platelet derived growth factor (PDGF) or 12-O-tetradecanoyl-phorbol-13-acetate (TPA), in fact, stimulated cells to produce active oxygens in quiescent cells (Shibanuma et al., 1987; Szatrowski & Nathan, 1991). From these observations, we examined the effect of hydrogen peroxide on DNA synthesis in quiescent Balb 3T3 cells by measuring the labeling index on autoradiography. As shown in Table I, hydrogen peroxide alone had essentially no effect on the start of DNA synthesis, but it increased the number of labeled nuclei in the presence of insulin (a progression factor). Number of grains in labeled muclei in cells treated with hydrogen peroxide plus insulin were comparable with those in serum-treated cells.

Effect of catalase on DNA synthesis at various periods during cell cycle
Above results indicated that active oxygens generated by mitogenic stimuli worked as a competence factor, but next question was how were their effects on cells progressing through cell cycle. To answer this question, we first examined the effect hydrogen peroxide on DNA synthesis added exogenously in Balb 3T3 cells during cell cycle, and found that hydrogen peroxide at non-toxic doses reversibly inhibited DNA synthesis in middle to late G1 phase (data not shown). Next, the effect of scavengers of active oxygens (catalase) on DNA synthesis was evamined at various time points during cell cycle. Resting Balb 3T3 cells were stimulated with TPA or PDGF, and DNA synthesis was measured by incorporation of [^{3}H]thymidine 24 to 36 hr

after stimulation. Catalase was added at various times following stimulation , and its effect on DNA synthesis was examined. As shown in Fig. 1, DNA synthesis in cells treated with catalase at 0 hr (G0 phase) was lower than that in cells treated at 3 to 12 hr (middle G1 phase) after stimulation, and it started to decrease by addition of catalase 15 to 18 hr following stimulation (late G1 phase). The effect of catalase in cells at G0 phase could be interpreted to indicate that hydrogen peroxide generated by growth stimuli worked as a positive signal for DNA synthesis, since DNA synthesis was lower in the presence of catalase at 0 hr following stimulation with serum compared to that at 6-9 hr.

Table I. Effect of H_2O_2 on induction of DNA synthesis in quiescent Balb 3T3 cells. Quiescent cells were stimulated by each agent, and labeling index (means + standard deviations) was estimated by autoradiography for cells incubated with [^{3}H]thymidine 12 to 24 hr after stimulation.

Addition		Labelling index (%)
–		0
Insulin	1μM	15.3 ± 3.5
H_2O_2	0.15 mM	0
Insulin + H_2O_2		29.1 ± 8.0
Serum	10%	33.5 ± 7.6
TPA	10ng/ml	18.0 ± 6.0

On ther other hand, the effect in late G1 phase seemed to suggest the negative effect of hydrogen peroxide, as DNA synthesis was higher at 9-12 hr in the presence of catalase than that at 12-15 hr. These indirect evidence suggested that active oxygens worked as a positive signal for quiescent fibroblasts, whereas they negatively control DNA synthesis in progressing cells, especially in late

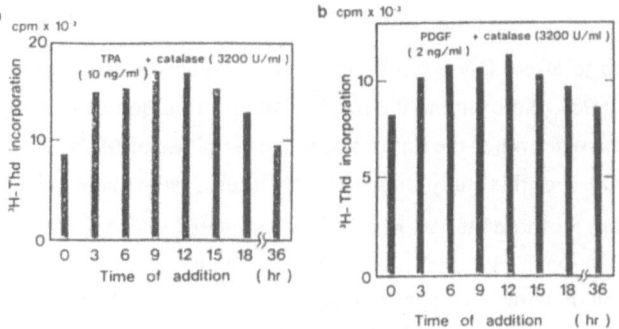

Fig. 1. Effect of catalase on DNA synthesis in Balb 3T3 cells stimulated with TPA or PDGF. Cells in quiescent state were stimulated either with (a) TPA or (b) PDGF at 0 hr. Catalase was added at various times after stimulation, and incorporation of [^{3}H]thymidine was measured 24 to 36 hr after the stimulation.

G1 phase of cell cycle. From screening of several mouse fibroblastic cells that are sensitive to negative regulation of DNA synthesis by hydrogen peroxide, mouse osteoblastic cells, MC3T3-E1, were found to show a more clear negative response by hydrogen peroxide compared to Balb 3T3 cells.

Cell cycle-dependent nhibition of DNA synthesis by TGFβ1 and H2O2 in MC3T3 cells

In cultures of MC3T3 cells in the logarithmic growth phase, DNA synthesis was inhibited significantly by one of the growth suppressive factor, TGF β1, and hydrogen peroxide at concentrations of 0.1 - 0.2 mM (at non-toxic levels) also inhibited DNA synthesis (Table II).

Table II. Inhibition of [^{3}H]thymidine incorporation by TGF β1 and H2O2 in growing MC3T3 cells. Logarithmically growing cells preincubated in DMEM containing 1 mg/ml BSA for 3 hr were treated with TGF β1 or H2O2. [^{3}H]Thymidine (0.1 μCi/ml) was added after treatment for 24 hr, and after further incubation for 12 hr, acid-precipitable radioactivity was measured. Values are means for duplicate samples + SE.

Addition		[^{3}H] Thymidine incorporation (cpm) (%)
None		12,370 ± 930 (100)
TGF β1	1 ng/ml	9,300 ± 770 (75)
	10 ng/ml	2,140 ± 130 (17)
H2O2	0.1 mM	8,680 ± 890 (70)
	0.2 mM	5,000 ± 340 (40)

TGF β1 is known to affect DNA synthesis in late G1 phase of cell cycle (Laiho et al., 1990; Pietenpol et al., 1990). To examine the mechanism of inhibition of [^{3}H]thymidine by TGF β1 and H2O2, we next determined the time point in the G1 phase of MC3T3 cells that is sensitive to TGF β1 and H2O2. For this study, subconfluent cultures were made quiescent by incubation in serum-free medium for more than 24 hr and then stimulated with 5% fetal bovine serum. Under this condition, DNA synthesis, measured as [^{3}H]thymidine incorporation, reached a peak about 24 hr after addition of serum. At appropriate times during the G1 phase, TGF β1 (10 ng/ml) or H2O2 (0.15 mM) was added, and the levels of DNA synthesis between 24 and 36 hr after stimulation were measured. Results shown in Fig. 2 indicate that both TGF β1 and H2O2 inhibited DNA synthesis when added in the late G1 phase (about 15 hr after stimulation with serum).

Effect of catalase on the inhibition of DNA synthesis by TGF β1.

The results of Fig. 2 suggested that H_2O_2 was involved in the inhibition of DNA synthesis caused by TGF β1. To examine this possibility, effect of catalase on the inhibition of DNA synthesis caused by TGF β1 was examined. Table III shows that TGF β1-induced inhibition of DNA synthesis in MC3T3 cells, as determined by the labeling index, was abolished at least in part in the presence of catalase at 3200 units/ml. Heat-inactivated catalase, superoxide dismutase, and mannitol had little effect (data not shown). These results indicate that hydrogen peroxide mediates the inhibitory effect of TGF β1.

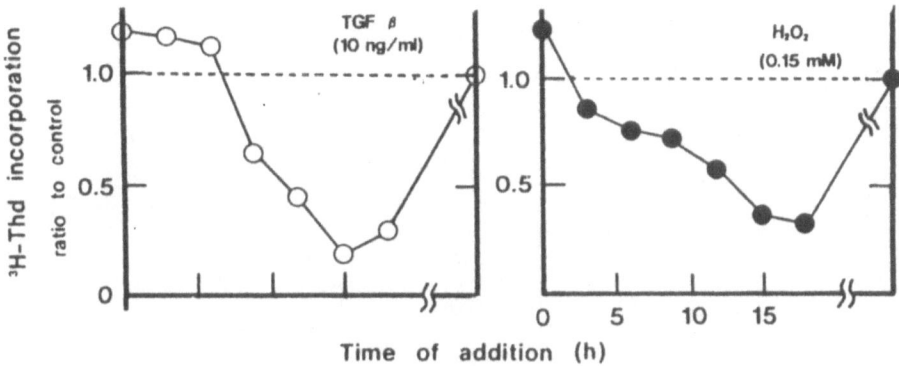

Fig. 2. Differential sensitivity to growth inhibition by TGF β1 and H_2O_2. Subconfluent cultures of MC3T3 cells were incubated in DMEM containing 1 mg/ml BSA for 24 hr, and then stimulated with 5% serum (0 hr). TGF β1 (10 ng/ml) (a) or H_2O_2 (0.15 mM) (b) was added at various times after stimulation, and [3H]thymidine incorporation between 24 and 36 hr was measured and is plotted versus the time of addition.

Table III. Effect of catalase on inhibition of DNA synthesis by TGF β1 in MC3T3 cells. Quiesent cells were stimulated with serum at 0 hr, and TGF β1 or H_2O_2 was added at 12 hr. Cells were labeled with [3H]thymidine (1 μCi/ml) from 12 to 36 hr and processed for autoradiography. The data are means of triplicated samples + SD.

Addition	Labelling (%)
None	6.0 ± 1.5
Serum (20%)	47.6 ± 5.7
+ TGF-β1 (5ng/ml)	27.6 ± 4.3
+ TGF-β1 + Catalase (3200 units/ml)	54.7 ± 6.2
+ Catalase	56.2 ± 9.0

28

Production of hydrogen peroxide from cells treated with TGFβ1

The above observations suggest that treatment with TGFβ1 may result in production of H_2O_2 from cells, leading to inhibition of DNA synthesis. To examine this possibility, we added TGFβ1 (5 ng/ml) to cells in the G0 phase (without serum stimulation), or late G1 phase (14 hr after serum stimulation), and measured the concentrations of H_2O_2 in the medium periodically by a fluorogenic method. Fig. 3 shows that the H_2O_2 concentration in the medium increased after TGF β1 treatment of cells in the late G1 phase, but not in the G0 phase, suggesting that H_2O_2 was released from cells by treatment with TGF β1 in the late G1 phase.

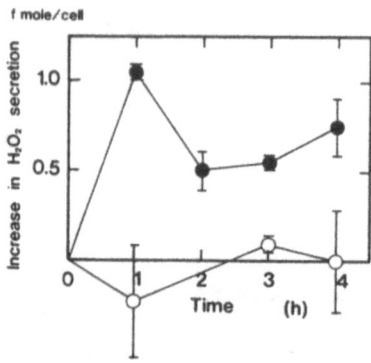

Fig. 3. Increase in H_2O_2 concentration in the medium of MC3T3 cells treated with TGF β1. Cells in the G1 phase (○) or late G1 (●) were incubated in Eagle's MEM containing 1 mg/ml BSA for 2 hr and then treated with TGF β1 (5 ng/ml) at 0 hr. After the indicated times, the H_2O_2 concentrations in the medium were determined. Values are means + SD.

Phosphorylation of HSP28 as a common target of TGF β1 and H2O2

To get molecular targets of the actions of TGF β1 and H_2O_2, we first examined the alterations in protein phosphorylation by these agents in MC3T3 cells. Cells were labeled with [^{32}Pi] for 6 hr and then treated with TGF β1 or H_2O_2, lysed and subjected to two-dimensional gel electrophoresis. Increased phosphorylation was detected in two proteins with molecular weights of about 30 kDa and pI values of 6.2-6.7, when the cells were treated either with TGF β1 or H_2O_2, (shown by arrows in Fig. 4). Increase in its phosphorylation became maximum 15 min after treatment with H_2O_2, while 2 hr was required in the case of phosphorylation by TGF β1. The 30 kDa spots phosphorylated in cells treated with H_2O_2 or TGF β1 gave the same peptide map with a small molecular weight heat shock protein (HSP28) (Shibanuma et al., 1992b), and hence HSP28 phosphorylation was one of the common molecular targets for TGF β1 and hydrogen peroxide. Catalase abolished, at least in part, the HSP28 phosphorylation increased by TGF β1 (Shibanuma et al., 1991).

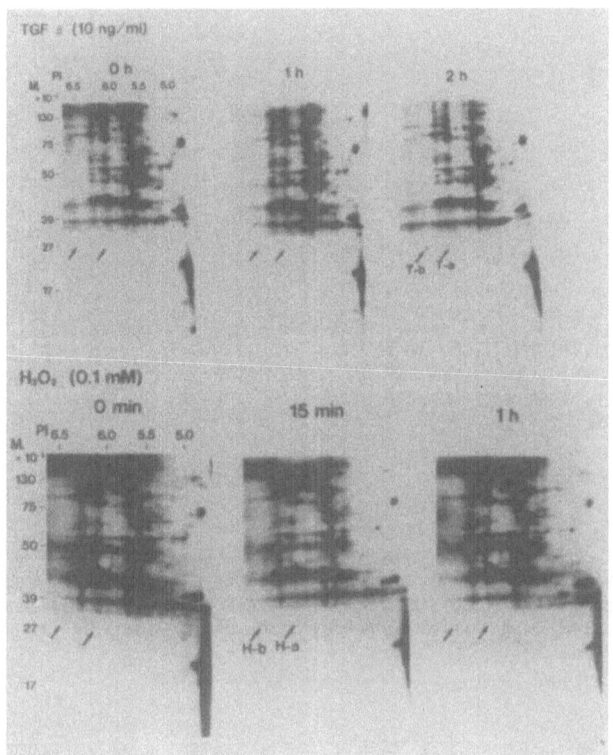

Fig. 4. Phosphorylation of HSP28 by hydrogen peroxide and TGFβ1. MC3T3 cells in late G1 phase were labeled with [^{32}Pi] (0.2 mCi/ml) for 46hr and treated either with 0.1 mM H_2O_2 or 10 ng/ml TGF β1 for indicated times. Then, the cells were lysed and subjected to two-dimensional electrophoresis. Arrows, position of HSP28.

Isolation of cDNA clones that are induced both by H_2O_2 and TGF β1.

As a common target molecule for H_2O_2 and TGF β1, we tried to isolate cDNA clones whose expression was increased by both agents using a differential screening method. Among six TGF β1-inducible clones, one (named HIC-5) clone was found to be induced by both agents (Fig. 5). HIC-5 seemed to encodes a polypeptide with a molecular weight of about 50 kDa from an open reading frame, and its amino acid sequence predicted from cDNA showed two domains; N-terminal half is rich in Pro and acidic amino acids, and C-terminal part contained 7 repeats of putative Zn-finger structures.(Fig.6) To examine the function of HIC-5 in growth regulation, we constructed expression vectors, and introduced into NIH 3T3 or MC3T3 cells. Number of colonies formed in the presence of geneticin was decreased significantly in cells transfected with sense-expression vector compared to that transfected with anti-sense-expression vector in

MC3T3 cells (Table IV). Numbers of colonies were not affected by both vectors. These results indicate that HIC-5 expression inhibited growth of MC3T3 cells but not of NIH 3T3 cells.

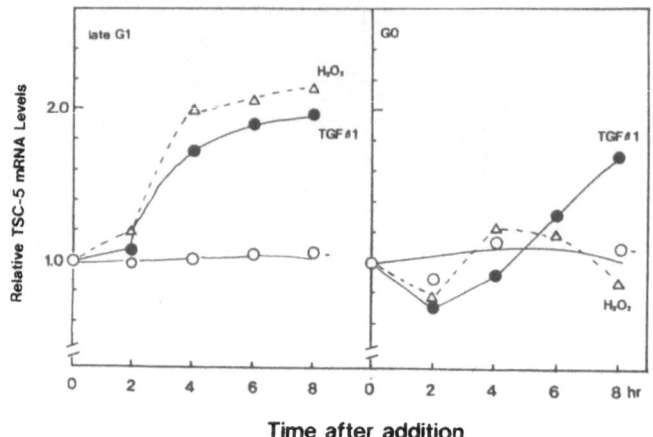

Fig. 5. Induction of HIC-5 mRNA. MC3T3 cells in G0 (without serum stimulation) or late G1 phase (12 hr after stimulation of quiescent cells) were treated either with TGF β1 (closed circle) or H2O2 (open circle) for indicated times, and RNA was extracted. Levels of HIC-5 mRNA were estimated by Northern blot hybridization, and the relative levels were calculated using α–tubulin mRNA as a reference.

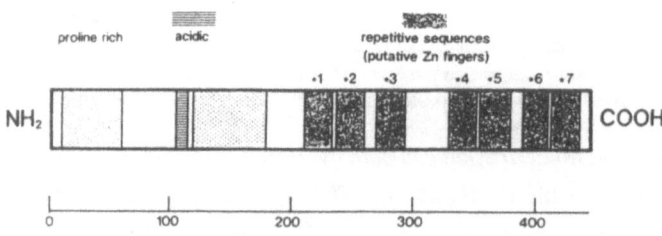

Fig. 6. Schematic presentation of the structure of HIC-5 protein predicted from the open reading frame of cDNA.

Table IV. Effect of HIC-5 expression vectors on colony formation. MC3T3 or NIH 3T3 cells were transfected with plasmids (total amount of 10 μg/dish), and cultured in the presence of geneticin. Numbers of colonies were counted 14 days later. CMV was a vector without cDNA, S5 was a sense-expression vector for HIC-5 in CMV plasmid, and AS5 was an antisense-expression vector.

Cells	Plasmids	number of colonies / dish	%
MC3T3			
	1 CMV	12.0	100
	2 CMV + S5 (2μg)	12.0	100
	3 CMV + S5 (5μg)	11.5	96
	4 S5 (10μg)	6.0	50
	5 AS5 (10μg)	12.8	107
NIH3T3			
	1 CMV	57.5	100
	2 CMV + S5 (2μg)	72.3	126
	3 CMV + S5 (5μg)	57.5	100
	4 S5 (10μg)	55.3	96
	5 AS5 (10μg)	68.8	120

Discussion

Active oxygens elicite various biological responses, most of which are hazordous in nature. Living cells produce active oxygen species under physiological conditions following stimulation by infection, growth factors or other environmental substances, and anti-oxidants in cells are undoubtly protecting damaging reactive oxygens. The present study raised a proposal that such active oxygens could have some biological important functions for the control of cell growth, thereby regulating gene expression and phosphorylating proteins. We found that natural polypeptide growth factors stimulated cells to produce hydrogen peroxide in cultured fibroblastic cells. Scavenging of the active oxygens modulated the progression of cell cycle, both positively and negatively depending on the period during cell cycle. The most striking phenomenon was that both hydrogen peroxide and TGF β1 inhibited DNA synthesis when added at late G1 phase in osteoblastic cells. TGF β1 really stimulated cells to produce hydrogen peroxide.

To show that hydrogen peroxide worked as a second messenger for TGFβ1 signaling, we tried to identify molecular changes that are induced by both agents. One of the important criteria for this notion is that the change caused by TGFβ1 should be quenched by specific scavengers for hydrogen peroxide. We have identified so far two such candidates; one was the phosphorylation of HSP28 and other was the cDNA clone (HIC-5). Role of these two gene products are not yet fully understood, but HSP28 is known to be associated with growth arrest in B cells (Spector et al., 1992), and its expression level was decreased upon transformation with adenovirus (Zantema et al., 1989). One of the biological functions of HSP28 seems to be connected with resistance to heat shock or DNA damaging agents (Rollet et al., 1992; Huot et al., 1991), but above results together with our findings suggest that HSP28 may have some roles in negative growth regulation.

From our preliminary experiments, HIC-5 seems to inhibit growth of at least some types of cultured fibroblasts. The HIC-5 gene seems to encode a new member of proteins that contain Zn-finger motifs, and could be a candidate that conveys TGF β1- and H_2O_2 signals to the inhibition of DNA synthesis.

Another important issue on the effect of hydrogen peroxide is what kinds of protein kinases are involved in the above signaling. Induction of early response genes by H_2O_2 was inhibited by a kinase inhibitor H-7. TGFβ1 activates its receptor kinase that phorphorylates Ser or Thr (Lin et al., 1992). Recent reports describe that radiation stimulates src-tyrosine kinase activity (Devary et al., 1992), and that a certain protein phosphatase is inactivated by oxidation (Guy et al., 1993). Identification of protein kinase(s) and phosphatases that are directly regulated by reactive oxygen species and participate in signaling cascades needs further extensive study.

References

Cerutti, P.S. (1985) Prooxidant state and tumor promotion. *Science* 227: 375-381.

Crawford, D., Zbinden, I., Amstad, P., and Cerutti, P. (1988) Oxidant stress induces the proto-oncogenes c-fos and c-myc in mouse epidermal cells. *Oncogene* 3: 27-32.

Devary, Y., Gottlieb, R.A., Smeal, T., and Karin, M. (1992) The mammalian ultraviolet response is triggered by activation of src tyrosine kinases. *Cell* 71: 1081-1091.

Graham, F.L., and Van der Eb, A.J. (1974) *Virology* 52: 456-457.

Guy, G.R., Cairns, J., Ng, S.B., and Tan, Y.H. (1993) Inactivation of redox-sensitive protein phosphatase during the early events of tumor necrosis factor/interleukin-1 signal transduction. *J. Biol. Chem.*, 268 : 2141-2148.

Huot, J., Roy, G., Lambert, H., Chretien, P., and Landry, J. (1991) Increased survival after treatments with anticancer agents of Chinese hamster cells expressing the human Mr27,000 heat shock protein. *Cancer Res.* 51: 5245-5252.

Hyslop, P.A., Hungshaw, D.B., Halsey, W.A., Jr., Schraufstatter, I.V., Sauerherber, R.D., Spragg, R.G., Jackson, J.H., and Cochrane, C.G. (1988) Mechanisms of oxidant-mediated cell injury. *J. Biol. Chem.* 263: 1665-1675.

Imlay, J.A., and Linn, S. (1988) DNA damage and oxygen radical toxicity. *Science* 240: 1302-1309.

Kurihara, N., Ishizuka, S., Kiyoki, M., Hatake, Y., Ikeda, K., and Kumegawa, M. (1986) Effects of 1,25-dihydroxyvitamin D3 on osteoblastic MC3T3-E1 cells. *Endocrinology* 118: 940-947.

Laiho, M., DeCapiro, J.A., Ludlow, J.W., Livingston, D.M., and Massague, J. (1990) Growth inhibition by TGF β linked to suppression of retinoblastoma protein phosphorylation. *Cell.* 62: 175-185.

Lin, H.Y., Wang, X-F., Ng-Eaton, E., Weinberg, R.A., and Lodish, H.F. (1992) Expression cloning of the TGF-β type II receptor, a functional transmembrane serine/threonine kinase. *Cell* 68: 775-785.

Lewis, S.A., Lee, M.G-S., Cowan, N.J. (1985) Five mouse tubulin isotypes and their regulated expression during development. *J. Cell Biol.* 101: 852-861.

May, J.M., and De Haen, C. (1979) The insulin-like effect of hydrogen peroxide on pathways of lipid synthesis in rat adipocytes. *J. Biol. Chem.* 254: 9017-9021.

Maniatis, T., Fritsch, E.F., and Sambrook, J. (1982) Molecular Cloning: *A Laboratory Manual.* Cold Spring Hrbor, NY, Cold Spring Harbor Laboratory.

Mukhwejee, S.P., Land, R.H., and Lynn, W.S. (1978) Endogenous hydrogen peroxide and peroxidative metabolism in adipocytes in response to insulin and sulfhydryl reagents. *Biochem. Pharmacol.* 27: 2589-2594.

O'Farrell, P.H. (1975) A high resolution two-dimensional electrophoresis of proteins. *J. Biol. Chem.* 250: 4007-4021.

Pardee, A.B. (1989) G1 events and regulation of cell proliferation. *Science* 246: 603-608.

Pietenpol, J.A., Stein, R.W., Moran, E., Yaciuk, P., Schlegel, R., Lyons, R.M., Pittelkow, M.R., Munger, K., Howley, P.M., and Moses, H.L. (1990) TGF β1 inhibition of c-myc transcription and growth in keratinocytes is abrogated by viral transforming proteins with pRB binding domains. *Cell* 61: 777-785.

Robertson, F.M., Beavis, A.J., Oberyszyn, T.M., O'Connell, S.M., Dokidos, A., Laskin, D.L., Laskin, J.D., and Reiner, Jr., J.J. (1990) Production of hydrogen peroxide by murine epidermal keratinocytes following treatment with the tumor promoter 12-O-tetradecanoylphorbol-13-acetate. *Cancer Res.* 50: 6062-6067.

Rollet, E., Lavoie, J.N., Landry, J., and Tanguay, R.M. (1992) Expression of drosophila's 27 kDa heat shock protein into rodent cells confers thermal resistance. *Biochem. Biophys. Res. Commun.* 185: 116-120.

Shibanuma, M., Kuroki, T., and Nose, K. (1987) Effect of the protein kinase C inhibitor H-7 and calmodulin antagonist W-7 on superoxide production in growing and resting human histiocytic leukemia cells (U937). *Biochem. Biophys. Res. Commun.* 144: 1317-1323.

Shibanuma, M., Kuroki, T., and Nose, K. (1988) Induction of DNA replication and expression of proto-oncogene c-myc and c-fos in quiescent Balb/3T3 cells by xanthine/xanthine oxidase. *Oncogene* 3: 17-21.

Shibanuma, M., Kuroki, T., and Nose, K. (1990) Stimulation by hydrogen peroxide of DNA synthesis, competence family gene expression and phosphorylation of a specific protein in quiescent Balb/3T3. *Oncogene* 5: 1025-1032.

Shibanuma, M., Kuroki, T., and Nose, K. (1991) Release of H_2O_2 and phosphorylation of 30 kilodalton proteins as early responses of cell cycle-dependent inhibition of DNA synthesis by transforming growth factor β1. *Cell Growth and Diff.* 2: 583-591.

Shibanuma, M., Kuroki, T., and Nose, K. (1992a) Isolation of a gene encoding a putative leucine zipper structure that is induced by transforming growth factor β1 and other growth factors. *J. Biol. Chem.* 267: 10219-10224.

Shibanuma, M., Kuroki, T., and Nose, K. (1992b) Cell-cycle dependent phosphorylation of HSP28 by TGFβ1 and H_2O_2 in normal mouse osteoblastic cells (MC3T3-E1), but not in their ras-transformants. *Biochem. Biophys. Res. Commun.* 187: 1418-1425.

Spector, N.L., Samson, W., Ryan, C., Gribben, J., Urba, W., Welch, W.J., and Nadler, L.M. (1992) Growth arrest of human B lymphocytes is accompanied by induction of the low molecular weight mammalian heat shock protein (Hsp28). *J. Immunol.* 148: 1668-1673.

Szatrowski, T.P. and Nathan, C.F. (1991) Production of large amount of hydrogen peroxide by human tumor cells. Cancer Res. 51: 794-798.

Yamada, T. (1988) Hydrogen peroxide generation in whole rat pancreatic islets: synergistic regulation by cytoplasmic free calcium and protein kinase C. *Biochem. Biophys. Res. Commun.* 155: 569-575.

Zaitsu, K., and Ohkura, Y. (1980) New fluorogenic substrates for horseradish peroxidase; rapid and sensitive assays for hydrogen peroxide and the peroxidase. *Anal. Biochem.* 109: 109-113.

Zantema, A., de Jong, E., Landenoije, R., and van der Eb, A.J. (1989) The expression of heat shock protein hsp27 and a complexed 22-kilodalton protein is inversely correlated with oncogenicity of adenovirus-transformed cells. *J. Virol.* 63: 3368-3375.

Oxidative Stress, Cell Activation and Viral Infection
C. Pasquier et al. (eds)
© 1994 Birkhäuser Verlag Basel/Switzerland

Oxidative Stress and Growth Factor-Mediated Signal Transduction

A. Stern

New York University Medical Center, Department of Pharmacology, New York, NY 10016 USA

Summary

Reversible enzymatic phosphorylation of protein tyrosine residues is considered to be an important mechanism of cellular regulation. Such processes have been associated with the control of cellular proliferation, differentiation and transformation. Enhanced protein tyrosine phosphorylation is essential to the stimulation of cell proliferation by peptide growth factors. Growth factors bind to their specific receptors which contain protein tyrosine kinase activity and initiate a cascade of events which include receptor autophosphorylation on tyrosine residues and tyrosine phosphorylation of other proteins of the signal transduction pathway. Dephosphorylation of tyrosine residues on tyrosine phosphorylated proteins occurs by the action of protein tyrosine phosphatases. Oxidative stress may also play a role in growth factor-mediated signal transduction as evidenced by its effect on cellular growth. We have recently demonstrated that diamide or ascorbic acid has inhibitory effects on protein tyrosine phosphatases in intact murine fibroblasts transfected with the human EGF-receptor. This inhibition can be prevented by EGF, which implies that an interactional link exists between protein tyrosine phosphatase and protein tyrosine kinase activities in the signal transduction pathway of cell growth induced by EGF when cells are exposed to redox agents.

Introduction

Reactive oxygen species and other oxidants in subtoxic concentrations have been implicated as stimulating either pathophysiological or physiological processes (Cerutti, 1991). Sublethal levels of reactive oxygen species stimulate growth of human fascial fibroblasts in cell culture (Murrell et al., 1990). Reactive oxygen species also causes cell growth (Muehlematter et al., 1988) and the expression of the proto-oncogenes c-fos and c-myc in mouse epidermal cells (Crawford et al., 1988). This is accompanied by an increase in DNA strand breaks and poly-ADP ribosylation of chromosomal proteins (Muehlematter et al., 1988) and ribosomal S-6 phosphorylation (Cerutti et al., 1989). Oxidative stress may induce c-jun expression in vanadate treated C127 cells (Yin et al., 1992), the activation of a transcriptional regulator of genes associated with the inflammatory response in T cells (Schreck et al. 1991) and the induction of a gene encoding for a protein tyrosine phosphatase in human skin fibroblasts (Keyse and Emslie, 1992). Redox processes have also been demonstrated for maintaining the activity of a protein kinase in the endoplamic reticulum of a B lymphocyte line (Bauskin et al., 1991), for attenuating the activation of c-src and inducing of c-jun expression in Hela S3 cells by uv light (Devary et al., 1992), for the regulation of gene expression in S. typhimurium (Storz et al., 1990), for maintaining DNA binding to Fos and Jun heterodimer proteins from E. coli (Abate et al., 1990), for regulating protein-RNA interactions in human K562 cell cytoplasmic extracts (Hentze et al., 1989) and for regulating tyrosine phosphorylation in T-cell receptor signaling (Kanner et al., 1992). These studies imply that oxidative stress and redox processes may contribute to growth factor-mediated signal transduction at all levels of the pathway from the growth factor receptor to the nucleus.

Growth Factor Receptors

Growth-related events are generally associated with increased protein tyrosine phosphorylation. This phosphorylation is stimulated by polypeptide growth factors (ex. epidermal growth factor, insulin-like growth factor) that bind at cell surface receptors to initiate signal transduction mechanisms resulting in cell proliferation. Growth factor receptors have a tyrosine kinase activity which is activated by binding of the growth factor to the receptor. All growth factor receptors are composed of an extracellular ligand binding domain, a transmembrane domain for anchoring the receptor to the cell surface and a cytoplasmic domain that contains the tyrosine kinase catalytic region. The ligand binding domain, following binding of the growth factor, undergoes an allosteric dimerization process that results in transmission of the biological signal to the cytoplasmic domain leading to autophosphorylation of tyrosine residues in this domain. The enhancement of tyrosine phosphorylation of the internal domain leads to further transmission of the signal along the signal transduction pathway to initiate nuclear events (Ullrich and Schlessinger, 1990 and References therein).

Tyrosine phosphorylated residues in the cytoplasmic domain are critical and specific sites of interaction for cytoplasmic target proteins that comprise the growth factor-mediated signal transduction pathway. In particular, proteins such as phospholipase C-g, an enzyme that generates diacylglycerol and inositol phosphates that both play a role in the signal transduction pathway, interact at certain phosphorylated tyrosine residues located in the cytoplasmic domain of growth factor receptors. The association of the cytoplasmic proteins with the autophosphorylated tyrosines of the cytoplasmic domain of growth factor receptors results in tyrosine autophosphorylation of these proteins and functional regulation of their activity. The interaction of tyrosine-phosphorylated regions in growth factor receptors and the cytoplasmic proteins is governed by a conserved region of approximately 100 amino acids on the cytoplasmic proteins and is referred to as the src homology 2 domain. These domains are considered as recognition motifs for specific tyrosine-phosphorylated peptide sequences and may contribute to the specificity of receptor signaling (Schlessinger and Ullrich, 1992 and References therein).

Dephosphorylation of autophosphorylated protein tyrosine kinases is catalysed by a family of specific protein tyrosine phosphatases (Alexander, 1990, Fischer et al., 1991). These enzymes are either found intracellularly in the cytosol or as transmembrane proteins with an external domain composed of immunoglobulin-like and fibronectin sequences, a transmembrane anchoring domain and an internal domain with tandem protein tyrosine phosphatase activity. The specific inhibition of protein tyrosine phosphatases can result in increased tyrosine phosphorylation in specific proteins (Alexander, 1990 and References therein), implying that a dynamic relationship exists in cellular tyrosine phosphorylation/dephosphorylation processes.

Oxidative Stress and Insulin Receptor-Mediated Signal Transduction

Hydrogen peroxide exhibits insulinomimetic actions. It partially resembles insulin action by increasing glucose transport and utilization. In contrast to insulin, it only increased glucose metabolism through the hexose monophosphate shunt but not the glycolytic pathway (Czech et al., 1974). Treatment of intact adipocytes or H-35 rat hepatoma cells with hydrogen peroxide resulted in increased tyrosine and serine phosphorylation of the b subunit of the insulin receptor (Koshio et al., 1988). Insulin sensitive Fao hepatoma cells exposed to hydrogen peroxide showed an increase in tyrosine phosphorylation of a number of cellular proteins as determined by use of a antiphosphotyrosine antibody (Heffetz and Zick, 1989). Hydrogen peroxide alone stimulated protein tyrosine phosphorylation, but this took place at later time intervals (20-30 min) than with insulin (45 sec) or insulin plus hydrogen peroxide (5-15 min). Vanadate, an insulinomimetic agent which may have potential as an antidiabetic agent (Heyliger et al., 1985, Brichard et al., 1990), in combination with hydrogen peroxide caused a 10 fold increase in protein tyrosine phosphorylation in Fao cells (Heffetz et al., 1990), an accumulation of insulin-like growth factor II receptor in rat adipocyte membranes (Kadota et al., 1987) and an increase in insulin receptor tyrosine kinase activity in rat adipocytes (Fantus et al., 1989). Peroxo-complexes of vanadium may occur in the reaction of hydrogen peroxide and vanadate (Mimoun et al., 1983, Kadota et al., 1987, Stankiewicz et al., 1991), but which species is responsible for its biological actions is not easily identifiable. This may complicate understanding of how these complexes effect the insulin receptor-mediated signal transduction pathway.

Stimulation of the insulin receptor by insulin results in stimulation of phosphoinositol metabolism via the hydrolysis of inositol containing phospholipids by phospholipase C. Hydrogen peroxide treatment of Fao cells caused minor changes in inositol phosphate production. However, the combination of hydrogen peroxide and vanadate resulted in a large increase in inositol phosphate formation (Zick and Sagi-Eisenberg, 1990). The specific mechanism for this stimulation is not apparent. The enhancement of protein tyrosine phosphorylation by hydrogen peroxide and vanadate may possibly result from the inhibitory action of each of these agents on protein tyrosine phosphatases. Hydrogen peroxide inhibited protein tyrosine phosphatase in Fao cells by 50 % at concentrations used in the above experiments (Heffetz et al., 1990). The protein tyrosine phosphatase activity is fully inhibited in Fao cells with the combination of hydrogen peroxide and vanadate, although vanadate itself had no significant effect. Increased protein tyrosine phosphorylation may result from inhibition of protein tyrosine phosphatases (Bernier et al., 1987 et al., Garcia-Morales et al., 1990). Nevertheless, vanadate exhibits complex behavior in cells, in that it can inhibit protein tyrosine phosphatase activity while also activating protein tyrosine kinase activity of the insulin receptor (Tamura et al., 1984).

Redox-Cycling Agents and Growth Factor-Mediated Signal Transduction

Redox-cycling agents which generate reactive oxygen species are known mitogens and tumor promotors (Cerutti, 1985, Troll and Wiesner, 1985). Among these agents, redox-cycling naphthoquinones have received particular attention regarding their cytotoxic effects in hepatocytes which are manifest by changes in calcium homeostasis (Thor et al., 1982, Bellomo et al., 1982), sulfhydryl oxidation (Thor et al., 1982, Gant et al., 1988) and alterations in the redox status of pyridine nucleotides (Frei et al., 1986, Stubberfield and Cohen, 1989).

The cytosolic activity of rat hepatocyte protein kinase C, a component of growth factor-mediated signal transduction pathways is increased by redox-cycling naphthoquinones (Kass et al., 1989). This stimulation in activity may be related to a delicate balance in critical thiol/disulfide groups in the enzyme. Menadione, a naphthoquinone vitamin K analog, did not result in a change in inositol phosphates in rat hepatocytes, but inhibited an increase in inositol phosphates commonly seen following stimulation by angiotensin II (Bellomo et al., 1987), an activator of protein tyrosine phosphorylation (Huckle et al., 1992).

Tyrosine-specific phosphorylation was increased in an artificial substrate, poly-glutamate-tyrosine (4:1) by several naphthoquinones, including menadione (Chan et al., 1986). Vitamin K_5 increased phosphorylation of a partially purified protein from rat hepatocyte cell membranes that was also phosphorylated by insulin. Followup studies, using thin layer chromatography on membrane fractions from rat hepatocytes exposed to naphthoquinones revealed both increased phosphotyrosine residues and increased phosphoinositol phosphorylation. Antiphosphotyrosine antibodies were used to identify a substantial increase in phosphoinositol kinase activity in rat hepatocyte membranes (Chen et al., 1990). These observations indicate that redox-cycling may play a contributory role in the regulation of growth factor-mediated signal transduction pathways and suggests a potential role for oxidative metabolism in this process.

Redox Regulation of Growth Factor-Mediated Signal Transduction

Diamide (Monteiro et al., 1991) and ascorbic acid at physiological concentrations (manuscript in preparation) inhibited protein tyrosine phosphatase activity in murine fibroblasts transfected with epidermal growth factor receptors (HER 14 cells). The substrate for measuring protein tyrosine phosphatase activity was immunoprecipitated tyrosine phosphorylated epidermal growth factor receptor. Incubation of intact HER 14 cells with EGF prior to addition of ascorbic acid or diamide prevented their inhibitory effect on protein tyrosine phosphatase activity implying that activation of the EGF receptor could have consequences on protein tyrosine phosphatase activity. The reversal of diamide-inhibited protein tyrosine phosphatase activity by b-mercaptoethanol indicates that sulfhydryl groups are important in the intracellular behavior of the enzyme. This is consistent with the observation that protein tyrosine phosphatase activity depends on specific cysteine residues in its catalytic domain (Tonks et al., 1988).

Conclusion

Growth factor-mediated signal transduction appears to be affected by redox processes and oxidative stress. Reactive oxygen species and redox cycling agents have mitogenic activity. They alter protein tyrosine phosphorylation through actions on both protein tyrosine kinase and protein tyrosine phosphatase activities. Oxidative stress and redox processes may contribute to growth factor-mediated signal transduction pathways by a complex mechanism involving tyrosine phosphorylation /dephosphorylation reactions.

40

Acknowledgements:

Supported by the National Institutes of Health grant ES 03425.

References:

Abate, C., Patel, L., Rauscher III, F.J., and Curran, T. (1990) Redox regulation of Fos and Jun DNA-binding activity in vitro. *Science* 249: 1157-1160.

Alexander, D.R. (1990) The role of phosphatases in signal transduction. *The New Biologist* 2: 1049-1060.

Bauskin, A.R, Alkalay, I., and Ben-Neriah, Y. (1991) Redox regulation of a protein tyrosine kinase in the endoplasmic reticulum. *Cell* 66: 685-696.

Bellomo, G., Jewell, S.A., and Orrenius, S. (1982) The metabolism of menadione impairs the ability of rat liver mitochondria to take up and retain calcium. *J. Biol. Chem.* 257: 11558-11562.

Bellomo, G., Thor, H., and Orrenius, S. (1987) Alterations in inositol phosphate production during oxidative stress in isolated hepatocytes. J. Biol. Chem. 262: 1530-1534.

Bernier, M., Laird, D.M.,and Lane, M.D. (1987) Insulin-activated tyrosine phosphorylation of a 15-kilodalton protein in intact 3T3-L1 adipocytes. *Proc. Natl. Acad. Sci.* (USA) 84: 1844-1848.

Brichard, S.M., Bailey, C.J., and Henquin, J.-C. (1990) Marked improvement of glucose homeostasis in diabetic ob/ob mice given oral vanadate. *Diabetes* 39: 1326-1332.

Cerutti, P. (1985) Prooxidant states and tumor promotion. *Science* 227: 375-381.

Cerutti, P. Larsson, R. Krupitza, G., Muehlematter, D., Crawford, D., and Amstad, P. (1989) Pathophysiological mechanisms of active oxygen. *Mutation Res.* 214: 81-88.

Cerutti, P.A. (1991) Oxidant stress and carcinogenesis. Eur. J. Clin. Invest. 21: 1-5.

Chan, T.M., Chen, E., Tatoyan, A., Shargill, N.S., Pleta, M. and Hochstein, P. (1986) Stimulation of tyrosine-specific protein phosphorylation in the rat liver plasma membrane by oxygen radicals. *Biochem. Biophys. Res. Comm.* 139: 439-445.

Chen, Y., Yang, D.-C., Brown, A.B., Jeng, Y., Tatoyan, A. and Chan, T.M. (1990) Activation of a membrane-associated phosphatidylinositol kinase through tyrosine-protein phosphorylation by naphthoquinones and orthovanadate. *Arch. Biochem. Biophys.* 283: 184-192.

Crawford, D. Zbinden, I., Amstad, P. and Cerutti, P. (1988) Oxidant stress induces the proto-oncogenes c-fos and c-myc in mouse epidermal cells. *Oncogene* 3: 27-32.

Czech, M.P., Lawrence, Jr., J.C. and Lynn, W.S. (1974) Evidence for electron transfer reactions involved in the Cu^{2+}-dependent thiol activation of fat cell glucose utilization. *J. Biol. Chem.* 249: 1001-1006.

Devary, Y., Gottlieb, R.A., Smeal, T. and Karin, M. (1992) The mammalian ultraviolet response is triggered by activation of *src* tyrosine kinases. *Cell* 71: 1081-1091.

Fantus, I.G., Kadota, S., Deragon, G., Foster, B., and Posner, B.I. (1989) Pervanadate [Peroxide(s) of Vanadate] mimics insulin action in rat adipocytes via activation of the insulin receptor tyrosine kinase. *Biochemistry* 28: 8864-8871.

Fischer, E.H., Charbonneau, H. and Tonks, N.K. (1991) Molecular characterization of protein tyrosine phosphatases. *Science* 253: 401-406.

Frei, B., Winterhalter, K.H., and Richter, C. (1986) Menadione (2-methyl-1,4- naphthoquinone)-dependent chemistry, enzymatic redox cycling and calcium release by mitochondria. Biochemistry 25: 4438-4443.

Gant, T.W., Rao, D.N.R., Mason, R.P. and Cohen, G.M. (1988) Redox cycling and sulfhydryl arylation. Their relative importance in the mechanism of quinone cytotoxicity to isolated hepatocytes. *Chem. Biol. Interact.* 65: 157-163.

Garcia-Morales, P., Minami, Y., Luong, E., Klausner, R.D., and Samelson, L.E. (1990) Tyrosine phosphorylation in T cells is regulated by phosphatase activity: Studies with phenylarsine oxide. *Proc. Natl. Acad. Sci.* (USA) 87: 9255-9259.

Heffetz, D., Bushkin, I., Dror, R. and Zick, Y. (1990) The insulinomimetic agents H_2O_2 and vanadate stimulate protein tyrosine phosphorylation in intact cells. *J. Biol. Chem.* 265: 2896-2902.

Heffetz, D., and Zick, Y. (1989) H_2O_2 potentiates phosphorylation of novel putative substrates for the insulin receptor kinase in intact Fao cells. *J. Biol. Chem.* 264: 10126-10132

Hentze, M.W., Rouault, T.A., Harford, J.B., and Klausner, R.D. (1989) Oxidation-reduction and the molecular mechanism of a regulatory RNA-protein interaction. *Science* 244: 357-359

Heyliger, C.E., Tahiliani, A.G., and McNeill, J.H. (1985) Effect of vanadate on elevated blood glucose and depressed cardiac performance of diabetic rats. *Science* 227: 1474-1477.

Huckle, W.R., Dy, R.C. and Earp, H.S. (1992). Calcium-dependent increase in tyrosine kinase activity stimulated by angiotensin II. *Proc. Natl. Acad. Sci.* (USA) 89: 8837-8841

Kadota, S., Fantus, I.G., Deragon, G., Guyda, H.J., and Posner, B.I. (1987) Stimulation of insulin-like growth factor II receptor binding and insulin receptor kinase activity in rat adipocytes: Effects of vanadate and H_2O_2. *J. Biol. Chem.* 262: 8252-8256

Kadota, S., Fantus, I.G., Deragon, G., Guyda, H.J., Hersh, B. and Posner, B.I. (1987) Peroxide(s) of vanadium: A novel and potent insulin-mimetic agent which activates the insulin receptor kinase. *Biochem Biophys. Res. Commun.* 147: 259-266

Kanner, S.B., Kavanagh, T.J., Grossmann, A., Hu, S.-L., Bolen, J.B., Rabinovitch, P.S. and Ledbetter, J.A. (1992) Sulfhydryl oxidation down-regulates T-cell signaling and inhibits tyrosine phosphorylation of phospholipase C1. *Proc. Natl. Acad. Sci.* (USA), 89:300-304

Kass, G.E.N., Duddy, S.K., and Orrenius, S. (1989) Activation of hepatocyte protein kinase C by redox-cycling quinones. *Biochem. J.* 260: 499-507

Keyse, S.M. and Emslie, E.A. (1992). Oxidative stress and heat shock induce a human gene encoding a protein-tyrosine phosphatase. *Nature* 359: 644-647

Koshio, O., Akanuma, Y. and Kasuga, M. (1988). Hydrogen peroxide stimulates tyrosine phosphorylation of the insulin receptor and its tyrosine kinase activity in intact cells. *Biochem J.* 250: 95-101.

Mimoun, H., Saussine, L., Daire, E., Postel, M., Fischer, J. and Weiss, R. (1983). Vanadium (V) peroxo complexes. New versatile biomimetic reagents for epoxidation of olefins and hydroxylation of alkanes and aromatic hydrocarbons. *J. Am. Chem. Soc.* 105: 3101-3110.

Monteiro, H.P., Ivaschenko, Y. Fischer, R. and Stern, A. (1991). Inhibition of protein tyrosine phosphatase activity by diamide is reversed by epidermal growth factor in fibroblasts. *FEBS Lett.* 295: 146-148

Muehlematter, D., Larsson, R. and Cerutti, P. (1988). Active oxygen induced DNA stand breakage and poly ADP-ribosylation in promotable and non-promotable JB6 mouse epidermal cells. *Carcinogenesis* 9: 239-245.

Murrell, G.A.C., Francis, M.J.O and Bromley, L. (1990) Modulation of fibroblast proliferation by oxygen free radicals. *Biochem. J.* 265: 659-665.

Schlessinger, J. and Ullrich, A. (1992) Growth factor signaling by receptor tyrosine kinases. *Neuron* 9: 383-391.

Schreck, R., Rieber, P. and Baeuerle, P.A. (1991) Reactive oxygen intermediates as apparently widely used messengers in the activation of the NF-kB transcription factor and HIV-1. *EMBO J.* 10: 2247-2258.

Stankiewicz, P.J., Stern, A., and Davison, A.J. (1991) Oxidation of NADH by vanadium: Kinetics, effects of ligands and role of H_2O_2 or O_2. *Arch. Biochem. Biophys.* 287: 8-17.

Stubberfield, C.R. and Cohen, G.M. (1989) Interconversion of NAD(H) and NADP(H). A cellular response to quinone-induced oxidative stress in isolated hepatocytes. *Biochem. Pharm.* 38: 2631-1637.

Storz, G., Tartaglia, L.A. and Ames, B.N. (1990) Transcriptional regulator of oxidative stress-inducible genes: Direct activation by oxidation. *Science* 248: 189-194.

Tamura, J., Brown, T.A., Whipple, J.H., Fugita-Yamaguchi, Y., Dubler, R.F., Cheng, K. and Larner, J. (1984) A novel mechanism for the insulin-like effect of vanadate on glucogen synthase in rat adipocytes. *J. Biol. Chem.* 259: 6650-6658.

Thor, H., Smith, M.T., Hartzell, P., Bellomo, G., Jewell, S.A. and Orrenius, S. (1982) The metabolism of menadione (2-methyl-1,4-naphthoquinone) by isolated hepatocytes *J. Biol. Chem.* 257: 12419-12425.

Tonks, N.K., Diltz, C.D., and Fischer, E.H. (1988) Purification of the major protein-tyrosine-phosphatases of human placenta. *J. Biol. Chem.* 263: 6722-6730.

Troll, W. and Wiesner, R. (1985) The role of oxygen radicals as a possible mechanism of tumor promotion. *Ann. Rev. Pharm. Toxicol.* 25: 509-528.

Ullrich, A.., and Schlessinger, J. (1990) Signal transduction by receptors with tyrosine kinase activity. *Cell* 61: 203-212.

Yin, X., Davison, A.J., and Tsang, S-S (1992) Vanadate-induced gene expression in mouse C127 cells: Roles of oxygen derived active species. *Biochem. Cell Biol.* 115: 85-96.

Zick, Y. and Sagi-Eisenberg, R. (1990) A combination of H_2O_2 and vanadate concomitantly stimulates protein tyrosine phosphorylation and polyphosphoinositide breakdown in different cell lines. *Biochemistry* 19: 10240-10245.

Oxidative Stress, Cell Activation and Viral Infection
C. Pasquier et al. (eds)
© 1994 Birkhäuser Verlag Basel/Switzerland

Cellularly generated active oxygen species as signals in the activation of tumour cell growth

R.H. Burdon

University of Strathclyde, Dept. of Bioscience & Biotechnology, Glasgow G4 0NR, Scotland, UK

Summary

It has become clear that superoxide and hydrogen peroxide at low levels can stimulate growth, or growth responses, in a variety of normal and tumour cell types when added exogenously. It is also evident that these active oxygen species are released from, or generated within, most of these cell types. Whereas 'normal' cells appear to require the stimulus of serum factors or phorbol esters, tumour cells seem to release superoxide constitutively. Experiemnts with extracellularly added superoxide dismutase and catalase suggest that superoxide radicals and hydrogen peroxide may have important biological roles in intra- and intercellular 'messengers', or 'signals', promoting cell proliferation and maintaining cell viability.
With regard to possible mechanisms, suggestions have been made that superoxide, and hydrogen peroxide, might function as mitogenic stimuli through biochemical processes common to cell growth factors. It is possible that they 'signal' by covalent modification of key cellular growth regulatory proteins on the basis of redox potenital, thus setting redox states appropriate fo cell growth responses. Altenatively they may oxidatively inactivate extracellular proteases thereby facilitating normal growth factor signalling. In the case of tumour cells reduced levels of antioxidant enzymes may more readily permit them to adjust their redox status in such a way that growth signal pathways are continuously in an up-regulated state.

The cellular regulatory redox paradigm

Previous results of ours, and others, have indicated that *low concentrations* (10 nM-1µM) of superoxide anions and hydrogen peroxide, can stimulate growth, or growth responses, in a variety of mammalian cell types when added exogenously to the culture medium. These include oncogene transformated hamster (BHK-21/PyY) and rat (RFAGT1) fibroblasts (Burdon & Rice-Evans, 1989; Burdon et al, 1990), mouse epidermal cells (JB6) (Crawford et al, 1988; Amstad et al, 1992), Balb/3T3 cells (Shibanuma et al, 1988; Shibanuma et al, 1990, mouse osteoclastic cells (MC3T3) (Nose et al, 1991) human primary fibroblasts (Murell et al, 1990), HeLa cells, human amnion cells (Ikebuchi et al, 1991) and human histiocytic leukaemia cells (U937) (Shibanuma et al, 1988). These derived oxygen species stimulate the expression of early growth regulated genes such as *c-fos* and *c-myc* . Such observations have led to suggestions that superoxide and hydrogen peroxide may function as, or augment, mitogenic stimuli through biochemical processes common to natural growth factors. The growth related effects of superoxide are extremely rapid. Intracellular pH is increased in human amnion cells within 10 sec (Ikebuchi et al, 1991) and this is followed by an increase in intracellular Ca^{2+} in 20-40 sec.

Studies on c-fois induction have suggested that protein kinase C is important and in mouse epidermal cells it is activated and translocated following exposure of the cells to superoxide (Amstad et al, 1992). In vitro activation of protein kinase C can be achieved by selective oxidative modification of the regulatory domain (Gopalakrishna and Anderson, 1989). Besides the possibility of redox modification of protein kinase C, redox regulation of a protein kinase of the endoplasmic reticulum has also been observeed (Buskin et al, 1991). In the context of early growth response gene activation it may also be significant that the binding to DNA of the transcription fractors Fos and Jun, as well as NF-κB, can be modulated by oxidation-reduction in vitro (Toledano & Leonard, 1991; Abate et al, 1990; Schreck et al, 1990). Importantly the transcription factor NF-κB can also be activated and translocated within intact cells exposed to hydrogen peroxide (Baeurle & Schreck, 1992). Such observations raise the possibility of a cell regulatory redox paradigm whereby the activity of components of signal transduction pathways can depend on their oxidation-reduction status.

Extracellularly released superoxide and hydrogen peroxide as autocrine signals

A growing number of cell types, including those already mentioned, release low levels of superoxide of hydrogen peroxide. In the case of human fibroblasts release involves the activity of a plasma membrane NADPH-oxidase and requires the stimulus of cytokines (Meier et al, 1989; Meier et al, 1990). Superoxide release from human epithelial cells also requires cytokines (Matsuyama & Ziff, 1986). Rabbit articular chondrocytes release hydrogen peroxide when stimulated by cytokines (Tiku et al, 1990) and Balb/3T3 cells require the presence of the growth factor PDGF (Shibanuma et al, 1990). Thus the potential for superoxide, or hydrogen peroxide to act in an autocrine fashion appears closely tied to the ability of cytokines, or growth factors, to stimulate NADPH-oxidase catalysed extracellular release.

In contrast a variety of human tumour cells (melanoma, colon carcinoma, pancreatic carcinoma, neuroblastoma, ovarian and breast carcinoma) are all reported to release large amounts of hydrogen peroxide (50 nmol/hr/10^6 cells) without any specific growth stimulus (Szatrowski & Nathan, 1991). This constitutive release also appears to involve a form of NADPH-oxidase. The fact that human tumour cells do not specifically require growth factor, or cytokine stimulation for hydrogen peroxide release may be an important phenotypic distinction between 'normal' and tumour cells. We also find that oncogene transformed hamster (BHK-21/PyY) (Burdon, 1992a) and rat (RFAGT1) cells, as well as HeLa cells, all release superoxide constitutively.

That an autocrine growth regulatory mechanism involving superoxide and hydrogen peroxide is operational is suggested from experiments in which the enzymes superoxide dismutase and catalase are added exogenously (at 100 µg/ml) to the culture medium of oncogene transformed hamster (BHK-21/PyY) (Burdon, 1992a &b; Burdon et al, 1992) or rat (RFAGT1) fibroblasts (Burdon, 1992a). These enzymes both very significantly reduce cell proliferation (superoxide dismutase by around 50-60 % , and catalase by around 80-90 %). We have observed similar outcomes when these enzymes were added to cultures of human melanoma cells, human ovarian carcinoma cells as well as HeLa cells, although the entent of down-regulation can vary a little with cell type. Importantly the growth inhibitory effects of superoxide dismutase in combination with catalase were not additive suggesting that both superoxide and hydrogen peroxide are each individually imortant as autocrine 'signals'.

Intracellular generation of superoxide

Uni-electron reduction of oxygen generates superoxide anions and an important source is cellular electron transport chains. Although components of these chains pass the bulk of electrons onto the next component of the chain, some appear to 'leak' electrons to oxygen (Boveris & Chance, 1973). In order to assess the level of any intracellularly generated superoxide we investigated the use of a tetrazolium salt, 3-(4,5-dimethylthiazol-2-yl)-2,5-diphenylformazanbromide (MTT). Superoxide release from cells has often been detected by following the reduction of nitroblue tetrazolium (Meier et al, 1989) On the other hand MTT is able to enter cells and by virtue of intracellular reduction to blue formazan offers a means of assessing intracellular generation of superoxide. MTT however is also known to interact directly with components of the mitochondiral respiratory chain (Slater et al, 1963) and therefore its reduction in part also reflects normal mitochondiral electron transport. Moreover it could be argued that any superoxide generated in the mitochondria would be rapidly disproportionated by the Mn-superoxide dismutase of mitochondria. Despite these limitations it is significant that an in vivo inhibitor of the cytosolic Cu,Zn-superoxide dismutase, the copper chelator diethyldithocarbamate (DDC) (Heikkila et al, 1976), greatly increases the level of MTT reduction within oncogene transformed hamster fibroblasts (BHK-21) (Burdon & Rice-Evans, 1989) and human carcinoma cells, HeLa (Table I). That a part of the MTT reduction is due to intracellular superoxide generation is supported by the observation that a low molecular weight lipophilic mimic of superoxide dismutase, copper II-(3,5 diisopropylsalicylate)$_2$ (CuDIPS), (Enger & Kensler, 1985), could also inhibit intracellular superoxide generation in BHK-21 and HeLa cells, another source of the intracellular superoxide leading to intracellular reduction of MTT may be xanthine oxidase (Halliwell & Gutteridge, 1989). Studies with inhibitors of xanthine oxidase such as allopurinol and oxypurinol (Chalmers et al, 1968), like CuDIPS, lead to over 30 % reduction in levels of intracellular reduction of MTT in HeLa cells (Table I).

Table I : Intracellular generation of MTT-formazan in HeLa cells.
Triplicate monolayer cultures of HeLa (0.5 x 106 cells per 3.5 cm petri dish) were established in 2 ml Glasgow modification of Eagle's minimal essential medium supplemented with 10 % fetal calf serum by growth at 37°C for 24 hr. The medium was then replaced, with the indicated additions and after 2 hr, 0.25 mg MTT [3-(4,5-dimethylthizol-2-yl)-2,5-diphenyltetrazolium bromide] added and incubation continued at 37°C for 30 min. The media was removed and the remaining monolayers extracted with 2 ml DMSO. The absorbance of DMSO extracts containing the blue formazan (6) from triplicate cultures was determined at 570 nm and the results presented are means ± s.d. (n = 3).

Expt	Additions	Blue formazan accumulation within cells (A_{570}) ± s.d.
1	0.1 % ethanol	0.201 ± 0.010
	0.1 % ethanol ± CuDIPS (10µM)	0.157 ± 0.007
2	none	0.407 ± 0.042
	diethyldithiocarbamate (1mM)	0.606 ± 0.020
	oxypurinol (10 µM)	0.232 ± 0.029
	allopurinol (10 µM)	0.180 ± 0.037

Whilst tumour necrosis factor-α (TNF-α) has no toxic effect towards HeLa cells, it is highly toxic to murine L-929 cells. In these sensitive cells, it is clear that TNF-α specifically induces extensive mitochondrial superoxide generation (Heennet et al, 1993). Whilst this may contribute to the toxicity TNF-α in sensitive cells, it highlights the possibility of considerable variation between cell types with regard to the source and extent of intracellular superoxide generation, excessive amounts being potentially lethal.

Redox mechanisms and the source of active oxygen species

The possibility that specific oxygen radicals and radical derived species function as growth regulatory agents for a considerable variety of normal and tumour cell types may have wide significance. The observed effects on hamster and rat fibroblasts of exogenously added superoxide dismutase and catalase suggest that superoxide and hydrogen peroxide not only have growth stimulatory effects but their presence, at least at low levels, appears necessary for cell viability as judged by trypan blue staining (Burdon, 1992a).

Whilst exogenously added superoxide dismutase (or catalase) brings about the 'down-regulation' of growth responses, these experiments however do not unequivocally support the involvement of only extracellularly released superoxide (or hydrogen peroxide) as growth regulators. Intracellularly generated superoxide could also be an important growth signal. For example, there are reports of the endocytosis of extracellularly added superoxide dismutase in numerous cell types (Li et al, 1992; Saez et al, 1987; Kyle et al, 1988; Beckman et al, 1988). Indeed the pathways followed by superoxide dismutase within cells after uptake has been followed and it appears that the enzymes can remain in lysosomes for a considerable time. Beckman et al (1988) have suggested that superoxide anions generated within cells may have access through anion channels present on lysosomal membranes. As Table II shows, exposure of HeLa cells to the low molecular weight lipophilic superoxide dismutase mimic, CuDIPS, like superoxide dismutase itself, leads to the down- regulation of cell proliferation. Thus both extracellular and intracellular superoxide could be involved in growth promoting processes. In this context it is perhaps significant that in HeLa cells use of the xanthine oxidase inhibitors oxypurinol and allopurinol, which depress the intracellular generation of superoxide as discussed in the previous section, can also reduce celll growth (see Table II). Clearly a role for intracellularly generated superoxide in cell growth promotion should not be overlooked. Recent data from mitogenically activated T-lymphocytes also implicate intracellular reactive oxygen species in the induction of responses associated with cell division (Hunt & Fragonas, 1992).

Table II : Growth of HeLa cells
Triplicate monolayer cultures fo HeLa (0.5 x 106 cells per 3.5 cm petri plate) were established as in Table I. The medium was replaced together with the additions indicated and incubation continued at 37°C for 3 days. The live cells in each plate were counted and the results presented are the means of triplicate experiments ± s.d. (n = 3).

Expt	Additions	Live cells/plate after 3 days x 10^6 ± s.d.
1	none	2.44 ± 0.02
	oxypurinol (10 μM)	1.61 ± 0.06
	allopurinol (10 μM)	2.18 ± 0.09
2	0.1 % ethanol	1.38 ± 0.02
	0.1 % ethanol + CuDIPS (10 μM)	1.10 ± 0.03
	0.1 % ethanol + CuDIPS (10 μM) + oxypurinol (10 μM)	1.17 ± 0.03

Growth control involving oxidative inactivation of extracellular proteinase inhibitors

Whilst a redox regulatory paradigm has attractions, it raises questions of specificity. A feature of growth factor - growth receptor interaction is its inherent specificity with regard to individual cell types. Hydrogen peroxide for example with its ability to traverse cell membranes and its potenital to modify oxidatively a number of regulatory proteins, does not, at first sight, appear to afford this level of specificity with regard to cell-type. In view of the many components that consittute the cellular pathways, and second messenger systems, whereby growth signals are normally transduced, it would be perhaps surprising that such delicately balanced multicomponent systems could be accurately activated by simple direct exposure to superoxide, or hydrogen peroxide. This of course may not be a critical issue in the context of cancer cell proliferation, but inappropriate proliferation of normal cells would be undesirable.

It is possible to approach this issue of specificity from another direction. This related to the extreme sensitivity of serum alpha-1 antitrypsin to oxidative inactivation (Wasil et al, 1987). It has been shown that cultured fibroblasts release a growth related proteinase which is believed to be involved in the local remodelling of the cell coat, or glycocalyx, to permit the release, or activity, of certain normal growth factors (Tse & Scott, 1991). Antibodies against this proteinase inhibit cell proliferation. Several macromolecular anti-proteinases are also active against this growth regulated proteinase and as a result block the action of several growth factors, but crucially not the intracellular signal transduction events linked to growth factor action. Whilst supplementation of serum with alpha-1 antitrypsin depresses the growth of BHK-21 cells, addition of hydrogen peroxide at low concentrations partly counters these effects (Burdon, 1992b).

Thus as an alternative, the signalling of growth responses involving released superoxide, or hydrogen peroxide, may be mediated through the oxidative inactivation of serum proteinase inhibitors allowing serum proteinases to remodel the cell surface, or glycocalyx, and thereby facilitate, or modulate, the action of normal growth factors. Thus the necessary growth specificity is retained. Such a mechanism may also have considerable relevance to the processes of metastasis by facilitating the action of tumour associated proteases.

In summary it is possible that redox growth regulation involving cellular regulatory molecules (such as protein kinase C and transcription factors) as well as extracellular antiproteinase activity, occur simultaneously. This is shown schematically in Figure 1A. On the other hand whilst extracellularly released active oxygen species are potentially important especially in relation to the redox inactivation of extracellular antiprotienases, the contribution to redox growth regulation of intracellularly generated active oxygen species should not be ignored. This alternative is shown schematically in Figure 1B. Further experimentation will be required to evaluate the relative importance for cell growth promotion of extracellularly released active oxygen species as compared with species generated intracellularly. This balance may vary considerably with cell type and circumstance.

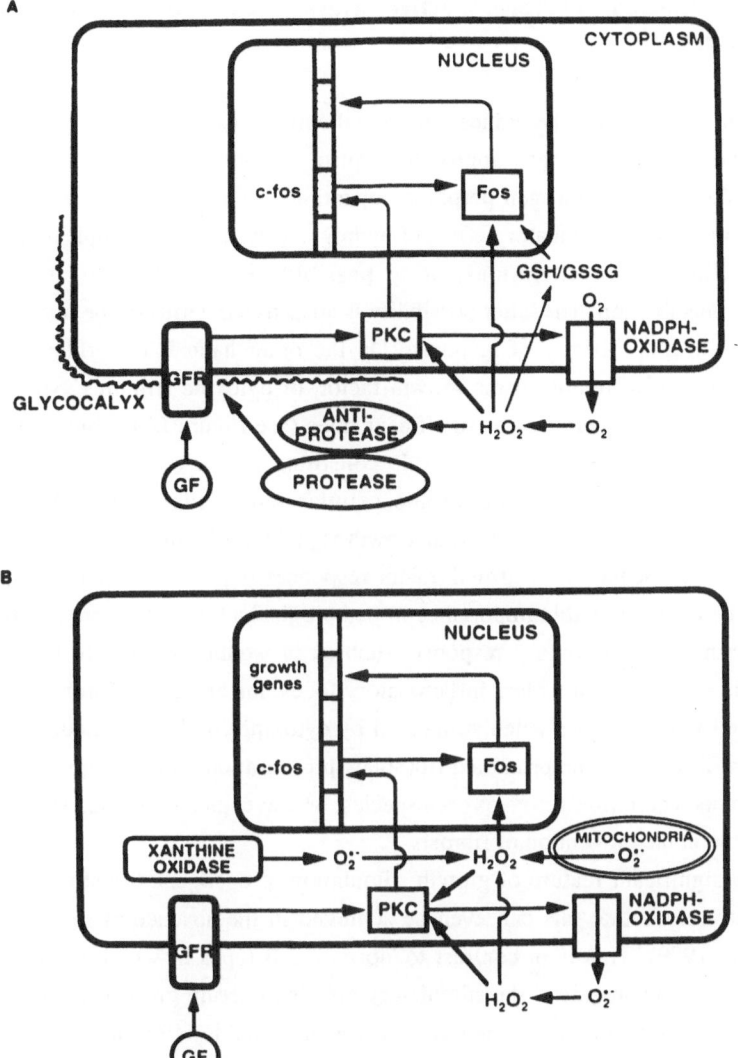

Figure 1 : Schematic diagrams illustrating (A) cellular release of active oxygen species and potential cell regulatory targets. (B) intracellular sources of active oxygen species and potential cell regulatory target [PKC, protein kinase C; growth factor; GFR, growth factor receptor]

A comparison of the possible roles of active oxygen species in the proliferation of 'normal' and tumour cell types

Whilst growth responses can be induced in normal cells by exogenous active oxygen species, a crucial observation is that these appear to require the additional presence of some serum components. Essentially hydrogen peroxide appears to augment the effects of natural growth factors. Thus the growth stimulatory effect of hydrogen peroxide for example may be ascribed to some intercellular redox alterations, or to possible extracellular effects at the level of antiproteinase inactivation. The latter possibility is attractive in terms of the necessity for normal cells to maintain a specificity of response. On the other hand it is perhaps significant that 'normal' cells appear to require some growth factor, or cytokine stimulus, before they release superoxide or hydrogen peroxide at significant rates (see Figure IA). Thus it may be that the released superoxide and hydrogen peroxide constitute a type of autocrine system with the general purpose of setting an appropriate cellular redox 'tone' that will permit optimal functioning of the cell proteins involved in growth signal transduction.

The concept of augmentation of growth factor responses in processes involving active oxygen species may be of considerable importance in pathological situations where cell proliferation is associated with an inflammatory response, such as in wound healing. In fibrotic conditions where there is an absence of a clear inflammatory focus, the ability of fibroblasts themselves to release active oxygen species when stimulated by cytokines may be particularly important for augmentation of the cytokine promoted fibroblast proliferation that is characteristic of fibrosis. Moreover agents that inhibit active oxygen species, or scavenge the released species, may have a therapeutic role in the prevention of fibrosis.

A potentially significant feature of growth stimulation of oncogene transformed fibroblasts by hydrogen peroxide is that this can even be achieved in the absence of serum growth factors (Burdon et al, 1989). This is in contrast to 'normal' cell-types in which it appears that active oxygen species act to 'augment' the stimulatory effects of serum growth factors. It may be that the constitutively released superoxide and hydrogen peroxide is sufficient in tumour cells to adjust the cellular redox status in such a way that growth signal pathways are continuously in an "up-regulated" state.

Thus reduced levels of intracellular superoxide dismutase, or catalase, that are a feature of many tumour cell types (Oberley & Oberley, 1988) may permit the accumulation of superoxide, or hydrogen peroxide, within tumour cells. Taken together with data suggesting that oncogene transformed cells respond significantly better to the growth promoting effects of low levels of superoxide or hydrogen peroxide, it may be the reduced levels of antioxidant enzymes contribute to the achievement of a cellular redox state which facilitates the advantageous growth of neoplastic cells in the presence of superoxide and hydrogen peroxide, either as part of a constitutively active autocrine system, or from adjacent inflammatory cells.

References

Abate, C., Patel, L., Rauscher, F.J. and Curran, T. (1990) Redox regulation for Fos and Jun DNA-binding activity in vitro. *Science* 249: 1157-1161.

Amstad, P.A., Krupitza, G. and Cerutti, P. (1992) Mechanisms of c-fos induction by active oxygen. *Cancer Res.* 52: 3952-3960.

Baeuerle, P.A. and Schreck, R. (1992) A role of reactive oxygen intermediates in the activation of the transcription factor NF-kappa B. *Biol. Chem. Hoppe-Seyler* 373: 740.

Beckman, J.S., Minor, R.L., White, C.W., Repine, J.E., Rosen, G.M. and Freeman, B.A. (1988) Superoxide dismutase and catalase conjugated to polyethylene glycol increases endothelial enzyme activity and oxidant resistance. *J. Biol. Chem.* 263: 6884-6892.

Boveris, A. and Chance B. (1973) The cellular production of hydrogen peroxide. *Biochem. J.* 128: 617-630.

Burdon, R.H. (1992a) Cell proliferation and oxidative stress: basis for anticancer drugs. *Proc. Roy. Soc. Edin.* 99B: 169-176.

Burdon, R.H. (1992b) Released active oxygen species as intercellular signals: their role in regulation of normal and tumour cell proliferation. *Biol. Chem. Hoppe-Seyler* 373: 739-740.

Burdon, R.H., Gill, V. and Rice-Evans, C. (1989) Cell proliferation and oxidative stress. *Free Rad. Res. Commun.* 7: 149-159.

Burdon, R.H., Gill, V. and Rice-Evans, C. (1990) Oxidative stress and tumour cell proliferation. *Free Rad. Res. Commun.* 11: 65-76

Burdon, R.H., Gill, V. and Rice-Evans, C (1992) Active oxygen species in the promotion and suppression of tumour cell growth. In: K.J.A. Davies (eds): *Oxidative Damage and Repair*, Pergamon Press, Oxford, pp. 791-795.

Burdon, R.H. and Rice-Evans C. (1989) Free radicals and the regulation of mammalian cell proliferation. *Free Rad. Res. Commun.* 6: 346-348.

Buskin, A.R., Alkalay, I. and Ben-Neriah, Y. (1991) Redox regulation of a protein kinase in the endoplasmic reticulum. *Cell* 66: 685-696.

Chalmers, R.A., Kroner, H., Palmer, G., Scott, J.T. and Watts, R.W.E. (1968) A comparative study of the xanthine oxidase inhibitors allopurinol in man. *J. Clin. Sci.* 35: 353-362.

Crawford, D., Zbinden, I., Amstad, P. and Cerutti, P. (1988) Oxidant stress induces the protooncogenes c-fos and c-myc in mouse epidermal cells. *Oncogene* 3: 27-32.

Enger P.A. and Kensler, T.W. (1985) Effects of a biomimetic superoxide dismutase on complete and multistage carcinogenesis in mouse skin. *Carrcinogenesis* 6: 1167-1172.

Gopalakrishna, R. and Anderson, W.B. (1989) Ca^{2+} and phospholipid-independent activation of protein kinase C by selective oxidative modification of the regulatory domain. *Proc. Natl. Acad. Sci. USA* 86: 6758-6762.

Halliwell, B. and Gutteridge, J.M.C. (1989) *Free Rad. Biol. Med.*, 2nd Ed., Clarendon Press, Oxford.

Heikkila, R.E., Cabbat, F.S. and Cohen, G. (1976) In vivo inhibition of superoxide dismutase in mice by diethyldithiocarbamate. *J. Biol. Chem.* 251: 2182-2185.

Hennet, T., Richter, C. and Peterhans, E. (1993) Tumour necrosis factor-α induces superoxide generation in mitochondria of 929 cells. *Biochem. J.* 289: 587-592.

Hunt, N.H. and Fragonas, J.C. (1992) Effects of antioxidants on ornithine decaboxylase in mitogenically activated T-lymphocytes. *Biochim. Biophys. Acta* 1133: 261-267.

Ikebuchi, Y., Masumoto, N., Tasaka, A., Koike, K., Kasahara, K., Miyake, A. and Tanizawa, O. (1991) Superoxide anion increases intracellular pH, intracellular free calcium and arachidonate release in human amnion cells. *J. Biol. Chem.* 266: 13233-13237.

Kyle, M.E., Nakae, D., Sakaida, I., Miccadei, S. and Farber, J.L. (1988) Endocytosis of superoxide dismutase is required in order for the enzyme to protect hepatocytes from cytotoxicity and hydrogen peroxide. *J. Biol. Chem.* 263: 3784-3789.

Li, L., Wattiaux-De Conick, S. and Wattiaux, R. (1992) Endocytosis of superoxide dismutase by rat liver. *Biochem. Biophys. Res. Commun.* 184: 727-732.

Matsuyama, T. and Ziff, M. (1986) Increased superoxide anion release from human endothelial cells in response to cytokines. *J. Immunol.* 137: 3295-3304.

Meier, B., Cross, A.R., Hancock, J.T., Kaup, F. and Jones, O.T.G. (1991) Identification of a superoxide generating NADPH oxidase system in human fibroblasts. *Biochem. J.* 275: 241-245.

Meier, B., Radeke, H.H., Selle, S., Younes, M., Seis, H., Resch, K. and Habermehl, G.G. (1989) Human fibroblasts release reactive oxygen species in response to interleukin-1 or tumour necrosis factor-α. *Biochem. J.* 263: 539-545.

Murell, G.A.C., Francis, M.J.O. and Bromley, L. (1990) Modulation of fibroblast proliferation by oxygen free radicals. *Biochem. J.* 265: 659-665.

Oberley, L.W. and Oberley, T.D. (1988) Role of antioxidant enzymes in cell immortalization and transformation. *Mol. Cell. Biochem..* 84: 147-153.

Saez, J.C., Kessler, J.A., Bennett, M.V.L. and Spray, D.C. (1987) Superoxide dimutase protects cultured neurones against death. *Proc. Natl. Acad. Sci. USA* 84: 3056-3059.

Schreck, R., Reiber, P. and Baeuerle, P.A. (1990) Reactive oxygen intermediates as apparently widely used messengers in the activation of the NF-kappa B transcription factor and HIV-1. *EMBO J.* 10: 2447-2258.

Shibanuma, M., Kuroki, T. and Nose, K. (1988) Induction of DNA replication and expression of proto-oncogenes c-myc and c-fos in quiescent Balb/3T3 cells by xanthine-xanthine oxidase. *Oncogene* 3: 17-21.

Shibanuma, M., Kuroki, T. and Nose, K. (1988) Superoxide as a signal for increase in intracellular pH. *J. Cell. Physiol.* 136: 379-383.

Shibanuma, M., Kuroki, T. and Nose, K. (1990) Stimulation by hydrogen peroxide of DNA synthesis competence family gene expression and phosphorylation of a specific protein in quiescent Balb/3T3 cells. *Oncogene* 5: 1025-1032.

Slater, T.F., Sawyer, B. and Strauli, V. (1963) Studies on succinate-tetrazolium reductase systems III Points of coupling of four different tetrazolium salts. *Biochim. Biophys. Acta* 77: 383-393.

Szatrowski; T.P. and Nathan, C.F. (1991) Production of large amounts of hyrogen peroxide by human tumour cells. *Cancer Res.* 51: 794-798.

Tiku, M.L., Liesch, J.B. and Robertson, F.M. (1990) Production of hydrogen peroxide by rabbit articular chondrocytes. Enhancement by cytokines. *J. Immunol* . 145: 690-697.

Toledano, M. and Leonard, W.J. (1991) Modification of transcription factor NF-kappa B binding activity by oxidation/reduction in vitro. *Proc. Natl. Acad. Sci. USA* 88: 4543-4547.

Tse, C.A. and Scott, G.K. (1991) Growth-related proteinase activity in a recently established human tumour cell culture. *Biochem. Soc. Trans.* 19: 285S.

Wasil, M., Halliwell, B., Hutchinson, D.C.S. and Baum, H. (1987) The antioxidant action of human extracellular fluids. The effect of human serum and its protein components on the inactivation of α-1 antiprotease by hypochlorous acid and by hydrogen peroxide. *Biochem. J.* 243: 219-223.

Oxidative Stress, Cell Activation and Viral Infection
C. Pasquier et al. (eds)
© 1994 Birkhäuser Verlag Basel/Switzerland

Effect of oxygen radicals on the IL-1 production by monocytes and IL-2 receptor expression in lymphocytes during primary and secondary immunodeficiency

I.B. Afanas'ev and L.G. Korkina

Russian Hematology Institute for Children, Moscow, Russia

Summary

Monocytes and lymphocytes of patients with agammaglobulinemia and ophtalmoherpes accompanied with immunodeficiency syndrom have been studied in the in vitro experiments. The effects of oxygen radicals on the producion and reception of IL-1 were evaluated by the addition of antioxidant enzymes, SOD and catalase, and the bioflavonoid rutin, a free radical scavenger and an iron chelator in culture medium. It was found that primary and secondary immunodeficiency were characterized by a decrease in IL-1 production by monocytes stimulated with latex particles and by the reduction of the amount of IL-1 receptors on the surface of T-lymphocytes. SOD, catalase, and rutin normalized these parameters as well as the inhibited lucigenin-amplified CL of stimulated monocytes. Thus, superoxide ion is apparently able to suppress cytokine production by monocytes and to modulate the sensitivity of lymphocytes to cytokines.

Introduction

It is now widely accepted that the formation of oxygen free radicals is an event that triggers many unfavorable processes in an organism leading to free radical pathologies including ischemia, inflammation, and others. However, it is now also clear although not frequently recognized that oxygen radicals are important mediators of many physiological processes including cell proliferation and differentiation.

Evidence of stimulating effects of oxygen radicals on cell proliferation have been obtained during the last decade for various types of cells, principally based on the inhibitory effects of antioxidants and free radical scavengers. For example, it was found that antioxidants, chelators, or antioxidant enzymes inhibited lymphocyte mitogenesis (Novogrosky et al., 1982), the proliferation of T-lymphocytes (Chaudri et al., 1988; Dornand and Gerber, 1989), the proliferation of cultured fibroblasts (Murrell et al., 1990), etc.. Therefore, it may be suggested that the mediation of cell proliferation and transformation by oxygen radicals is an universal mechanism typical for cells. The effects of a wide spectrum of the antioxidants, scavengers, and chelators (e.g. hydroxyl radical scavengers, α-tocopherol, NDGA, BHA, DOD, catalase) suggests that superoxide ion and possibly much more active hydroxyl radicals participate in this process.

For some years we have studied the intercommunication between oxygen radical production and the activity of immunocompetent cells. It is well known that major producers of oxygen radicals in the immune system are phagocytes; therefore, stimulatory effect of oxygen radicals on other immunological cells, and first of all, lymphocytes, must be realized through the interaction of phagocytes (monocytes) and lymphocytes. It is important that the same stimuli (e.g. PMA, phytohemagglutinin, concanavalin A) are responsible for the release of oxygen radicals by monocytes and the proliferation activity of lymphocytes.

In this investigation we have studied the effects of oxygen radicals on the cytokine production and their effects on lymphocytes of healthy donors and patients with primary and secondary imunodeficiency.

Material and Methods

Materials

Lucigenin, trypan blue, bovine superoxide dismutase, and catalase were from Sigma Chemical Co (St Louis, MO). Ficoll-Hypaque was from Flow Laboratories (McLean, VA). Fetal calf serum, human serum albumin, latex (1 μm), and L-glutamine were from Serva (Germany), and PHA and HEPES buffer were from Difco (USA). Recombinant human interleukin-2 was supplied by Institute of Organic Chemistry (Latvia). The other materials and solutions used were of high purity and of Russia production.

Leukocyte preparation

The mononuclear leukocyte fraction (more than 95 % viable as shown by trypan blue exclusion) was obtained from heparinized (10 U/ml) venous blood by the routine one-step ficoll-Hypaque separation procedure (Boyum, 1976). Residual red blood cells were lysed with 3 ml ice-cold 0.22 % sodium chloride solution for 50 sec and then with 3 ml 0.88 % sodium chloride solution. The fraction obtained (15-20 % monocytes and 80-85 % lymphocytes) was divided into the two parts containing adherent and nonadherent cells. Monocyte-depleted samples were prepared by allowing the cells to adhere to a tissue culture-grade plastic Petri plate for 45 minutes at 37°C (5% CO_2). After that, the supernatant containing nonadherent cells was collected, centrifuged (150 g x 5 min) and resuspended in the cold 199 medium. The preparation of nonadherent cells contained 90-95 % lymphocytes as confirmed by a negative nonspecific esterase stain. The adherent cell fraction (95-97 % monocytes) was dissolved in 1 ml ice-cold EDTA solution, centrifuged 150 g x 10 min at 4°C, washed twice with HBSS and finally resuspended in the 199 medium contained 10 % fetal calf serum.

Media with interleukin-1 activity preparation

Monocyte suspension (1 ml, 2×10^5 cells/ml) was incubated for 72 hours at 37°C with or without antioxidants. For cell stimulation, 0.1 ml 0.01 % latex suspension was added in each flask. Then the samples were cooled to 0°C, and the cells were removed by centrifugation, yielding a cell-free interleukin-1 containing medium, which was stored at -20°C until testing.

Cell sensitivity to interleukin-1 and interleukin-2

Cell response to interleukins was estimated using the blast-transformation reaction). Non adherent cells (0.1 ml, 10^5 cells) were added to 0.06 ml of 199 medium containing streptomycin (100 μm/ml), penicillin (100 U/ml), HEPES buffer (20 mM), L-glutamine (2mM), and PHA (0.02 ml, 10 μg/ml) in microvial immunological plates and cultivated with or without antioxidants for 72 hours at 37°C and 5 % CO_2. Four hours before the end of cultivation, [^{3}H] Thymidine was added. Then, the cells were collected on the glass filter, and fixed radioactivity was determined on a scintillation β-counter (Beckman, USA).

Leukocyte chemiluminescence

Purified monocyte suspensions (0.1 ml) were incubated in HBSS containing 0.1 % human serum albumin with or without antioxitants at 37°C under continuous mixing. Then lucigenin (0.65 mM) and 0.1 ml 0.01 % latex suspension were added and chemiluminescence intensity was measured on a LKB luminometer (model 1251, Sweden).

Results

It was found that SOD and catalase suppressed blast transformation reaction of lymphocytes in a concentration- dependent manner (Fig. 1), with SOD being a much more effective inhibitor than catalase.

The combined application of SOD and catalase produced the strongest effect (a decrease in the stimulation index (SI) of BT was 22 % (Catalase), 24 % (SOD), 33 % (SOD + Catalase).

(SI = [proliferation rate + PHA]/[proliferation rate without PHA]). Similarly SOD and catalase strongly inhibited IL-1 production by monocytes (46 %, 47 %, and 54 % inhibition) and the proliferation of lymphocytes in the PHA-induced blast transformation reaction stimulated by recombinant IL-2 (55 %, 62 %, and 68 % inhibition) for catalase, SOD, and Cat+SOD, respectively.

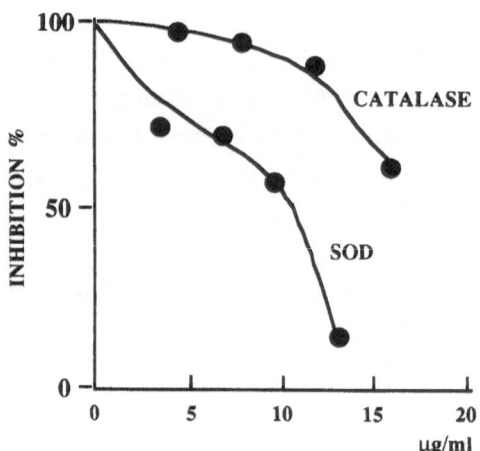

Fig. 1: Inhibition by SOD and catalase of PHA-induced respiratory burst of human lymphocytes.

Furthermore, spontaneous and Concanavalin A-stimulated luminol-amplified chemiluminescence by monocytes from healthy donors or patients with ophtalmoherpes was measured (Table I). It was found that there was a good correlation between chemiluminescence intensity and the level of IL-1 production by monocytes (r = 0.900) or chemiluminescence and the rate of lymphocyte proliferation in the presence of IL-1 (r = 0.950). Monocytes from the patients with diagnosed immunodeficiency states (herpes- induced uveit, lymphogranulomatosis, and Downs syndrome) exhibited a smaller capacity to produce oxygen radicals in the response to IL-1 and IL-2 than those of healthy donors (Table 1).

Tab.I: Decrease in activation index of patients with primary and secondary immunodeficiency.

Disease	Number of patients	Activation index for a patient Activation index for donors [x) Percent stimulation	
		IL-1	IL-2
Herpes-induced uveit	43	10.2	9.1
Lymphogranulomatosis	3	15.0	17.0
Downs syndrome	3	18.5	14.2

[x) Activation index is the ratio of the monocyte chemiluminescence response in the presence or absence of IL-1 or IL-2.

Discussion

Active oxygen species, superoxide ion and hydrogen peroxide, which are released by stimulated monocytes apparently mediate the production of IL-1 by monocytes, the proliferation of lymphocytes in PHA-induced respiratory burst reaction, and the reception of IL-2 on the surface of lymphocytes. However, the connection between the production of oxygen species and lymphocyte proliferation turns out to be more complicated for patients with primary and secondary immunodeficiency. For example, it was found that oxygen radical production by the monocytes of patients with ophtalmoherpes during the acute stage of the disease was 10-15 times greater than that for normal donors although the rate of lymphocyte proliferation in this stage was 3.2 times smaller. This phenomenon is apparently explained by an inflammatory reaction during virus infection. In remission, the chemiluminescence response approached that of healthy donors. However, a decrease in the oxygen radical production was observed for monocytes from patients with diagnosed immunodeficiency states especially accompanied by numerous relapses (Table 1).

Until now, the mechanism of stimulation of cell proliferation by oxygen radicals remains unclear. As it follows from preceeding works, two major mechanisms of the lymhocyte stimulation by oxygen radicals can be considered. The first one is the activation of protein kinase C, and the second one is the activation of lipooxygenase. Although Boscoboinik et al. (1991) reported that α-tocopherol inhibited protein kinase C during vascular smooth muscle cell proliferation, they also concluded that this effect is not connected with the antioxidant properties of α-tocopherol. Therefore, the second mechanism, i.e. lipoxygenase activation (supported by the strong inhibitory effect of NDGA, a lipoxygenase inhibitor and antioxidant on lymphocyte proliferation (Dornand and Gerber, 1989) is possibly of greater importance.

Our findings emphasize an importance of supporting normal levels of oxygen radical production in an organism to prevent the development of immunodeficiency states. Therefore, new drugs possessing high stimulatory activities to phagocytes are needed for combating diseases originating from primary and secondary immunodeficiency.

References

Boscoboinik, D., Szewczyk, A., and Azzi, A. (1991) α-tocopherol (vitamin E) regulates vascular smooth muscle cell proliferation and protein kinase-C activity. *Arch. Biochem. Biophys*. 286: 264-270.

Boyum, I.(1976) Isolation of lymphocytes, granulocytes and macrophages. *Scand. J. Immunol*. 5(suppl): 9-15.

Chaudri, G., Hunt, N.H., Clark, I.A., and Ceredic, R. (1988) Antioxidants inhibit proliferation and cell surface expression of receptors for interleukin-2 and transferrin in T lymphocytes stimulated with phorbol myristate acetate and ionomycin. *Cell. Immunol*. 115: 204-213.

Dornand, J., and Gerber, M.(1989) Inhibition of murine T-cell responses by antioxidants: the targets of lipooxygenase pathway inhibitors. *Immunology* 68: 384-391.

Murrell, G.A.C., Francis, M.J.O., and Bromley, L. (1990) Modulation of fibroblast proliferation by oxygen free radicals. *Biochem. J.* 265: 659-665.

Novogrosky, A., Ravid, A., Rubin, A.L., and Stenzel, K.H. (1982) Hydroxyl radical scavengers inhibit lymphocyte mitogenesis. *Proc. Natl. Acad. Sci. USA* 79: 1171-1179.

Oxidative Stress, Cell Activation and Viral Infection
C. Pasquier et al. (eds)
© 1994 Birkhäuser Verlag Basel/Switzerland

The effect of reactive oxygen species (ROS) on human T and B lymphoid cells

B.M. Hannigan, S. Ranjbar and L. Cromie

Department of Biological and Biomedical Sciences, University of Ulster, Coleraine, Northern Ireland BT52 1 SA.

Summary

T cells are known to use ROS as intracellular signals but also to exhibit an enhanced sensitivity, relative to other cell types, to the DNA damaging effects of various forms of radiation. These effects may be characteristic of DNA repair capacity and/or intracellular antioxidant levels. In a more physiologically-relevant situation T-lymphoid (Molt-3) cells have been shown to exhibit enhanced susceptibility to H_2O_2-induced cytogenetic damage and to have decrease intracellular catalase acitvity, relative to a B-lymphoid line (Raji) without any greater loss of cell viability. Such observations must be reconciled with the need for T cells to proliferate efficiently and accurately when initiating, regulating and participating in human immune responses of maintaining immunological memory throughut the lifespan of the individual.

Introduction

Studies on the interaction between reactive oxygen species (ROS) and non-phagocytic cells of the human immune system have, until very recently, been few. With the recognition that ROS produced in inflammatory responses can interact with adjacent host cells, it is necessary to understand how such species might influence the function of affected cells. The close packing of infiltrating cells at inflammatory foci means that the most likely host cell types to be exposed to ROS are leukocytes, including both T and B lymphocytes. The importance of lymphocyte exposure to ROS is two-fold. Firstly, these cells, and particularly T lymphocytes, are responsible for directing the specificity of immune responses and, secondly, lymphocytes do not die at inflammatory sites, as phagocytes are thought to do. Rather, lymphocytes re-circulate via the lymphatic and cardiovascular vessels to secondary lymphoid tissues and peripheral tissues according to the needs of the immune system at any particular time. While there is no current consensus on the *in vivo* life-span of lymphocytes (days or weeks ?) it is likely that T cells can replicate by division in peripheral tissues with B cells always originating from bone marrow precursors (Freitas and Rocha, 1993). Hence any genomic alteration may be maintained within the T cell population and any ROS-mediated changes incurred at one inflammatory reaction may have implications for the future integrity of immune responses. It is recognised that not all ROS-mediated events in lymphocytes are potentially deleterious, given current information on the widespread use of ROS as activating signals for T-lymphocytes (Schreck et al, 1992) and the apparent inhibition of T cell proliferation by antioxidaants (Kalsi et al, 1991; Hunt et al, 1992). Shifting the balance of ROS effects to those which are the most beneficial involves a range of factors of which the most important must be the dose of ROS generated relative to the target cell antioxidant and DNA repair capacities.

Lymphocyte exposure to ROS in vivo

Precise data on ROS levels at inflammatory foci are not yet available however we may speculate on the basis of data from other sites e.g. H_2O_2 levels in liver are in the sub-micromolar range while in blood plasma concentrations of 0.25 - 5 µM have been measured (Schreck et al, 1988). The maximum amount of H_2O_2 inducible from phagocytes (phorbol ester (PMA) - stimulated neutrophils) has been determined *in vitro* to be in the range of µmoles/10^6 cells in addition to other ROS e.g. superoxide anions which are convertible to H_2O_2. Cell crowding in the inflamed tissue would cause limited tissue oxygenation so maximal levels would be unlikely to be attained. On the other hand, however, crowding may eliminate any antioxidant molecules normally expected in extracellular fluids. The duration and frequency of an individual cell's ROS exposure *in vivo* is also difficult to estimate. Therefore, until more precise data are available, it will be important to consider the effects of a very wide (nM - mM) range of H_2O_2 concentrations in experiments designed to predict possible immunological outcomes.

Lymphocyte antioxidant enzyme activities

The antioxidant status of a cell is frequently measured as the activities of the enzymes superoxide dismutase (SOD), catalase (CAT) and glutathione peroxidase (GPx). Both CAT and GPx are primarily concerned with H_2O_2 removal however they differ in kinetic constants (Km) such that only GPx would be expected to be active under physiological conditions with only CAT active when H_2O_2 concentrations are raised. A "bystander" cell without sufficient CAT activity would thus be susceptible to ROS fluxes at inflammatory sites. The current literature does not detail SOD or CAT activities in mature human lymphocytes perhaps due to the relatively large numbers of cells necessary to make such measurements. In a study on rats Carville and Strain (1989) reported total lymphocyte SOD activity of 1.6 U/10^7 cells.

Lymphocyte DNA repair capacity

DNA repair capacities of lymphocytes have been assessed particularly in experiments using potentially-damaging ionising or non-ionising radiation. Using ultra-violet radiation (UV-C) resting T lymphocytes have been seen to be more susceptible to DNA damage and cell death than are B lymphocytes or mitogen-stimulated T lymphocytes (Arlett et al, 1992). These authors attributed such findings to a decreased DNA repair capacity in resting T cells with mitogen-stimulated T cells showing enhanced DNA excision repair. It has been shown that radio-resistant (R) and radio-sensitive (S) variants of the murine L5178Y lymphoma cell line exhibit different sensitivities to the physiologically-relevant xanthine+xanthine oxidase system of ROS generation (Hannigan et al, 1992). Both R and S cells are similarly deficient in DNA excision repair but the more ROS-sensitive S cells have lower intracellular activities of SOD, CAT and GPx. Thus differences in a cell's susceptibility to ROS-induced DNA damage cannot necessarily be explained on the basis of DNA repair capacity alone. The relative DNA excision repair capacities of Raji and Molt-3 are not yet known.

Material and Methods

Lymphoid cell antioxidant activities

Using the human cell lines Molt-3 and Raji, of human T and B lymphoblastoid origins respectively, the intracellular acitvities of SOD, CAT and GPx were measured by the methods of Jones and Suttle (1981), Aebi (1984) and Paglia and Valentine (1967), respectively. These data are shown in Table I.

Lymphoid cell susceptibility to H_2O_2-induced damage

Raji and Molt-3 cells were exposed to a range of H_2O_2 concentrations (0-250 μM) in the presence or absence of exogenous CAT (cells pre-incubated for 2 hours with 10 U/ml CAT). The extent of the subsequent damage was measured by a variety of cytogenetic end-points. These data are summarised in Table II. For clarity, only a single H_2O_2 dose (50 μM) is shown although similar patterns were observed across the range of H_2O_2 doses tested.

Results

Table I shows that SOD values were not dissimilar to those reported for rats. Molt-3 showed relatively low SOD and CAT activities while GPx activities were similar in both cell types. These data would suggest that both T and B lymphoid cells can tolerate equally well low (physiological) H_2O_2 concentrations while T cells woul be more susceptible to higher doses.

Table I: Intracellular antioxidant superoxide dismutase (SOD), catalase (CAT) and glutathione peroxidase (GPX) activities in T (Molt-3) and B (Raji) lymphoblastoid cells; U/10^6 cells; means ± s.d., n = 3.

	SOD	CAT	GPx
Molt-3	0.068 ± 0.004	5.10 ± 0.70	0.478 ± 0.065
Raji	0.136 ± 0.008	13.10 ± 1.10	0.385 ± 0.053

From Table II it is clear that the T cell line shows both higher baseline levels of DNA abnormalities and an increased susceptibility to the clastogenic effects of H_2O_2. Exogenous CAT could only prevent a fraction of such damage. This may be because CAT cannot enter live cells although H_2O_2 can cross intact membranes and hence gain access to DNA either directly or, indirectly, through the formation of intracellular derivatives. Thus the relatively low endogenous CAT levels of the T cells may still be considered to contribute significantly to the observed DNA damage.

Table II : H_2O_2-induced cytogenetic changes in human T and B lymphoblastoid cells. Cell survival (%), chromosome aberrations (aberr., per 50 metaphase cells), cloning efficiency (c.e., %), point mutations (pt. mut., at the thymidine kinase locus) and sister chromatid exchanges (SCE, per 30 metaphase cells, mean ± s.d.) in T (Molt-3) and B (Raji) lymphoblastoid cells in the presence and absence of H_2O_2 (50 µM) and catalase (CAT, 10 U/ml). *Raji cells do not incorporate bromodeoxyuridine (BrdU) so SCE cannot be detected.

	MOLT-3			RAJI		
		50 µM H_2O_2+			50 µM H_2O_2+	
	control	no antioxidant	CAT	control	no antioxidant	CAT
Survival	98	39.7	89.2	98	46.9	94.1
Aberr.	100	1750.0	n.d.	10	240.0	n.d.
c.e.	17.63	7.01	n.d.	13.0	6.1	n.d.
pt. mut.	8.5x10^{-5}	7.13x10^{-4}	3.47x10^{-4}	4.3x10^{-5}	9.8x10^{-5}	6.5x10^{-5}
SCE	4.1 ± 2	12 ± 2	7.1 ± 1	n.d.	n.d.	n.d.*

Conclusion

These experiments obviously suffer from the flaw of using cell lines rather than their normal in vivo counterparts. Further, the relative DNA excision repair capacities of Molt-3 and Raji cells are not yet known. Nonetheless such data provide support for the performance of experiments which evaluate the roles of both endogenous antioxidants and DNA repari processes in T lymphocytes. This system is likely to be one of the most highly sophisticated in mammalian cells because T cells must proliferate rapidly in response to ROS-mediated signalling systems yet avoid, as far as possible, the loss of immune responsiveness ensuing from T cell death and the potentially serious consequences of mutations which could influence the antigen-specificity of subsequent T cell generations.

Acknowledgements

This work was supported by DHSS (N.I.) and by the Ulster Cancer Foundation. L. Cromie was supported by a D.E.N.I. Postgraduate Studentship.

References

Aebi, H. (1984) Catalase in vitro. *Methods in Enzymology* 195, 121-124.

Arlett, C.F., Harcourt, S.A., Cole, J., Green, M.H.L. and Anstey, A.V. (1992). A comparison of the response of unstimulated and stimulated T-lymphocytes and fibroblasts from normal, xeroderma pigmentosum and trichothiodystrophy donors to the lethal action of UV-C. *Mutation Research, DNA Repair*: 273, 127-135.

Carville, D.G.M. and Strait, J.J. (1989) The effect of copper deficiency on blood anitoxidant enzymes in rats fed sucrose or sucrose and lactose diets. *Nut. Rep. Int.*: 39, 25-32.

Freitas, A.A. and Rocha, B.B. (1993) Lymphocyte lifespans: homeostasis, selection and competition. *Immunology Today* 14: 25-29.

Hannigan, B.M., Richardson, S.-A., M. and Mc Kenna, P.G. (1992) DNA damage in mammalian cell lines with different antioxidant levels and DNA repair capacities. In:Emerit, I. and Chance, B. (eds): *Free Radicals and Aging*, Birkhauser Verlag, Basel, pp. 247-250.

Hunt, N.H., Jeitner, T.M., Fragonas, J.C., Kneale, C.L. and van Reyk, D.M. (1992) Redox mechanisms in T cell activation. *Free Rad. Res. Commun.*: 16, suppl. 1, 17.2.

Kalski, J.K., Clay,K. and Hall, N.D. (1991) Antioxidant drugs as modulators of lymphocyte activation in vitro. *Molec. Aspects Med.*: 12, 161-167.

Paglia, D.E. and Valentine, W.N. (1967) Studies on the quantitative and qualitative characterization of erythrocyte glutathione peroxidase. *J. Lab. Clin. Med.* 70, 158-162.

Schreck, R., Rieber, P. and Baeuerle, P.A. (1992) Reactive oxygen intermediates as apparently widely used messengers in the activation of the NF-κB transcription factor and HIV-1. *The EMBO Journal* 10: 2247-2258.

Oxidative Stress, Cell Activation and Viral Infection
C. Pasquier et al. (eds)
© 1994 Birkhäuser Verlag Basel/Switzerland

Leukocyte Adhesion and Endothelial Cytokine Production in Hypoxia /Reoxygenation

G. Modat*, C. Bonne* and J. Dornand**

*Laboratoire de Physiologie Cellulaire, Université Montpellier I, 15 avenue Charles Flahault, 34060 Montpellier, France
**Laboratoire de Biologie Cellulaire, INSERM U65, USTL, place eugène Bataillon, 34060 Montpellier, France

Summary

We submitted cultured human umbilical vein endothelial cells to 5h hypoxia followed by 20 min to 24h reoxygenation. This resulted in an increased adhesion of human resting neutrophils to monolayers with two peaks after 20 min and 4h reoxygenation. This was correlated with the respective expression of the preformed granule membrane protein 140 (GMP-140) and of the *de novo* synthesized endothelial leukocyte adhesion molecule 1 (ELAM-1) on endothelial surface. We also found that the spontaneous endothelial production of interleukin 1 (IL-1) α and β, and interleukin 6 (IL-6) greatly increased after 5h hypoxia/19h reoxygenation. Superoxide dismutase, catalase or glutathione peroxidase added to culture before hypoxia efficiently prevented both effects. These results underline the crucial role played by endothelial-derived oxygen free radicals at reoxygenation in leukocyte adhesion and cytokine synthesis, which could lead to amplify the oxidative stress injury. The protection offered by free radical scavengers emphasizes the potential interest of antioxidants in post ischemic vascular disorders.

Introduction

Because of its key position between blood and tissues, vascular endothelium is a major site for oxidative stress in ischemia/reperfusion processes. Much attention has been devoted to the importance of the enhanced neutrophil adhesion to endothelial lining in microvessels during post ischemic periods (Go *et al.*, 1988; Hernandez *et al.*, 1987; Jaeschke *et al.*, 1990). However the mechanisms responsible for such complex cell/cell interaction which requires an increased expression of surface adhesion molecules either on neutrophils (Horgan *et al.*, 1990) or on endothelial cells (Schreeniwas *et al.*, 1992) remain to be defined more clearly. Although previous studies have underlined the involvement of oxygen-derived free radicals in the phenomenon (McCord, 1987; Suzuki *et al.*, 1991; Zimmerman *et al.*, 1990), the role of oxyradicals largely produced by endothelial cells at reperfusion (Arroyo *et al.*, 1990; Palluy *et al.*, 1991; Zweier *et al.*, 1988) is still unclear. Another function of endothelium is to produce and/or respond to various cytokines (Mantovani *et al.*, 1989) including the pleiotropic interleukin 1 (IL-1) and interleukin 6 (IL-6) (Jirik *et al.*, 1989; Modat *et al.*, 1990; Warner *et al.*, 1987) known to regulate inflammatory and immunological responses. In particular IL-1 stimulates endothelial cells to promote neutrophil adhesion by inducing the expression of two adhesive molecules, namely endothelial leukocyte adhesion molecule 1 (ELAM-1) (Bevilacqua *et al.*, 1987) and intercellular adhesion molecule 1 (ICAM-1) (Rothlein *et al.*, 1986). It has also been demonstrated that oxygen-derived free radicals activate monocytes and macrophages to produce IL-1 (Ansel *et al.*, 1984; Gougerot-Pocidalo *et al.*, 1989; Kasama *et al.*, 1989). Moreover, IL-1 primes free radical production in various cell types (Ozaki *et al.*, 1987; Radeke *et al.*, 1990; Warren *et al.*, 1988) including endothelial cells (Matsubara *et al.*, 1986). On the basis of these data, the aim of the present study was to investigate whether cultured human umbilical vein endothelial cells synthesize IL-1 and IL-6, and bind neutrophils in response to H/R with particular interest in the adhesion molecules involved, laying emphasis on the role played by endothelial-generated free radicals.

Materials and Methods

Cell culture and hypoxia/reoxygenation (H/R) procedure
Human umbilical vein endothelial cells (HUVEC) were cultured as previously described (Palluy *et al.*, 1992) in M199 Earle medium supplemented with 20% fetal calf serum, L-glutamine (4 mM), penicillin (100 U/ml), streptomycin (100 µg/ml), amphotericin (2.5 µg/ml) and heparin (10 U/ml). Confluent HUVEC (passages 3-4) grown in 24 multiwell plates for the adhesion assay (Palluy *et al.*, 1992) or in 25 cm^2 flasks for the interleukin assay (Ala *et al.*, 1992) were washed twice and covered with DMEM (300 µl/well) or RPMI 1640 (5ml/flask). Cells were then exposed to 5h hypoxia in a modular incubator chamber in a 95% N2/5% CO2 atmosphere. Oxygen tension in the culture medium was 20-30 mmHg and there was no change in medium pH. Reoxygenation was obtained by replacing HUVEC in normoxic conditions (95% air/5% CO2) for various times. Controls consisted in HUVEC incubated for the same duration of experiments in normoxia.

Adhesion assay
It was carried out as previously described (Palluy *et al.*, 1992). Briefly, human neutrophils from healthy donors were isolated from heparinized venous blood, resuspended in DMEM/HEPES 40 mM, pH 7.4, and added to twice-washed HUVEC monolayers at a concentration of 5.10^5 cells/well, for the last 15 min reoxygenation (or normoxia for controls). Non adherent neutrophils were then removed by washing and remaining well contents were solubilized with 0.5% hexadecyltrimethyl ammonium bromide in phosphate buffer 50 mM, pH. 6. Adhesion was evaluated by the spectrophotometric measure of myeloperoxidase activity in the extracts.

Interleukin assay

IL-1 and IL-6 were determined in HUVEC supernatants as previously described (Ala *et al.*, 1992). Briefly, IL-1α and IL-1β were measured by specific enzyme immunoassays using commercially available tets (Immunotech, Marseille, France). IL-6 was assayed by virtue of its capacity to stimulate the growth of the IL-6 dependent murine hybridoma cell line B9. Proliferation was measured by [^{3}H] methylthymidine incorporation. Samples (serial dilutions of HUVEC supernatants) were always related to a standard curve, one unit IL-6 corresponding to 1 pg/ml human recombinant IL-6.

Reagents

Bovine catalase, glutathione peroxidase and superoxide dismutase were obtained from Sigma Chemial Co. (St Louis, USA). mAb BBA2 was from British Biotechnology (Oxon, U.K.). mAb IOP62 was from Immunotech (Marseille, France). mAb F2 B 1.8 was given by Dr. P. Carayon (Sanofi, Montpellier, France). B9 cell line was obtained from Dr. L.A. Aarden (Amsterdam, Holland). mAb B-E8 was a gift of Dr. K. Klein (Montpellier, France). [^{3}H] methylthymidine was from the Commissariat à l'Energie Atomique (Saclay, France). Tissue culture materials were from ATGC (Paris, France). All other chemicals used were the highest grade available.

Statistics

Statistical differences between means were determined using analysis of variance followed by Student t test, Dunnett's test, or Newman-Keuls's test according to experiments. $p < 0.05$ was considered significant.

Results

Neutrophil/HUVEC adherence during H/R

Exposure of HUVEC to 5h hypoxia/20 min-24h reoxygenation resulted in a time-related increase in neutrophil adhesion to monolayers as compared to corresponding normoxic controls (Fig. 1). A first and transient peak was found after 20 min reoxygenation and maximal adhesion was observed after 4h. Return to basal adhesion occured after 24h reoxygenation. Hypoxia alone had no effect since HUVEC immediately fixed in 2.5% glutaraldehyde after hypoxia bound neutrophils as fixed controls. The fixation procedure used did not modify the adhesion assay (no difference was found between fixed and unfixed HUVEC after 20 min and 4h reoxygenation). Among molecules expressed by activated endothelial cells to bind neutrophils, some including granule membrane protein 140 (GMP-140) act rapidly and transiently after stimulation (Geng *et al.*, 1990). In contrast, those requiring *de novo* protein synthesis such as ELAM-1 and ICAM-1 appear after hours (Pober *et al.*, 1991). Regarding the kinetics of adhesion, studies undertaken to determine the molecules involved have been focused on the maxima observed, *i.e.* after 20 min and 4h reoxygenation. As shown in Fig. 2, the protein synthesis inhibitor cycloheximide (2.10^{-5} M) added to cultures before hypoxia did not prevent the former but abolished the latter, thus clearly indicating the participation of a at least two different processes or two different types of molecules. This was confirmed by the use of monoclonal antibodies (mAb) directed against endothelial adhesion molecules, added to cultures before the adhesion assay. The mAb anti-GMP-140 IOP62 (5 µg/ml) totally inhibited adhesion after 20 min reoxygenation (Fig. 2, A) and the mAb anti-ELAM-1 BBA2 (50 µg/ml) totally inhibited adhesion after 4h reoxygenation (Fig. 2, B).

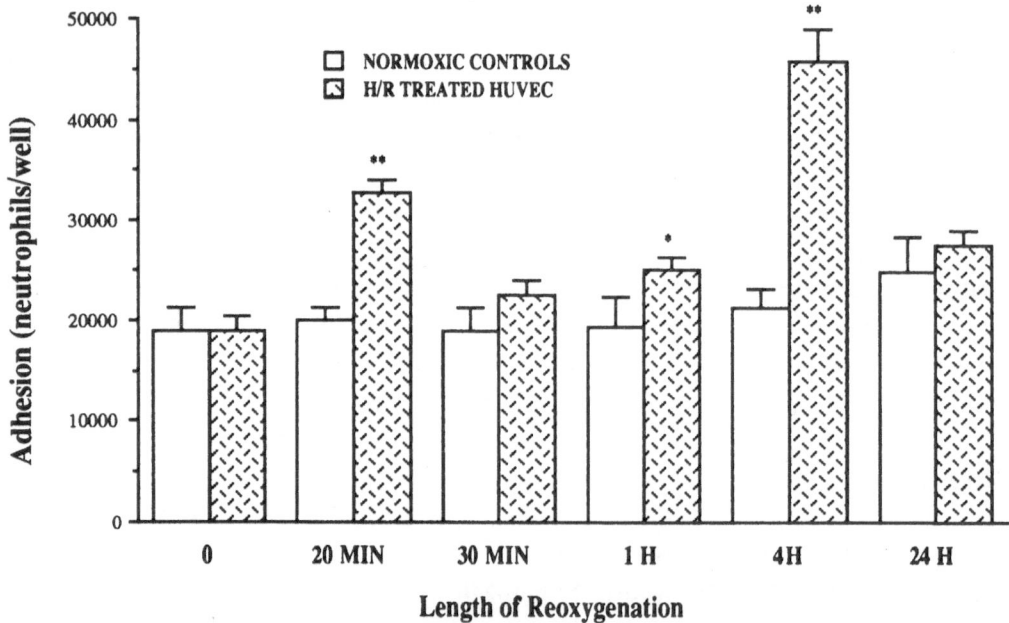

Figure 1. Kinetics of neutrophil adhesion to H/R-treated HUVEC monolayers. HUVEC were submitted to 5h hypoxia followed by 20 min to 24h reoxygenation as indicated in materials and methods. Adherence was evaluated by myeloperoxidase measurements. A rate of absorbance of 0.012/min corresponded to 10^4 neutrophils. Data are means ± SEM of 6 to 10 experiments, each in triplicate. * $p < 0.05$; ** $p < 0.01$ relative to corresponding controls.

IL-1 and IL-6 production by H/R-treated HUVEC

We compared the cytokine production by HUVEC under basal conditions (24h normoxia) and after 5h hypoxia/19h reoxygenation. This duration of experiments was retained in order to have measurable amounts of cytokines in HUVEC supernatants. Fig. 3 clearly shows that the basal production of IL-1 was effectively low, with a predominance of IL-1α over IL-1β. When cells were submitted to H/R, a significant increase in IL-1α (81.8 ± 8.9%) and IL-1β (138 ± 6%) was observed. Such increase was found cycloheximide-sensitive.

The proliferative response of the IL-6-dependent B9 cell line to serial dilutions of supernatants (down to the 1/40) from normoxic HUVEC allowed to detect an IL-6 like activity, reflecting a spontaneous endothelial production since we verified the absence of IL-6 in the culture medium used in experiments. Moreover, addition of the mAb anti-IL-6 B-E8 (5 µg/ml) to supernatants 1h before IL-6 bioassay led to a total inhibition of the B9 cell proliferation, thus precising the nature of the detected activity. As for IL-1, H/R resulted in a significant increase in IL-6 synthesis which was also inhibited by cycloheximide.

Effects of free radical scavengers on H/R-induced neutrophil adherence and cytokine synthesis
Addition of superoxide dismutase (300 U/ml) or Catalase (300 U/ml) to cultures before hypoxia prevented neutrophil adhesion with a maximal effect on the 4h reoxygenation peak of adhesion (Fig. 2). Superoxide dismutase was also able to prevent H/R-induced increase in IL-1 and IL-6 (Figs. 3 and 4). Glutathione peroxidase (200 U/ml) was also active, while to a lesser extent for IL-1. In all cases, heat-inactivated enzymes were ineffective, indicating that specific enzymatic activities were required for the protective effect.

Discussion

These studies first indicate that cultured human endothelial cells submitted to H/R express differential capacity to bind human resting neutrophils. By washing monolayers before the adhesion assay (otherwise limited to 15 min), we eliminated the potential effect of endothelial-derived chemoattractants produced during H/R (Farber *et al.*, 1987) on neutrophil activation and subsequent production of oxygen-derived free radicals by these cells. Since adhesion was not modified for fixed endothelial cells or neutrophils (data not shown), it can be concluded that the observed increase in adhesion at reoxygenation was only linked to modifications of the adhesive properties of the endothelial cell surface.

The 20 min reoxygenation peak of adhesion was transient and cycloheximide insensitive. According to its swiftness of expression, GMP-140 was a good candidate. Indeed, GMP-140 is known to be redistributed within minutes from endothelial secretory granules to the plasma membrane and serve as a ligand for neutrophils after various stimulations of endothelial cells (Geng *et al.*, 1990; Hattori *et al.*, 1991) including oxygen radicals (Patel *et al.*, 1991).

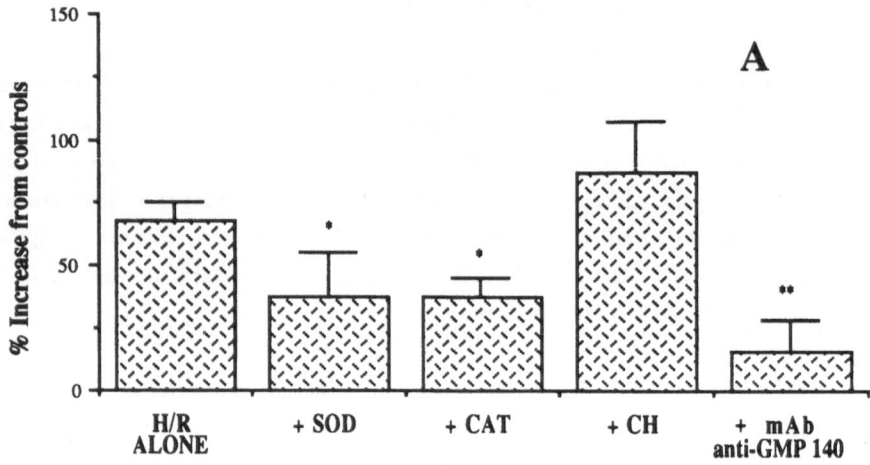

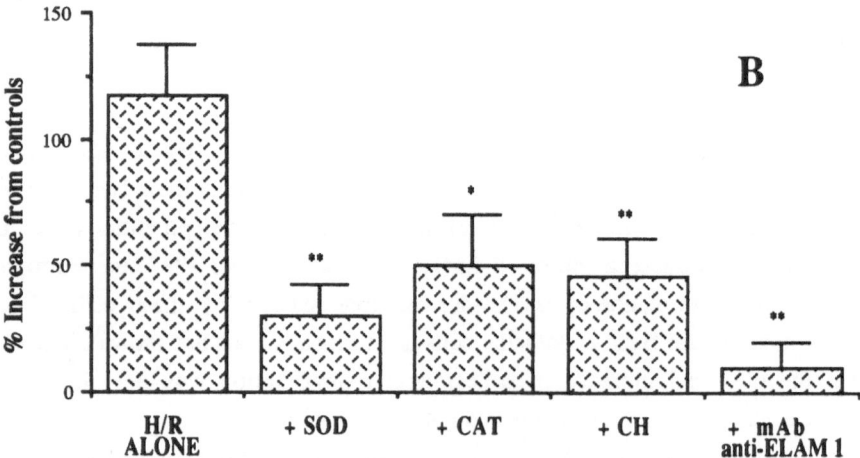

Figure 2. Effect of free radical scavengers, cycloheximide and mAbs anti-endothelail adhesive molecules on H/R-induced neutrophil adhesion after 20 min (A) and 4h (B) reoxygenation. Results are expressed as % increase in adhesion from normoxic controls. Data are means ± SEM of 6 to 10 experiments, each in triplicate. SOD, superoxide dismutase; CAT, catalase; CH, cycloheximide; mAb anti-GMP 140 (IOP62); mAb anti-ELAM 1 (BBA2). * p < 0.05; ** p < 0.01 relative to H/R alone.

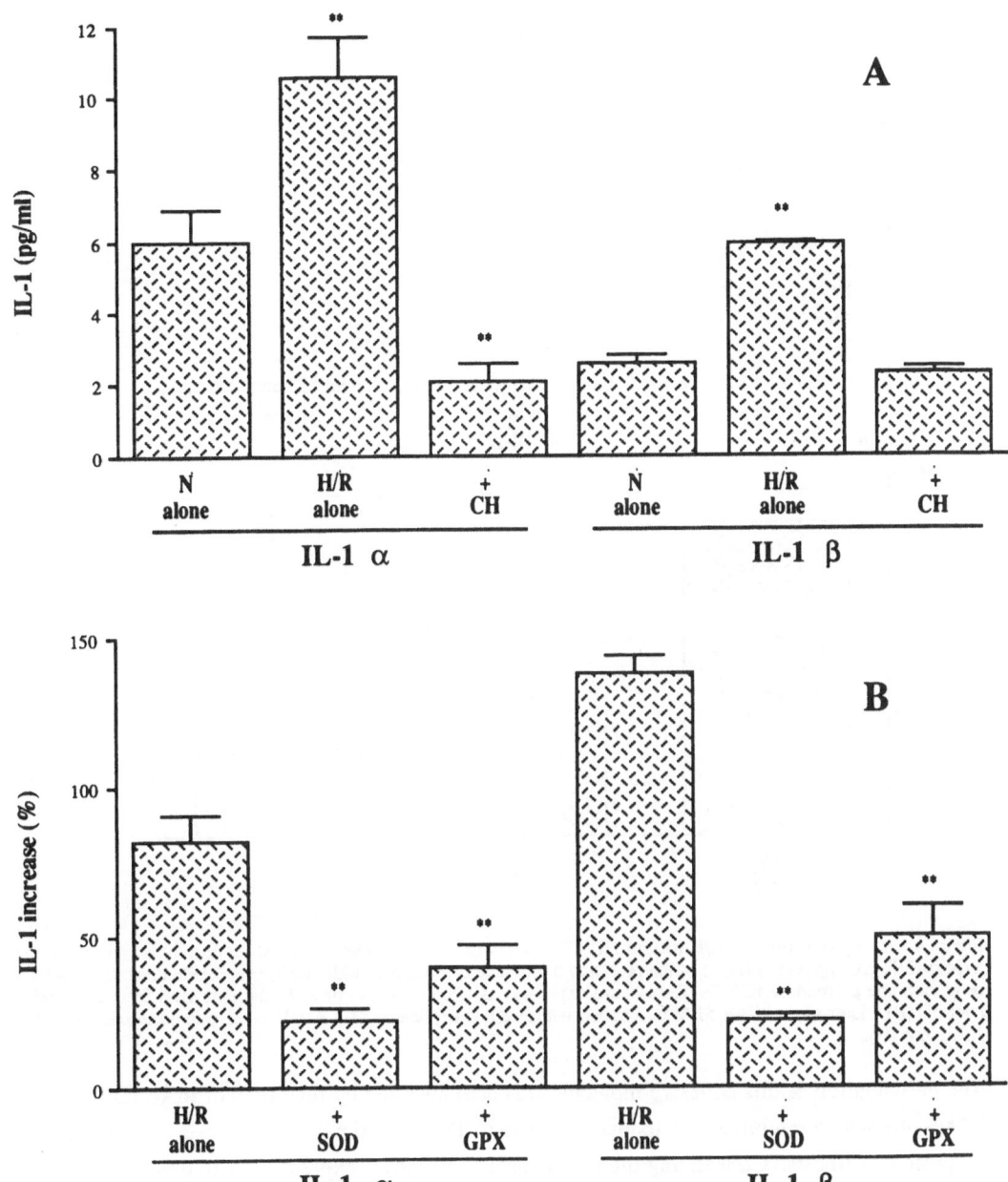

Figure 3. IL-1 production by H/R-treated HUVEC. HUVEC were incubated for 24h in normoxia (N) or for 5h hypoxia/19h reoxygenation (H/R) in the presence or absence of cycloheximide (CH), superoxide dismutase (SOD) or glutathione peroxidase (GPX). Results are expressed as pg/ml IL-1 (A) or as % increase IL-1 from normoxic controls (B). Data are means ± SEM of 3 experiments each in triplicate. ** $p < 0.01$ relative to normoxia (A) or to H/R alone (B).

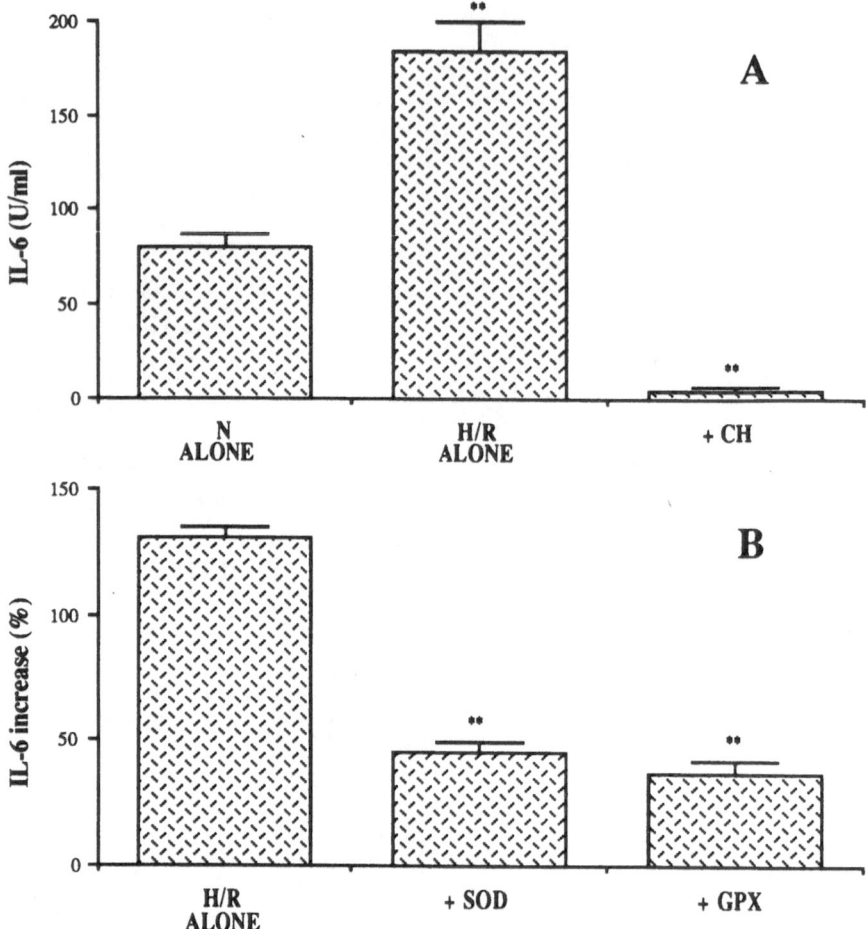

Figure 4. IL-6 production by H/R-treated HUVEC. HUVEC were incubated for 24h in normoxia (N) or for 5h hypoxia/19h reoxygenation (H/R) in the presence or absence of cycloheximide (CH), superoxide dismutase (SOD) or glutathione peroxidase (GPX). Results are expressed as U/ml IL-6 (A) or as % increase IL-6 from normoxic controls (B). Data are means ± SEM of 3 experiments each in triplicate. ** $p < 0.01$ relative to normoxia (A) or to H/R alone (B).

The participation of this adhesive molecule was demonstrated by the effect of mAb IOP62 anti-GMP-140, which inhibited H/R-induced adhesion. The second peak of adhesion was dependent on protein synthesis. Considering the time course of ELAM-1 expression with a maximum at 4-6h after endothelial stimulation (Bevilacqua et al., 1987; Pober et al., 1991) we studied the effect of the mAb anti-ELAM-1 BBA2, which effectively inhibited adhesion. The participation of free radicals generated by endothelial cells at the early phase of reoxygenation (Arroyo et al.,

1990; Zweier *et al.*, 1988) was indirectly demonstrated by the protective effect of scavenging enzymes. Results obtained with superoxide dismutase strongly suggest the involvement of superoxide anions. These radicals could be produced, at least in part, intracellularly through the xanthine oxidase pathway since exogenously added superoxide dismutase has been reported to penetrate cells (Chudej *et al.*, 1990; Markey *et al.*, 1990; Palluy *et al.*, 1991) and since it exists an important conversion of xanthine deshydrogenase to xanthine oxidase in H/R-treated HUVEC (Palluy *et al.*, 1992). Protection offered by catalase also implicate hydrogen peroxide.

In a second set of experiments, we further established that H/R greatly stimulates HUVEC to release more IL-1 and IL-6 molecules. By experiments using cycloheximide, we demonstrated that the observed cytokine enhancement corresponds to a *de novo* protein synthesis rather than a release of preformed activities, in particular membrane-associated IL-1 (Poubelle *et al.*, 1990). These results are in agreement with a very recent report indicating an increase in the level of mRNA for IL-1α in hypoxia-treated HUVEC (Shreeniwas *et al.*, 1992). However our data indicate that IL-1β was relatively more increased than IL-1α while this latter form was predominant in normoxia and has been reported to be the most inducible in endothelial cells (Poubelle *et al.*, 1990). Superoxide dismutase and glutathione peroxidase were able to prevent H/R-induced cytokine synthesis, again demonstrating the participation of superoxide anions and hydrogen peroxide in the phenomenon.

Given the variety of molecular targets for free radicals, it is difficult at present to assess how endothelial-derived reactive oxygen species are implicated in the biochemical events leading not only to the fusion of Weibel-Palade bodies with the plasma membrane and therefore expression of GMP-140 but also to ELAM-1, IL-1 and IL-6 protein synthesis. In particular, it is not clear whether IL-1, IL-6 and ELAM-1 are separately affected or IL-1 is only concerned, further inducing IL-6 (May et al., 1989; Sironi et al., 1989) and/or ELAM-1 (Bevilacqua et al., 1987; Pober et al., 1991; Shreeniwas et al., 1992) as previously reported. One possibility could be that oxygen-derived free radicals directly or indirectly modulate HUVEC gene expression through activation of transcriptional factors as elsewhere discutted in this volume.

Whatever the mechanism involved, our results provide additional evidence for the informative role played by endogenous free radicals on endothelial functionality during H/R, emphasizing the potential interest of anti-oxidants in ischemia/reperfusion processes.

Acknowledgements

The authors are grateful to Drs. L.A. Aarden and B. Klein for their kind gifts of lymphoid cells and mAb B-E8. They also wish to thank Miss H. Bennouar and Miss M. Souchon for their excellent secretarial assistance. These studies were supported by grants from the Institut National de la Santé et de la Recherche Médicale (INSERM 89 - 2008) and from the european Eureka comittee (EU-437).

References

Ala, Y., Palluy, O., Bonne, C., Modat, G. and Dornand, J. (1992) Hypoxia/reoxygenation stimulates endothelial cells to promote interleukin-1 and interleukin-6 production. Effects of free radical scavengers. *Agents & Actions* 37: 134-139.

Ansel, J., Luger, T.A., Kock, A., Hochstein, D. and Green, I. (1984) The effect of in vitro UV irradiation on the production of IL-1 by murine macrophages and P 388D cells. *J. Immunol.* 133: 1350-1355.

Arroyo, C. M., Carmichael, A. J., Bouscarel, B., Liang, J. H. and Weglicki, W. B. (1990) Endothelial cells as a source of oxygen-free radicals. An ESR study. *Free Radic. Res. Commun.* 9: 287-296.

Bevilacqua, M. P., Pober, J. S., Mendrick, D. L., Cotran, R. S. and Gimbrone, M. A. (1987) Identification of an inducible endothelial-leukocyte adhesion molecule. *Proc. Natl. Acad. Sci. USA* 84: 9238-9242.

Chudej, L. L., Koke, J. R. and Bittar, N. (1990) Evidence for transcytosis of exogenous superoxide dismutase and catalase from coronary capillaries into dog myocytes. *Cyto. Bios.* 63: 41-53.

Farber, H. W., Center, D. M. and Rounds, S. (1987) Effect of ambient oxygen on cultured endothelial cells from different vascular beds. *Am. J. Physiol.* 253: H878-H883.

Geng, J. G., Bevilacqua, M. P., Moore, K. L., Mc Intyre, T. M., Prescott, S. M., Kim, J. M., Bliss, G. A., Zimmerman, G. A. and Mc Ever, R. P. (1990) Rapid neutrophil adhesion to activated endothelium mediated by GMP-140. *Nature* 343: 757-760.

Go, L. O., Murry, C. E., Richard, V. J., Weischedel, G. R., Jennings, R. B. and Reimer, K. A. (1988) Myocardial neutrophil accumulation during reperfusion after reversible or irreversible ischemic injury. *Am. J. Physiol.* 255: H1188-H1198.

Gougerot-Pocidalo, M. A., Roche, Y., Fay, M., Perianin A. and Bailly, S.(1989) Oxidative injury amplifies interleukin-1-like activity produced by human monocytes. *Int. J. Immunopharmac.* 11: 961-969.

Hattori, R., Hamilton, K. K., Fugate, R. D., Mc Ever, R. P. and Sims, P. J. (1989) Stimulated secretion of endothelial Von Willebrand factor is accompanied by rapid redistribution to the cell surface of the intracellular granule membrane protein GMP-140. *J. Biol. Chem.* 264: 7768-7771.

Hernandez, L. A., Grisham, M. B., Twohig, B., Arfors, K. E., Harlan, J. M. and Granger, D. N. (1987) Role of neutrophils in ischemia/reperfusion-induced microvascular injury. *Am. J. Physiol.* 253: H699-H703.

Horgan, M. J., Wright, S. D. and Malik, A. B.(1990) Antibody against leukocyte integrin (CD18) prevents reperfusion-induced lung vascular injury. *Am. J. Physiol.* 259: L315-L319.

Jaeschke, H., Farhood, A. and Smith, C. W. (1990) Neutrophils contribute to ischemia/ reperfusion injury in rat liver in vivo. *FASEB J.* 4: 3355-3359.

Jirik, F. R., Podor,T. J, Hirano, T., Kishimoto, T., Loskutoff, D. J., Carson D. A. and Lotz, M.(1989) Bacterial lipopolysaccharide and inflammatory mediators augment IL-6 secretion by human endothelial cells. *J. Immnol.* 142: 144-147.

Kasama, T., Kobayashi, K., Fukushima, T., Tabata, M., Ohno, I., Negishi, M., Ide, H. Takahashi, T. and Niwa, Y. (1989) Production of interleukin-1-like factor from human peripheral blood monocytes and polymorphonuclear leukocytes by superoxide anion: the role of interleukin 1 and reactive oxygen species in inflamed sites. *Clin. Immunol. Immunopathol.* 53: 439-448.

Mantovani, A. and Dejana, E. (1989) Cytokines as communication signals between leukocytes and endothelial cells. *Immunol. Today* 10: 370-375.

Markey, B. A., Phan, S. H., Varani, J., Ryan, U. S. and Ward, P. A. (1990) Inhibition of cytotoxicity by intracellular superoxide dismutase supplementation. *Free Radic. Biol. Med.* 9: 307-314.

Matsubara, T. and Ziff, M. (1986) Increased superoxide anion release from human endothelial cells in response to cytokines. *J. Immunol.* 137: 3295-3298.

May, L. T.,Torcia, G., Cozzolino, F., Ray, A., Tatter, S. B., Santhanam, U., Sehgal, P. B.and Stern, D. (1989) Interleukin-6 gene expression in human endothelial cells: RNA start sites, multiple IL-6 proteins and inhibition of proliferation. *Biochem. Biophys. Res. Commun.* 159, 991-998.

Mc Cord, J. M. Oxygen-derived radicals: a link between reperfusion injury and inflammation. *Fed. Proc.* 46: 2402-2406.

Modat, G., Dornand, J., Bernad, N., Junquero, D., Mary, A., Muller, A. and Bonne, C. (1990) LPS-stimulated bovine aortic endothelial cells produce IL-1 and IL-6 like activities. *Agents Actions* 30: 403-411.

Ozaki, Y., Ohashi, T. and Kume, S. (1987) Potentiation of neutrophil function by recombinant DNA-produced interleukin 1α. *J. Leukocyte Biol.* 42: 621-627.

Palluy, O., Bonne, C. and Modat, G. (1991) Hypoxia/reoxygenation alters endothelial prostacyclin synthesis. Protection by superoxide dismutase. *Free Radic. Biol. Med.* 11: 269-275.

Palluy, O., Morlière, L., Gris, J. C., Bonne, C. and Modat, G. (1992) Hypoxia/reoxygenation stimulates endothelium to promote neutrophil adhesion. *Free Radic. Biol. Med.* 13: 21-30.

Patel, K. D., Zimmerman, G. A., Prescott, S. M., Mc Ever, R. P. and Mc Intyre, T. M. (1991) Oxygen radicals induce human endothelial cells to express GMP-140 and bind neutrophils. *J. Cell Biol.* 112: 749-759.

Pober, J. S. and Cotran, R. S. (1991) What can be learned from the expression of endothelial adhesion molecules in tissues ? *Lab. Invest.* 64: 301-305.

Poubelle, P. E., Grassi, J., Pradelles, P. and Marceau, F. (1990) Pharmacological modulation of interleukin 1 production by cultured endothelial cells from human umbilical veins *Immunopharmacol.* 19: 121-130.

Radeke, H. H., Meier, B., Topley, N., Flöge, J., Habermehl, G. G. and Resch, K. (1990) Interleukin 1-α and tumor necrosis factor-α induce oxygen radical production in mesangial cells. *Kidney Intern.* 37: 767-775.

Rothlein, R., Dustin, M. L., Martin, S. D. and Springer, T. A. (1986) A human intercellular adhesion molecule (ICAM-1) distinct from LFA-1. *J. Immunol.* 137: 1270-1275.

Shreeniwas, R., Koga, S., Karakurum, M., Pinsky, D., Kaiser, E., Brett, J., Wolitzky, B.A., Norton, C., Plocinski, J., Benjamin, W., Burns, D. K., Goldstein, A. and Stern, D. (1992) Hypoxia-mediated induction of endothelial cell interleukin-1α. An autocrine mechanism promoting expression of leukocyte adhesion molecules on the vessel surface. *J. Clin. Invest.* 90: 2333-2339.

Sironi, M., Breviario, F.,Proserpio, P., Biondi, A., Vecchi, A., Van Damme, J., Dejana, E. and Mantovani, A. (1989) IL-1 stimulates IL-6 production in endothelial cells. *J. Immunol.* 142: 549-553.

Suzuki, M., Grisham, M. B. and Granger, D. N. (1991) Leukocyte-endothelial cell adhesive interactions: role of xanthine oxidase-derived oxidants. *J. Leukocyte Biol.* 80: 488-494.

Warner,S. J. C., Auger, K. R. and Libby, P. (1987) Interleukin 1 induces interleukin 1. II Recombinant human interleukin 1 induces interleukin 1 production by adult human vascular endothelial cells. *J. Immunol.* 139: 1911-1917.

Warren, J. S., Kunkel, S. L., Cunningham, T. W, Johnson, K. J. and Ward, P. A. (1988) Macrophage-derived cytokines amplify immune complex-triggered O2·- responses by rat alveolar macrophages. *Am. J. Pathol.* 130: 489-495.

Zimmerman, B. J., Grisham, M. B. and Granger, D. N. (1990) Role of oxidants in ischemia/reperfusion-induced granulocyte infiltration. *Am. J. Physiol.* 258: G185-G 190.

Zweier, J. L., Kuppusamy, P. and Lutty, G. A. (1988) Measurement of endothelial cell free radical generation: evidence for a central mechanism of free radical injury in post ischemic tissues. *Proc. Natl. Acad. Sci. USA* 85: 4046-4050.

Oxidative Stress, Cell Activation and Viral Infection
C. Pasquier et al. (eds)
© 1994 Birkhäuser Verlag Basel/Switzerland

Effects of Antioxidants on IL-6 Secretion Induced by IL-1 in Human Cultured Lung Fibroblasts. Involvement of NFκB

M. Raes, P. Renard, E. Bosmans*, E. Delaive and J. Remacle

*Laboratoire de Biochime Cellulaire, Facultés Universitaires Notre-Dame de la Paix, Rue de Bruxelles, 61, 5000 Namur, * Eurogenetics, 3980 Tessenderlo, Belgium.*

Summary

IL-6 secretion was investigated in human cultured lung fibroblasts, stimulated with IL-1β. IL-1β stimulated the IL-6 secretion in a dose- and time-dependent manner. The transcriptional factor NFκB has been claimed to be involved in the expression of the IL-6 gene after IL-1 stimulation and its activation could be mediated, at least partially, by free radicals produced as second messengers in response to the IL-1 signal. So, we first checked if the NFκB factor was indeed activated in our model of human lung fibroblasts after IL-1 stimulation, using the electrophoretic mobility shift analysis (EMSA). As this was indeed the case, we further investigated whether oxidative mechanisms could be involved in the activation of NFκB. A indirect approach was chosen : cells were stimulated with IL-1 in the presence of various antioxidant molecules and their effect tested on the secretion of IL-6. Two thiol-reacting molecules, pyrrolidine dithiocarbamate (PDTC) and N-acetyl-cystein (NAC) were able to completely inhibit the IL-6 secretion in a dose-dependent manner. Trolox C, an hydrosoluble analogue of vitamin E also completely inhibited the IL-6 secretion, while desferrioxamine had no effect at least at the concentration used. These data strongly support the involvement of oxidative mechanisms in the activation of NFκB leading to IL-6 secretion, in human lung fibroblasts stimulated with IL-1.

Introduction

IL-1 is a pleiotropic cytokine, acting on a large spectrum of cells and inducing various cellular responses such as the arachidonic acid cascade, protease secretion, cytokine secretion, ... (Dinarello et al., 1990). IL-1 also induces the production of intracellular oxygen radicals, not only in inflammatory cells, but also in mesenchymal fibroblasts (Meier et al., 1989) and has been shown recently to induce nitric oxide synthase in cultured rat Langerhans islets (Corbett et al, 1993). However, despite a lot of progress in our knowledge of cytokine receptors and signal transduction (for a recent review, Taga and Kishimoto, 1993), the post-receptor events involved in the IL-1 induced responses remain unclear. Fortunately, the recent discovery and characterization of the transcription factors, and particularly of the transcription factor NFκB has shed some new light on the genetic regulation of some of the genes induced by IL-1. NFκB is constitutively present in the cytoplasm as an inactive complex. IL-1, but also TNF, PMA and other activating factors, lead to its activation followed by its translocation to the nucleus, where it recognizes specific sequences (for a review, see Baeuerle, 1991; Grimm and Baeuerle, 1993).

For instance, NFκB has been involved in the regulation of the gene coding for IL-6 (Shimizu et al, 1990; Zhang et al, 1990) and the *gro* gene encoding a protein with cytokine-like properties (Anisowicz et al., 1991). The gene encoding for E-selectin, a highly tissue specific adhesion molecule, transiently and exclusively expressed on cytokine and in particular IL-1 treated endothelial cells, has also a NFκB binding site, although other promoter elements are needed and modulate NFκB function (Hooft van Huijsduijnen et al., 1992). The VCAM-1 (vascular cell adhesion molecule 1), expressed at the surface of IL-1 stimulated endothelial cells, also displays a NFκB-element (Neish et al., 1992). Recently, the chicken mitogen and IL-1 inducible prostaglandin G/H synthase (cyclooxygenase-2) gene, has been cloned ans sequenced; the promoter region contained various responsive elements and amongst them, the NFκB recognition site (Xie et al., 1993). However, both the signal transduction molecules elicited by IL-1 leading to the activation of NFκB and the exact molecular mechanisms involved in this activation, remain a matter of controversy. It has been hypothesised recently that oxygen radicals could play a role in both phenomena, acting as second messengers for the IL-1 signal and activating the NFκB factor either directly or indirectly (Shreck and Baeuerle; 1991, Baeuerle, 1992; Grimm and Baeuerle, 1993; Schreck et al, 1993). Two types of experiments argue in favour of this hypothesis. First, it has been shown at least in some cellular models such as the T-like Jurkat cells, that antioxidants block some of the IL-1 induced responses. Secondly, there is some evidence that oxidative conditions that generate intracellular oxygen radicals can mimic some of the IL-1 induced responses. For instance, the gene encoding the heme oxygenase (HO) is induced by IL-1 in mouse liver Rizzardini et al., 1993) and by H_2O_2 in human skin fibroblasts (Keyse et Tyrrel, 1989). Rizzardini et al. (1993), by sequence alignment have found that the HO gene contains recognition sites for NFκB. Reactive oxygen intermediates (ROI) have also been shown to favour the HIV replication in cultured infected T cells and monocytes, through the activation of the NFκB factor (for a recent review, Müller, 1992).

In this investigation, we wanted to confirm the above-mentioned hypothesis in a model of human lung fibroblasts in vitro, focusing the attention on one particular response induced by IL-1, the IL-6 secretion.

Materials and Methods

Cell culture

Human lung fibroblasts were purchased from the NIA Aging Cell Repository (Camden, New Jersey, USA). The cells were serially cultured in OPTI-MEM (Gibco, Great Britain) with 10 % fetal bovine serum and plated in T75 flasks at 20,000 cells/cm^2 as previously described (Raes et al., 1993). When stimulated with rhIL-1β (Janssen Biochimica, Beerse, Belgium), cells were plated in OPTI-MEM + 10 % serum for 24 hours at the right cell density and in the required culture vessel for 24 hours, then rinsed and further cultured in the same medium without serum, in the presence of lactalbumin hydrolysate, as serum substitute, enriched or not with IL-1. For the IL-6 assay, cells were plated in 96-well plates at ± 3,000 cells/well. For the EMSA experiments, cells were plated in 35 mm Petri dishes at 10^6 cells /dish.

IL-6 assay

IL-6 was assayed using an ELISA kit, based on the avidin-biotin affinity (Eurogenetics, Tessenderlo, Belgium). The presence of IL-6 is detected thanks to HRP (horse radish peroxidase) coupled to the secondary antibodies, in the presence of H_2O_2 and tetramethylbenzidine. The enzymatic reaction is stopped after 20 min with 2N H_2SO_4 and the plates are read at 450 nm in a BioRad 3550 Multiplate Reader (Richmond, Cal, USA).

Electrophoretic mobility shift analysis

Cells were stimulated or not for different periods of time (0, 15, 30 and 45 min) in the presence or not of IL-1 (20 U/ml). Cells were then rinsed twice with cold PBS and centrifuged for 20 min at 1100 rpm in a table centrifuge (Hettich, Tüttlingen, Germany). The cell pellet was rinsed with 1 ml PBS and centrifuged a second time. The second cell pellet was further treated as described by Urban and Baeuerle (1991), with the modifications of Patestos et al. (in the press). Cellular proteins were extracted in ± 0.2 ml of extracting buffer containing 0.02 M Hepes at pH 7.5, 0.35 M NaCl, 20 % glycerol, 1 % NP-40, 1 mM $MgCl_2$, 0.5 EDTA pH 7.5, 0.1 mM EGTA pH 7.5, leupeptin at 10 μg/ml, 5 mM DTT, 0.0005 % Pefabloc and aprotinin at 100 μl/ml. The protease inhibitors are added just before use. After 10 min on ice, the cell lysate was centrifuged to eliminate the cell debris at 14,000 rpm (Janetzki) for 20 min at 4 °C; the supernatant is collected, fractionated and stored at -70 °C. Proteins were assayed with the Bradford's assay (1976), using bovine serum albumin (BSA) as a standard. The cell lysate was then incubated in the presence of the P-32-labelled double-stranded oligonucleotide probe specific for the activated form of NFκB factor. The labelled oligonucleotide was kindly provided by Dr. N. Patestos (Laboratory of Molecular Biology, University of Ghent, Belgium). It is a 26-mer oligonucleotide containing the murine κ light chain enhancer motif. The binding was performed in eppendorf tubes as described by Patestos et al. (in the press), in 20 μl of binding buffer containing 20 mM Hepes pH 7.5, 5 % Ficoll 400, 60 mM KCl, 2mM DTT, enriched with 2 μg of poly d[I-C] (resuspended in 10 mM TRIS, 10 mM EDTA and 100 mM NaCl at 1 mg/ml), 2 μl of BSA (at 10 mg/ml), 10,000 to 15,000 cpm of the purified labelled oligonucleotide and 8 to 12 μg of the extracted proteins, added last to the binding mixture. The binding mixture was then incubated for 30 min at room temperature. Before electrophoresis, 1 ml of 0.3 % bromophenol blue was added per tube. The mixtures were then subjected to electrophoresis on a native 4 % acrylamide gel (30:1 acrylamide:bis-acrylamide) according to the method developped by Patestos et al. (in the press). After electrophoresis, the gel was dried under vacuum and stored in a KP63375 Kodak Cassete, in the presence of a pre-flashed film. After 1 day, the film was processed as usual, dried and ready for quantitative anaysis with the Visage 101 (Millipore, Bedford, USA) image analysis system The gels were then scanned with a high resolution camera (Videk) and the corresponding images recorded and analysed with the Whole Band software (Millipore) on a SUN station. After image processing and calibration, the intensity of the different bands was given as I.O.D. (integrated optical density, i.e. O.D. X band surface (mm^2)). After image analysis of the gels, bands were cut and counted individually for radioactivity in 5 ml of Aqua-Luma (Lumac, Netherlands) in a β-scintillation counter.

Results

The effects of IL-1β incubation on lung fibroblasts were first tested on the secretion of IL-6. Cells were incubated for 24 hours, in the presence of IL-1, at increasing concentrations ranging from 0 to 80 U/ml (Fig. 1). IL-6 production was already induced at the lowest concentration used (1 U/ml), increased drastically for concentrations ranging from 5 to 10 U/ml, but seemed to reach a plateau value from 20 U/ml to the higher concentrations.

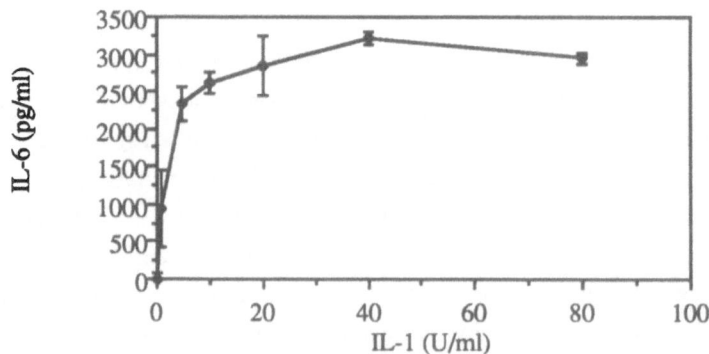

Fig. 1. Effect of IL-1β concentration on the secretion of IL-6. Cells were seeded as described in the Materials and Methods, and stimulated with IL-1 for 24 hours, at increasing IL-1 concentrations ranging fron 0 to 80 U/ml. The cell supernatants were then collected, diluted 15X and assayed for IL-6. Each value represents the mean of 3 assays ± S.D.

Cells were then incubated in the presence of IL-1 at 20 U/ml for different periods of time. As shown in Fig. 2, the induction of IL-6 is very rapid, since IL-6 is already detected for the shortest incubation times (less than 6 hours); however, IL-6 production increases over the longer incubation times (6 and 24 hours).

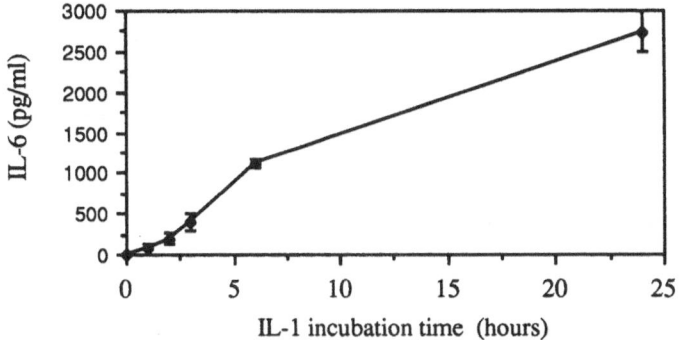

Fig. 2. IL-6 secretion as a function of IL-1 incubation time. Cells were incubated in the presence of IL-1 at 20 U/ml for increasing periods of time (0, 1, 2, 3, 6 and 24 hours). Each value represents the mean ±S.D.

As the transcriptional factor NFκB has been shown to be involved in the induction of the IL-6 gene expression in different cell types stimulated with IL-1 (Zhang et al., 1990; Shimizu et al., 1990), we checked whether IL-1 was indeed able to activate NFκB in the human lung fibroblasts used in this investigation. Therefore, the electrophoretic mobility shift analysis (EMSA) was applied on both unstimulated and IL-1-stimulated cells. The EMSA gels were scanned and analysed with an image analysis system. Some of the resulting integrated optical density (I.O.D.) profiles are illustrated in Fig. 3. In the blank lane (Fig. 3A), the O.D. profile reveals a major peak (peak D) corresponding to the free uncomplexed P32-labelled probe, in the absence of any cell lysate. Fig. 3B shows the profile obtained for unstimulated (control) cells : the D peak corresponding to the free probe is still present, but two other peaks appear (peaks B and C) . In Fig. 3C, the probe was incubated with lysates prepared from IL-1-stimulated cells : the peaks present in the control cells (C, B and D) are visible, but a fourt heavier peak (peak A) appears, revealing the complex between the activated form of NFκB and the associated labelled probe.

As shown in Table I, the activated NFκB was not detectable after 15 min of IL-1 stimulation, but appeared and increased from 30 to 45 min of IL-1 stimulation. Longer stimulation times were not performed. As shown in Table I, these conclusions can be drawn either by analysing the EMSA gels by the image analysis system or by counting the radioactivity associated with the individual peaks in each lane.

82

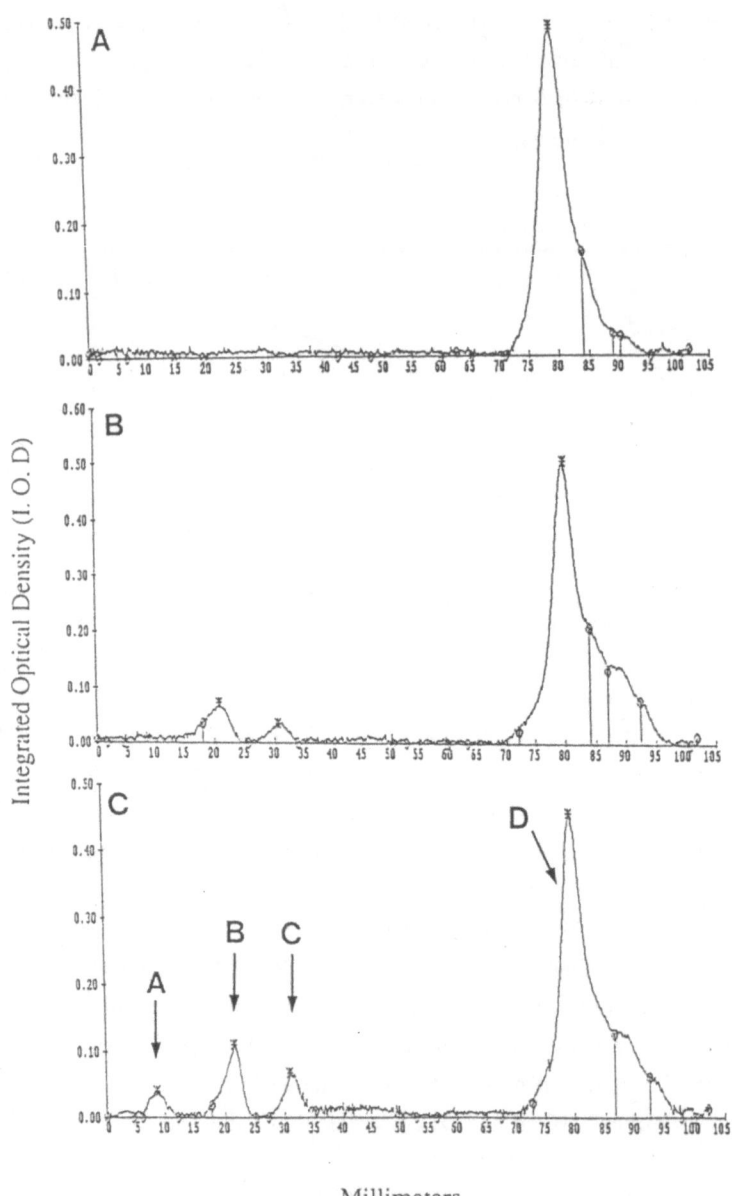

Fig. 3. I.O.D. profiles obtained by scanning and analysing the EMSA gels with the Visage 101 Image Analysis System, for the blank lane (oligo-probe without cell lysate), control cells (without IL-1) and cells stimulated with IL-1 for 30 min. Band A: probe/activated NFκB complex; D: uncomplexed oligo-probe. The EMSA experiments were performed as desribed in the Materials and Methods.

Table I. Estimation of NFκB activation by the EMSA method by image analysis of the overall gels or by radio-activity counting of the individual bands.

	Band A		Band B		Band C		Band D	
Lanes	% (IOD)	% (dpm)	% (IOD)	% (dpm)	% (IOD)	% (dpm)	% (IOD)	% (dpm)
Blank (1)	N.D	N.D	N.D	N.D	N.D	N.D	100	100
Control (2)	N.D.	N.D	8,9	6,7	4,7	4,9	86,4	88,4
IL-1 15 min (3)	N.D.	N.D	3,5	2,8	2,2	2,5	94,3	94,7
IL-1 30 min (4)	3,8	4,2	10,6	7,4	7,1	6,0	78,5	82,4
IL-1 45 min (5)	5,3	5,5	15,1	10,0	7,1	7,0	72,5	77,5

Cells were incubated in the presence of IL-1 for different periods of time. Control cells were incubated without IL-1. The blank lane corresponds to the oligo-probe alone, in the absence of cell lysate. After the EMSA, the gels were scanned and analysed with the Bio-Image system as described in the Materials and Methods. The IOD for each band was reported as % of total IOD per lane (Bands A+B+ C+D). The different bands were then eluted and their respective radioactivity (dpm) estimated in a scintillation counter; the radioactivity for each band was then reported as % of total radioactivity of the corresponding lane.

Since free radicals seems to be involved, at least partially, in the activation of NFκB and since this activation leads amongst other cellular responses to IL-6 secretion, we tested the effects of various antioxidant molecules on the IL-6 secretion. PDTC and NAC are thiol-reacting antioxidant molecules. Cells were first incubated with PDTC 0,1 mM. However, free radicals are probably not necessarily the only second messengers induced in response to the IL-1 signal. Therefore to avoid any interference with other second messengers pathways that could mask the free radical effect, we thought it would be useful to test PDTC in the presence or not of a mixture of molecules inhibiting the major second messenger pathways (Fig.4) : verapamil to block the calcium channels, neomycin sulfate to inhibit the phospholipase C cascade and 2'5'-dideoxyadenosine that inhibits adenylate cyclase. Surprisingly, the inhibitor mixture did not affect the IL-1 induced IL-6 secretion, at leat at the concentrations used. On the contrary, PDTC, whether in the presence or not of the mixture, clearly abolished completely the IL-6 secretion.

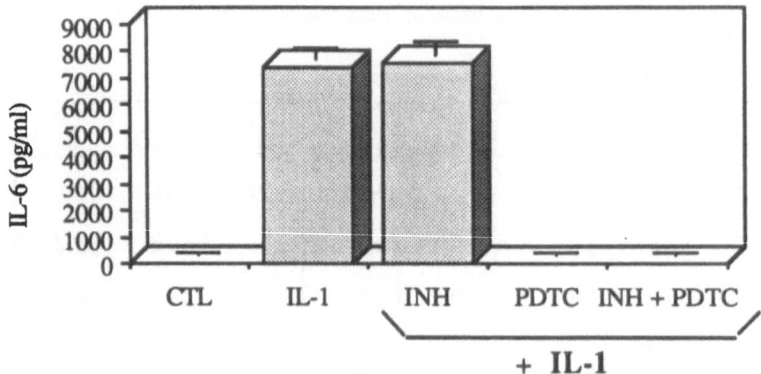

Fig. 4. Effect of PDTC on IL-6 secretion induced by IL-1. Cells were incubated for 6 hours in the presence of IL-1 at 20 U/ml alone, of IL-1 combined with PDTC at 0,1 mM , of IL-1 combined with a mixture of various inhibitors (verapamil 0,1 mM + neomycin sulfate 1 mM + 2'5'-dideoxyadenosine monophosphate 0,1 mM) or of IL-1 combined with both PDTC and the inhibitor mixture. IL-6 was assayed as described in the Materials and Methods. Each value represents the mean of 3 assays ± S.D.

Table II. Inhibitory effect of NAC on the IL-6 secretion induced by IL-1, as a function of a preincubation or not

NAC treatment	Inhibition (%)
IL-1 + NAC	30,1
1 hour preincubation with NAC followed by NAC + IL-1	100,0

Cells were preincubated or not, for 1 hour with NAC (30 mM) before the incubation for 6 hours in the presence of both IL-1(20 U/ml) and NAC (30 mM). The % of inhibition was calculated as follows :

$$\left\{ 1 - \left(\frac{IL6.produced.in.the.presence.of.IL6 + NAC}{IL6.produced.in.the.presence.of.IL6.alone} \right) \right\} \times 100\%$$

These experiments were repeated with NAC 30 mM (Table II). To our surprise, NAC was unable to completely inhibit the IL-6 secretion (30.1 % inhibition). Some authors recommend to preincubate the cells with NAC. We thus repeated the experiments, by preincubating the cells for 1 hour in the presence of NAC, before the IL-1 stimulation in the presence of NAC. In these conditions, NAC was able as well as PDTC, to inhibit the IL-6 secretion (Table II). Lower concentrations of both PDTC and NAC were also tested (Table III).

Table III. Estimation of the inhibitory potential of various antioxidants on the IL-6 production induced by IL-1β

Antioxydant	Concentration (mM)	Inhibition (%) *
PDTC	0.100 0.50 0.10	100 100 86
NAC	30 15 5 1	96 68 13 0
Desferrioxamine	0.1	0
Trolox C	0.1	98

Cells were incubated for 6 hours in the presence of both IL-1 and the antioxidants. *The % of inhibition was calculated as described in Table II. Each value represents the mean of 3 assays.

PDTC at 0, 05 mM was still fully inhibitory while at 0,01 mM the scored inhibition was about 86 %. The possible cytotoxicity of PDTC was tested (results not shown). PDTC clearly becomes toxic at 0,5 mM; at 0,1 mM it induces some cell mortality (about 20 %) at leat for 24 hour incubations. At 0,05 mM the molecule was not cytotoxic and fully inhibitory. NAC was also tested at different concentrations and shown not to be cytotoxic (cytotoxicity tests not shown) : NAC exerted a clear dose-dependent inhibition from 30 mM (96 % inhibition) to 5 mM (13 % inhibition), as shown in Table III. At 1 mM, NAC was no more effective in inhibiting the IL-6 secretion. Finally, 2 other antioxidant molecules, acting through completely different molecular mechanisms were tested (Table III) : desferrioxamine, a metal ion chelator, generally considered to limit OH° production and Trolox C, a hydrosoluble analog of vitaminE, considered to display the numerous antioxidant properties of its liposoluble natural counterpart. At the concentrations used (0,1 mM), Trolox C completely abolished the IL-6 secretion, while desferrioxamine had no effect.

Discussion

Using a model of human lung fibroblasts in culture, IL-1 was shown to induce the IL-6 secretion in a dose- and time-dependent manner. These results are in agreement with previous data obtained on other human mesenchymal strains in culture, such as human skin fibroblasts (FS4 cells) (Carty et al., 1990), chondrocytes (Bender etal., 1990; Bunning et al., 1990), and synovial cells (Harigai et al., 1989; Bender et al., 1990). However as observed by Carty et al. (1990) on skin fibroblasts, unstimulated lung fibroblasts do not constitutively secrete IL-6, while the joint cells do. The human lung fibroblasts used in this study also release arachidonic acid and prostaglandins in response to IL-1 (data not shown).

Since the transcriptional factor NFκB has been involved in the induction of the IL-6 gene in IL-1 treated cells (Zhang et al., 1990; Shimizu et al;, 1990), we first checked if NFκB was indeed activatable by IL-1 in the cellular model used in this study. Using the EMSA, we demonstrated both by image analysis and by radioactivity counting, the presence of a well-defined NFκB/probe complex, exclusively in total cell extracts of IL-1 treated cells. The appearance of this complex required 30 to 45 min of IL-1 stimulation.

In the last part of this investigation, we investigated the possible involvement of reactive oxygen intermediates (ROI) in the IL-6 secretion induced by IL-1. Therefore different antioxidant molecules were tested. We first tested 2 different thiol compounds. These molecules are well-known as protective agents against oxidative stress (see for instance, Udupi et Rice-Evans, 1992). They restore or protect sulphydryl groups and some of them, such as NAC, also enhance the availability for cystein, a substrate for glutathione synthesis. PDTC, a thiol-reacting molecule has already been shown to block the activation of NFκB (Schreck et al., 1991) at 0.1 mM. The molecule was thus tested at this concentration on the IL-6 secretion. In human fibroblasts, PDTC showed however a limited cytotoxicity at this concentration, at least for the long incubation times (24 hours). Since the molecule was still active at lower concentrations, it should be recommended to use it at 0,05 mM. NAC gave similar results, but required higher concentrations, which is in agreement with Schreck et al. (1991). However, full inhibition of IL-6 secretion by NAC was achieved, only if cells were preincubated with this antioxidant, before the IL-1 treatment. The fact that NAC not only is a thiol-reacting molecule, but also can raise intracellular GSH, could explain the need for some pre-incubation in some cell types.

Preliminary experiments were also performed with Trolox C and desferrioxamine. From our data, it appears that Trolox C was also successful in inhibiting the IL-6 secretion induced by IL-1, while desferrioxamine was not, at least at the concentration used (0,1 mM). Trolox C is an hydrophilic analogue of vitamin E that shares the same antioxidant properties, mainly the ability to scavenge peroxyl radicals, to interrupt the propagation of free radical induced reactions in polyunsaturated fatty acids and thereby to limit lipid peroxidation in cellular membranes.

Vitamin E has been shown by Israël et al. (1992) to suppress PMA and TNF induced, but also constitutive HIV enhancer activity in lymphoblastoid T cells and monocytic cells (U937), which is under the control of NFκB. However, given the extreme lipophilicity of this vitamin and its relative slow uptake by the cells, we preferred Trolox C. This molecule apparently has excellent cytoprotective properties both in vivo and in vivo (Wu et al., 1990; Zeng et al., 1990; Rubinstein et al., 1992). Desferrioxamine, a well-known iron chelator, has a very high affinity for iron (III) and is therefore thought to reduce iron-induced lipid peroxidation through inhibition of the Fenton reaction. According to Morel et al. (1992), the cytoprotective properties of desferrioxamine not only rely on its iron-chelating properties, but also on its hydroxyl-radical and peroxyl-radical scavenging activity. In our experiments, desferrioxamine did not block the IL-6 secretion induced by IL-1. However dose-response experiments in the presence of desferrioxamine, as well as the use of other antioxidants (other iron chelators (Morel et al., 1992), α-lipoic acid (Suzuki et al., 1992), butylated hydroxyanisole, nordihydroguairetic acid (Israël et al., 1992), ... are needed before drawing definite conclusions on the involvement of ROI and on their nature, in the IL-1 activation of NFκB. The intracellular sites of ROI production in activated cells also remain an open question, although at least in TNF-L929 treated cells, there is some evidence in favour of the mitochondria as possible source of ROI (Schultze-Osthoff et al., 1992; Hennet et al., 1993).

Finally, we were unable to induce IL-6 secretion by moderate oxidative stresses using either H_2O_2 or nitrofurantoin, at various concentrations and for different incubation times (not shown). Israël et al. (1992) also failed to increase the HIV enhancer activity by incubating their lymphoblastoid T and monocytic cells with H_2O_2. However, H_2O_2 potentiated the PMA effects. According to these authors, the inhibiting effects of various antioxidants on NFκB regulated cellular responses, is the consequence of a refractory state induced in cells by an imbalanced redox status, rather than by their scavenging of ROI acting as second messengers.

Human lung fibroblasts have been up to now poorly investigated regarding the NFκB activation by IL-1, despite the physiological importance of IL-1 in some inflammatory situations affecting the lung in vivo. In a preliminary study, we have shown that IL-1 indeed activates NFκB in these cells and that different antioxidant molecules completely block the secretion of IL-6. These results suggest the importance of ROI in the activation by IL-1 of cells not belonging to the immune system, such as lung fibroblasts, whatever the exact molecular mechanisms involved.

Acknowledgements

M.R. is a Research Associate of the F.N.R.S. (Brussels, Belgium). The authors gratefully thank Dr. N. Patestos, G. Haegeman and Prof. W. Fiers (University of Ghent, Belgium) for providing us the labelled oligonucleotide-probe as well as their technical advice for the EMSA experiments.

References

Anisowicz, A., Messineo, M., Lee, S.W. and Sager, R. (1991). An NF-κB-like transcription factor mediates IL-1/TNF-α induction of gro in human fibroblasts. *J. Immunol.* 147: 520-527.

Baeuerle, P.A. (1991). The inducible transcription activator NF-κB : regulation by distinct protein subunits. *Biochim. Biophys. Acta*, 1072: 63-80.

Bender, S., Haubeck, H.-D., Van de Leur, E., Dufhues, G., Schiel, X., Lauwerijns, J., Greiling, H. and Heinrich, P.C. (1990). Interleukin-1β induces synthesis and secretion of interleukin-6 in human chondrocytes. *FEBS Lett.* , 263: 321-324.

Bradford, M.M. (1976) A rapid and sensitive method for the quantitation of microgram quantities of protein utilizing the principle of protein-dye binding. *Anal. Biochem.* , 72: 248-254.

Bunning, R.A.D., Russell, R.G.G. and Van Damme, J. (1990). Independent induction of interleukin 6 and prostaglandin E2 by interleukin 1 in human articular chondrocytes. *Biochem. Biophys.Res.Comm.* 166: 1163-1170

Carty, S.E., Buresch, C.M. and Norton, J.A. (1991). Decreased IL-6 secretion by fibroblasts following repeated doses of TNF α or IL1α : post transcriptional gene regulation. *J. Surg. Res.*, 51 : 24-32.

Corbett, J.A., Sweetland, M.A., Lancaster, J.R. and McDaniel, M.L. (1993). A 1-hour pulse with IL-1β induces formation of nitric oxide and inhibits insulin secretion by rat islets of Langerhans : evidence for a tyrosine kinase signaling mechanism. *FASEB J.* 7: 369-374.

Dinarello, A., Burton, D., Ikejima, T., Adrian, P.D., Puren, J., Savage, N. and Rosoff, P.M. (1990). Interleukin 1 receptors and biological response. *Yale J. Biol. Med.*, 63: 87-93.

Grimm, S; and Baeuerle, P.A. (1993). The inducible transcription factor NF-κB : structure-function relationship of its protein subunits. *Biochem J.* 290: 297-308.

Harigai, M., Hara, M., Norioka, K., Kitani, A., Hirose, T., Suzuki, K., Kawakami, M., Masuda, K., Shinmei, M., Kawagoe, M. and Nakamura, H. (1989). Stimulation of interleukin 6-like B-cell differentiation factor production in human adherent synovial cells by recombinant interleukin 1. *Scand. J. Immunol.* 29: 289-297.

Hennet, T., Richter, C. and Peterhans, E. (1993). Tumour necrosis factor-α induces superoxide anion generation in mitochondria of L929 cells. *Biochem J.* 289: 587-592.

Hooft van Huijsduijnen, R., Whelan, J., Pescini, R., Becker-Andre, M., Schenk, A.M. and De Lamarter, J.F. (1992). A T-cell enhancer cooperates with NF-κB to yield cytokine induction of E-selectin gene transcription in endothelial cells. *J.Biol.Chem.* 267: 22385-22391.

Israel, N., Gougerot-Pocidalo, M.A., Aillet, F. and Virelizier, J.L. (1992). Redox status of cells influences constitutive or induced NF-kappa B translocation and HIV long terminal repeat activity in human T and monocytic cell lines. *J. Immunol.* 149: 3386-3393.

Keyse, S.M. and Tyrrell, R.M. (1989). Induction of the heme oxygenase gene in human skin fibroblasts by hydrogen peroxide and UVA (365 nm) radiation : evidence for the involvement of the hydroxyl radical. *Carcinogenesis* 11: 787-791.

Libermann, T.A., and Baltimore, D. (1990) Activation of interleukin-6 gene expression through the NF-kB transcription factor. *Mol. Cell. Biol.* 10: 2327-2334.

Lilienbaum, A. and Paulin, D. (1993). Activation of the human vimentin gene by the Tax human T-cell Leukemia virus I. (Mechanisms of regulation by the NF-κB transcription factor). *J. Biol.Chem.* 268: 2180-2188.

Meier, B., Radeke, H.H., Selle, S., Younes, M., Sies, H., Resch, K. and Habermehl, G.G. (1989). Human fibroblasts release reactive oxygen species in response to interleukin-1 or tumor necrosis factor-α. *Biochem. J.* 263: 539-545.

Morel, I., Cillard, J., Lescoat, G., Sergent O., Pasdeloup, N., Ocaktan, A.Z., Abdallah, M.A., Brissot, P. and Cillard, P. (1992). Antioxidant and free radical scavenging activities of the iron chelators pyoverdin and hydroxypyrid-'-ones in iron-loaded hepatocyte cultures : comparison of their mechanism of protection with that of desferrioxamine. *Free Rad. Biol. Med.* 13: 499-508.

Müller, F. (1992). Reactive oxygen intermediates and human immunodeficiency virus (HIV) infection. *Free Rad. Biol. Med.* 13: 651-657.

Neish, A.S., Williams, A.J., Palmer, H.J., Whitley, M.Z. and Collins, T. (1992) Functional analysis of the human vascular cell adhesion molecule 1 promoter. *J. Exp. Med.* 176: 1583-93.

Patestos, N. P., Haegman, G., Vandevoorde, V. and Fiers, W. NFκB activation may be required, but is not sufficient for TNF-induced IL-6 gene expression. *EMBO J.* (in the press).

Raes, M. Burton, M., Knott, I., Vanderbeck, S., Lecomte, V. and Remacle, J. (1993). Activation by interleukin-1 of human synovial cell, human skin and lung fibroblast proliferation, interaction with the arachidonic acid-derived metabolites. In : S. Nigam, K.V. Honn, L.J. Marnett and Walden, T.L., Jr. (eds) : *Eicosanoids and other bioactive lipids in cancer, inflammation and radiation injury,* Kluwer Academic Publishers, Boston, pp. 525-529.

Rizzardini, M., Terao, M., Falciani, F. and Cantoni, L. (1993). Cytokine induction of haem oxygenase mRNA in mouse liver. Interleukin 1 transcriptionally activates the haem oxygenase gene. *Biochem J.* 290 343-347.

Rubinstein, J.D., Lesnefsky, E.J., Byler, R.M., Fennessey, P.V. and Horwitz, L.D. (1992). Trolox C, a lipid-soluble membrane protective agent, attenuates myocardial injury from ischemia and reperfusion. *Free Rad. Biol. Med.* 13: 627-634.

Schreck, R. and Baeuerle, P.A. (1991). A role for oxygen radicals as second messengers. *Trends Cell Biol.* ,1: 39-42.

Schreck, R., Rieber, P. and Baeuerle, P.A. (1991) Reactive oxygen intermediates as apparently widely used messengers in the activation of the NF-kB transcription factor and HIV-1. *EMBO J.* 10: 2247-2258.

Schreck, R., Albermann, K. and Baeuerle, P.A. (1993) Nuclear factor kappa B : an oxidative stress-responsive transcription factor of eukaryotic cells (a review). *Free Rad. Res. Commun.* 17: 221-37.

Schulze-Osthoff, K. and Bakker, A.C., Vanhaesebroeck, B., Beyaert, R., Jacob, W.A. and Fiers, W. (1992). Cytotoxic activity of tumor necrosis factor is mediated by early damage of mitochondrial functions. Evidence for the involvement of mitochondrial radical generation. *J. Biol. Chem.* 267: 5317-5323.

Shimizu, H., Mitomo, K., Watanabe, T., Okamoto, S. and Yamamoto, R. (1990) Involvement of a NF-κB-like transcription factor in the activation of the interleukin-6 gene by inflammatory lymphokines. *Mol. Cell. Biol.* 10: 561-568.

Suzuki, Y.J., Aggarwal, B.B. and Packer, L. (1992) α-Lipoic acid is a potent inhibitor of NF-κB activation in human T cells *Biochem. Biophys. Res.Comm.* 189: 1709-1715.

Taga, T. and Kishimoto, T. (1993). Cytokine receptors and signal transduction. *FASEB J.*, 7 : 3387-3396.

Udupi, V. and Rice-Evans, C. (1992). Thiol compounds as protective agents in erythrocytes under oxidative stress. *Free Rad. Res. Comm.* 16: 315-323.

Urban, M.B. and Baeuerle, P.A. (1990). The 65 kD subunit of NF-κB is a receptor for IκB and a modulator of DNA-binding specificity. *Genes Dev.* 4: 1975-1984.

Wu, T.-W., Hashimoto, N., Wu, J., Carey, D., Li, R.-K, Mickle, D.A.G. and Weisel, R.D. (1990) The cytoprotective effect of Trolox demonstrated with three types oh human cells. *Biochem. Cell Biol.* 68: 1189-1194.

Xie, W., Merrill, J.R., Bradshaw, W.S. and Simmons, D.L. (1993) Structural determination and promoter analysis of the chicken mitogen-inducible prostaglandin G/H synthase gene and genetic mapping of the murine homolog. *Arch. Biochem. Biophys.* 300: 247-52.

Zeng, L-H., Wu, J., Carey, D.and Wu, T.W. (1991). Trolox and ascorbate : are they synergistic in protecting liver cells in vitro and in vivo ? *Biochem. Cell Biol.* 69: 198-201.

Zhang, Y., Lin, J.-X. and Vilcek, J. (1990) Interleukine-6 induction by tumor necrosis factor and interleukin-1 in human fibroblasts involves activation of a nuclear factor binding to a κB-like sequence. *Mol. Cell. Biol.* 10: 3818-3828.

Oxidative Stress, Cell Activation and Viral Infection
C. Pasquier et al. (eds)
© 1994 Birkhäuser Verlag Basel/Switzerland

Hydrogen peroxide-induced synthesis of the 32kDa stress protein (HO-1) in endothelial cells is serum dependent

V.R. Winrow, A. Watson, S.L. Harley and D.R. Blake

The Inflammation Research Group, ARC Bone and Joint Research Group, The London Hospital Medical College, 25-29 Ashfield Street, London E1 2AD, UK

Summary

The human 32kDa stress protein is synthesised following cellular redox imbalance and has been identified as a specific marker of oxidative stress; 32kDa stress protein-inducing agents include heavy metals, oxidising species and sulphydryl reactive compounds. As well as inducing the 32kDa protein, the oxidant H_2O_2 has been shown to induce transcription of the immediate-early genes and this activation is serum dependent. Our evidence that hypoxia/ reperfusion events occur in mobile inflamed joints prompted us to examine the vasculature. Therefore, human umbilical vein endothelial cells (HUVEC) were challenged with cadmium chloride, hydrogen peroxide (H_2O_2) and disodium aurothiomalate in the presence and absence of serum. The results show that a 32kDa protein can be induced in HUVEC, that the response to H_2O_2 is serum dependent and that it occurs at the level of transcription. Thus, two mechanisms of induction are possible: oxidants (eg H_2O_2) require serum while heavy metals and sulphydryl-reactive compounds may use a different pathway.

Introduction

Stress (heat-shock) proteins are highly conserved molecules which are essential to cell function, development and survival (Lindquist, 1986; Subjeck and Shyy, 1986; Carper, Duffy and Gerner, 1987; Riabowol, Mizzen and Welch, 1988; Welch, 1990; Ellis and van der Vies, 1991). In recent years, stress proteins have been linked to a number of autoimmune, infectious and inflammatory diseases (Young *et al*, 1988; Winrow, McLean, Morris and Blake, 1990; Cohen, 1991; Latchman, 1991; van Eden, 1991; Winrow and Blake, 1991; Welch, 1992) and inflammatory mediators including eicosanoids, cytokines and reactive oxygen species (ROS) have been shown to be potent regulators of stress protein induction (Ferris *et al*, 1988; Polla, 1988; Kaur, Welch and Saklatvala, 1989; Koizumi *et al*, 1991).

Physiological stress results in the upregulation of stress protein genes which are translated to produce the stress (heat shock) proteins and commonly include members of the 60kDa, 70kDa and 90kDa families. However, some stress protein genes are specifically activated by only certain agents. One of these is the 32kDa protein, now characterised as the microsomal enzyme haem oxygenase-1 (HO-1; Keyse and Tyrrell, 1989), which functions as the rate limiting step in the conversion of haem to bilirubin. A second non-inducible NADH-dependent enzyme, HO-2, exists in the mitochondria and is encoded by a separate gene (Maines, 1988). HO-1, first observed in cells exposed to heavy metals (Levinson, Oppermann and Jackson, 1980), has been induced in a number of different cell types using, as well as its substrate haem, UVA radiation, hydrogen peroxide (H_2O_2) and sulphydryl-reactive compounds (Caltabiano, Koestler, Poste and Greig, 1986; Keyse and Tyrrell, 1989); all of these agents alter the redox balance of the cell. Hence, HO-1 has been defined as an oxidation-specific protein (Donati, Slosman and Polla, 1990) and is believed to constitute a generalised response to oxidant stress (Keyse, Applegate, Tromvoukis and Tyrrell, 1990; Applegate, Luscher and Tyrrell, 1991).

In patients with reumatoid rathritis, the inflamed joint is hypoxic, hypercapnic and acidotic whilst also being oxidatively stressed (reviewed in Stevens, Williams, Farrell and Blake, 1991) and we have detected increased levels of 60kDa and 70kDa stress proteins in the rheumatoid synovium (Winrow *et al*, 1990). Furthermore, our group has evidence of ROS-mediated damage (Lunec *et al*, 1987; Allen, Blake, Nazhat and Jones, 1989; Winyard et al, 1991) and we have demonstrated that this occurs following pressure-induced hypoxia/ reperfusion cycles induced by movement (Blake *et al*, 1989). The immediate local response to hypoxia emanates from the blood vessels. Therefore, we chose to examine the response of human umbilical vein endothelial cells (HUVEC) to the ROS, H_2O_2, with particular reference to the 32kDa protein.

Protein synthesis is dependent, in part, on the appropriate activation of transcription factors. Using different cell lines, Nose and colleagues (1991) and Amstad, Krupitza and Cerutti (1992) showed that the induction by H_2O_2 of the immediate-early gene c-*fos* (coding a component of the AP-1 transcription factor) is mediated through a serum response element. To further elucidate the mechanism of HO-1 induction in HUVEC, we have assessed the requirement for serum following exposure to H_2O_2, cadmium chloride (Cd^{2+}) and the anti-rheumatic compound, disodium aurothiomalate (AuTM), another known inducer of the 32kDa protein (Caltabiano, Poste and Greig, 1988).

Materials and Methods

Endothelial Cell Isolation

Endothelial cells were isolated from human umbilical veins by the method of Jaffe and colleagues (1973). Cells were grown in Medium 199 (Gibco BRL, Paisley) containing 30µg/ml endothelial cell growth supplement (Sigma Chemical Co Ltd, Poole), 20% foetal calf serum (Biological Industries, Glasgow), 17U/ml heparin (CP Pharmaceuticals, Wrexham), 50U/ml penicillin (Sigma), 50µg/ml streptomycin (Sigma) and maintained in a humidified atmosphere of 5%CO_2/ 95% air. HUVEC, characterised by morphology and von Willebrand factor positivity, were used at passage 2 or 3; for protein studies, the cells were 8transferred to 24 well tissue culture plates (Nunc-Gibco BRL, Paisley) and for RNA studies, 35mm diameter tissue culture dishes (Falcon-Becton Dickinson, Cowley, Oxford) were used.

Treatment with 32kDa Stress Protein-Inducing Agents

All reagents were prepared immediately before use. Stock solutions were prepared in deionised water and diluted accordingly with medium. H_2O_2 (Sigma) was used at concentrations of 5µM, 25µM and 100µM, disodium aurothiomalate (AuTM; Aldrich Chemical Co Ltd, Gillingham) at $10^{-4}M$, $10^{-6}M$, $10^{-8}M$ and $10^{-10}M$ and cadmium chloride (Cd^{2+}; Sigma) at 20µM. H_2O_2 was applied for 30 minutes and washed off while AuTM and cadmium chloride remained in contact with the cells for the duration of the experiment.

Measurement of 32kDa Stress Protein Synthesis

HUVEC were grown on 24 well tissue culture plates (Nunc) until almost confluent. The cells were washed three times with complete Dulbecco's phosphate buffered saline (cD-PBS) and allowed to recover for 30-60 minutes in Eagles minimum essential medium with Earle's salts but without L-methionine (Gibco) and with or without 10% foetal calf serum (FCS). The 32kDa protein-inducing agents were added and HUVEC maintained in culture for a total of 8 hours from the addition of the stressor, 5µl (approx 50µCi) of ^{35}S-labelled methionine (Du Pont NEN Research Products, Stevenage) being added to each well after 6 hours. Following three washes in cD-PBS, 100µl of endothelial cell lysis buffer (Demolle, Lecompte and Boeynaems, 1988) was added to each well and the lysates stored at -70°C.

Sodium dodecyl sulphate polyacrylamide gel electrophoresis (SDS-PAGE) was carried out by the technique of Laemmli (1970) using a 12.5% separating gel and a 4% stacking gel. Proteins were separated under a constant current of 15mA/ gel, the gels dried at 80°C for 1 hour (BioRad gel dryer; Model 543) and autoradiographs prepared by exposure to X-OMAT S film (Kodak) at -70°C. Rainbow™ coloured or ^{14}C-methylated molecular weight markers (Amersham International plc, Amersham) were included on each gel.

Measurement of HO-1 mRNA Levels

HUVEC were grown on 35mm dishes until almost confluent, rinsed three times with cD-PBS and treated with H_2O_2 with or without 10% FCS. After 4 hours, RNA was isolated by the one step guanidinium thiocyanate/ phenol/ chloroform method of Chomczynski and Sacchi (1987). RNA samples (20µg) were run on agarose gels in the presence of glyoxal and blotted on to GeneScreen (Du Pont NEN Research Products). The HO-1 cDNA probe was prepared by EcoR1 digestion of a Bluescript plasmid containing the coding region of the HO-1 gene (Clone 2/10;

gift of Dr SM Keyse, ICRF, Edinburgh; Keyse and Tyrrell, 1989), yielding two fragments of 500bp and 1kb. The fragments were labelled with [_^{32}P] dCTP by random priming, Northern blots probed with either fragment using standard techniques and autoradiographs prepared.

Results

Measurement of 32kDa Stress Protein Synthesis

Figures 1, 2 and 3 are representative autoradiographs of the SDS-PAGE gel patterns obtained from lysates of HUVEC treated with 32kDa stress protein-inducing agents in the presence and absence of serum. Molecular weight markers are indicated for each gel. In figures 1 and 3, lysates of passage 2 cells were used and figure 2 displays passage 3 cells.

Figure 1a shows that HUVEC treated with H_2O_2 in the absence of serum do not synthesise the 32kDa stress protein but, when serum is added (figure 2a), an efficient response is observed. Both AuTM (figures 1b, 2b and 3) and Cd^{2+} (figures 2a and 3) induced a strong 32kDa protein response and this was not dependent on the addition of serum. No differences in expression of the 32kDa protein were noted with respect to passage number. However, some differential expression of other stress proteins occurred between passages with Cd^{2+} (figures 2a and 3).

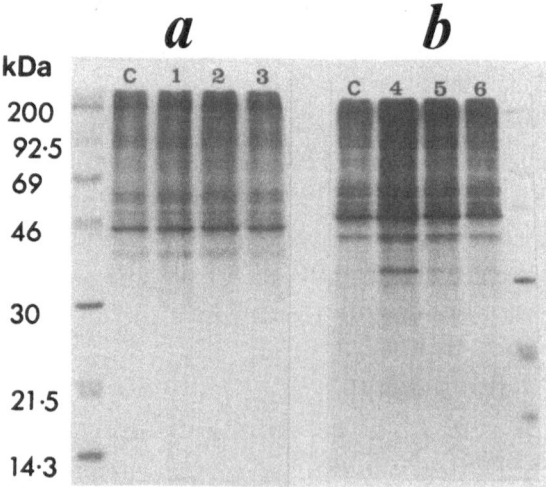

Figure 1. AuTM but not H_2O_2 induces the 32kDa stress protein in HUVEC (passage 2) in the absence of serum.
(a) H_2O_2 at 5μM (lane 1), 25μM (lane 2) and 100μM (lane 3).
(b) AuTM at 10^{-6}M (lane 4), 10^{-8}M (lane 5) and 10^{-10}M (lane 6). Untreated (control) cells are marked C.

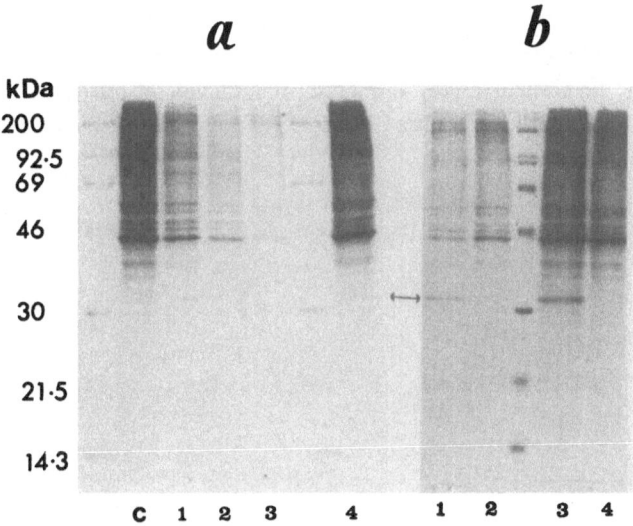

Figure 2. Induction of the 32kDa stress protein (arrowed) in HUVEC (passage 3).
(a) H_2O_2 and Cd^{2+} in the presence of serum.
Lanes 1, 2 and 3 show the gel patterns obtained after treatment with H_2O_2 at concentrations of 5μM, 25μM and 100μM respectively. Lane 4 shows the effect of 20μM Cd^{2+}. Untreated cells are marked C.
(b) AuTM in the absence (lanes 1 and 2) and presence (lanes 3 and 4) of serum.
Lanes 2 and 4 are untreated cells and lanes 1 and 3 are HUVEC treated with 10^{-6}M AuTM.

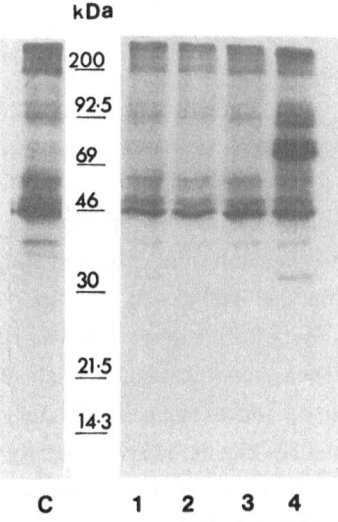

Figure 3. Induction of the 32kDa stress protein in HUVEC (passage 2) in the presence of serum.
Lanes 1, 2 and 3 show the patterns obtained following treatment of HUVEC with 10^{-4}M, 10^{-6}M and 10^{-8}M AuTM respectively Lane 4 shows the effect of 20μM Cd^2. Untreated cells are marked C.

Identification of HO-1 mRNA

As figure 4 shows, H_2O_2-induced transcription of HO-1 is clearly serum dependent. HUVEC in the presence of serum contained very low background levels of mRNA, often at the limits of detection, which were highly upregulated by H_2O_2 (Figure 3b). However, HUVEC cultured in serum free medium appeared to have higher resting levels of mRNA and no increased transcription was apparent following addition of H_2O_2.

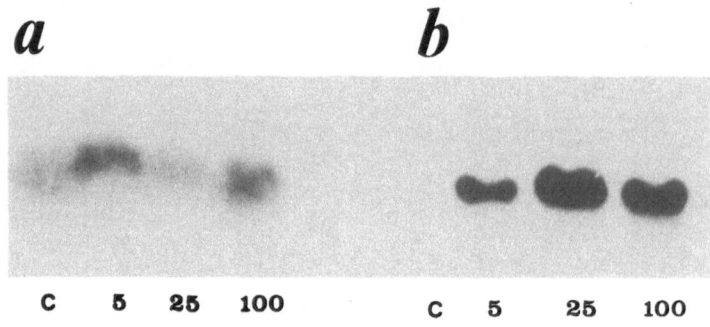

Figure 4. Induction of HO-1 mRNA by H_2O_2 in HUVEC. HUVEC were treated with 5μM, 25μM or 100μM H_2O_2, as indicated, in the (a) absence or (b) presence of serum. Untreated cells are marked C.

Discussion

Stress protein synthesis has been measured in numerous mammalian cell lines including epithelial cells, monocytic and lymphoblastoid cells and fibroblasts but comparatively little attention has been focused on the endothelial cell. This may be due to the fact that cultured endothelial cells are comparatively short-lived, becoming functionally deficient with increasing passage number. Indeed, it has been shown that endothelial cells vary in their response to heat shock, not only between different organisms (rat and calf) and at different sites (aorta and brain) but also at different passage number (Ketis, Hoover and Karnovsky, 1989). More recently, Jornot, Mirault and Junod (1991) showed that 70kDa proteins were differentially expressed in HUVEC exposed to either heat shock or H_2O_2; fewer 70kDa isoforms were seen following H_2O_2 treatment. Many of the published studies have been concerned only with the 70kDa response which is induced rapidly following many insults and is not specific to oxidation.

Our results presented here show that, following H_2O_2 treatment, the 32kDa human HO-1 protein is expressed in HUVEC only in the presence of serum, thereby underlining the need to re-appraise cell treatments. Published methods of stress protein induction vary between laboratories. Thus, Caltabiano, Koestler, Poste and Greig (1986) used AuTM and arsenite in the absence of serum while Keyse and Tyrrell (1989) used arsenite and H_2O_2 in the presence of serum. Here, we show that AuTM induces the 32kDa protein in the presence or absence of serum (figures 2 and 3) but have no data using arsenite; however, Cd^{2+} induction of HO-1 was also serum independent. Jornot, Mirault and Junod (1987) have shown that hyperoxia induces a translational defect in endothelial cells, thereby blocking 70kDa protein synthesis. This provides an important caveat since cells cultured *in vitro* are in a hyperoxic state in comparison with their tissue origin and thus, any abnormalities in stress protein expression may be a function of their hyperoxic state. It must be stated, however, that the degree of hyperoxia used to affect the translational defect (ie 95%) was excessive although prolonged culture at lower oxygen concentrations (40%) achieved similar effects.

A number of groups, using different cell types, have suggested that different mechanisms of 32kDa stress protein synthesis may operate. Thus, Mitani, Fujita, Sassa and Kappas (1990) showed differential expression of HO-1 by human hepatoma cell lines, being induced in only 1 of 3 lines tested. Keyse and Tyrrell (1990) demonstrated that induction of the 32kDa protein by H_2O_2 or UVA-radiation was inhibitable by iron chelators suggesting, at least in part, a role for the hydroxyl radical. Further, they showed transcriptional activation to be the major mechanism by which oxidants and sulphydryl reagents induced HO-1 expression (Keyse, Applegate, Tromvoukis and Tyrrell, 1990). Kantengwa and Polla (1991), using erythrophagocytosis by monocytes as a physiologic inducer of stress proteins, demonstrated that flavonoids, but not other protein kinase C (PKC) inhibitors, prevented 32kDa stress protein synthesis and linked the synthesis of stress proteins to ROS production by phagocytes; the inhibition by flavonoids was attributed to their radical scavenging activity, thus complementing the hypothesis of Burdon, Gill and Rice-Evans (1987) who suggested a role for ROS in stress protein induction following heat shock.

H_2O_2 is mitogenic at low concentrations and mitogenic activity is under the control of the immediate-early genes like c-*fos*, c-*jun* and *egr*-1. Amstad, Krupitza and Cerutti (1992), using a mouse epidermal cell line and three inducers of c-*fos*, namely serum, phorbol ester and the ROS-generating xanthine/ xanthine oxidase system, demonstrated that all three pathways converge to the same 5' regulator sequence of c-*fos*. Nose and colleagues (1991), using a mouse osteoblast line, also showed that H_2O_2 induction of c-*fos* was under the control of the serum response element and that this could be blocked using inhibitors of the major cell activator, PKC. Moreover, they further demonstrated c-*fos* induction involving protein kinases distinct from

PKC. HO-1 expression, which is known to be induced in mouse myeloleukaemia cells by a fos-AP1 complex (Kurata and Nakajima, 1990), was expressed following activation of either kinase system.

Recently, Applegate, Luscher and Tyrrell (1991) divided 32kDa stress protein-inducing agents into two groups, the first comprising oxidants or agents able to generate active intermediates and the second including agents which are known to interact with or modify cellular glutathione levels. Transcriptional activation of stress protein genes is controlled by a family of heat shock transciption factors (HSFs). Although HSF1 is activated by heat shock, oxidative stress and heavy metals, it is now known that other factors may play a role (Morimoto, 1993). Our results would suggest that the first group of HO-1 inducers, the oxidants, are under the control of a serum response element, perhaps in c-*fos*, while the second group of glutathione regulators may, in the absence of serum, use a different mechanism, perhaps analagous to the two kinase systems of Nose and colleagues (1991). Interestingly, the baseline mRNA levels in our studies appeared to be differentially expressed with higher levels in HUVEC cultured in the absence of serum. One possible explanation is that this results from activation of the second pathway of HO-1 induction although we observed no significant increase in mRNA levels when H_2O_2 was added.

Conclusion

Here, we have demonstrated that the 32kDa stress protein can be induced by two different pathways and that one of these, which appears to be oxidant-induced, is controlled at the transcriptional level by a serum response element.

Acknowledgements

We wish to thank the Arthritis and Rheumatism Council, London and Italfarmaco SpA, Milan, for financial assistance.

References

Allen, R.E., Blake, D.R., Nazhat, N.B., and Jones, P. (1989) Superoxide radical generation by inflamed human synovium after hypoxia. *Lancet* ii: 282-283.

Amstad, P.A., Krupitza, G., and Cerutti, P.A. (1992) Mechanism of c-fos induction by active oxygen. *Cancer Res* 52: 3952-3960.

Applegate, L.A., Luscher, P., and Tyrrell, R.M. (1991) Induction of heme oxygenase: A general response to oxidant stress in cultured mammalian cells. *Cancer Res* 51: 974-978.

Blake, D.R., Merry, P., Unsworth, J., Kidd, B.L., Outhwaite, J.M., Ballard, R., Morris, C.J., Gray, L., and Lunec, J. (1989) Hypoxic-reperfusion injury in the inflamed human joint. *Lancet* i: 289-293.

Burdon, R.H., Gill, V.M., and Rice-Evans, C. (1987) Oxidative stress and heat shock protein induction in human cells. *Free Radic Res Commun* 3: 129-139.

Caltabiano, M.M., Koestler, T.P., Poste, G., and Greig, R.G. (1986) Induction of 32- and 34-kDa stress proteins by sodium arsenite, heavy metals, and thiol-reactive agents. *J Biol Chem* 261: 13381-13386.

Caltabiano, M.M., Poste, G., and Greig, R.G. (1988) Induction of the 32-kD human stress protein by auranofin and related triethylphosphine gold analogs. *Biochem Pharmacol* 37: 4089-4093.

Carper, S.W., Duffy, J.J., and Gerner, E.W. (1987) Heat shock proteins in thermotolerance and other cellular processes. *Cancer Res* 47: 5249-5255.

Chomczynski, P., and Sacchi, N. (1987) Single-step method of RNA isolation by acid guanidinium thiocyanate-phenol-chloroform extraction. *Anal Biochem* 162: 156-159.

Cohen, I.R. (1991) Autoimmunity to chaperonins in the pathogenesis of arthritis and diabetes. *Annu Rev Immunol* 9: 567-589.

Demolle, D., Lecompte, M., and Boeynaems, J-M. (1988) Pattern of protein phosphorylation in aortic endothelial cells. Modulation by adenine nucleotides and bradykinin. *J Biol Chem* 263: 18459-18465.

Donati, Y.R., Slosman, D.O., and Polla, B.S. (1990) Oxidative injury and the heat shock response. *Biochem Pharmacol* 40: 2571-2577.

Ellis, R.J., and van der Vies, S.M. (1991) Molecular chaperones. *Annu Rev Biochem* 60: 321-347.

Ferris, D.K., Harel-Bellan, A., Morimoto, R.I., Welch, W.J., and Farrar, W.L. (1988) Mitogen and lymphokine stimulation of heat shock proteins in T lymphocytes. *Proc Natl Acad Sci USA* 85: 3850-3854.

Jaffe, E.A., Nachmann, R.L., Becker, C.G., and Minick, C.R. (1973) Culture of human endothelial cells derived from umbilical veins. Identification by morphologic and immunologic criteria. *J Clin Invest* 52: 2745-2756.

Jornot, L., Mirault, M.E., and Junod, A.F. (1987) Protein synthesis in hyperoxic endothelial cells: Evidence for translational defect. *J Appl Physiol* 63: 457-464.

Jornot, L., Mirault, M.E., and Junod, A.F. (1991) Differential expression of hsp70 stress proteins in human endothelial cells exposed to heat shock and hydrogen peroxide. *Am J Respir Cell Mol Biol* 5: 265-275.

Kantengwa, S., and Polla, B.S. (1991) Flavonoids, but not protein kinase C inhibitors, prevent stress protein synthesis during erythrophagocytosis. *Biochem Biophys Res Commun* 180: 308-314.

Kaur, P., Welch, W.J., and Saklatvala, J. (1989) Interleukin 1 and tumour necrosis factor increase phosphorylation of the small heat shock protein. Effects in fibroblasts, Hep G2 and U937 cells. *FEBS Lett* 258: 269-273.

Ketis, N.V., Hoover, R.L., and Karnovsky, M.J. (1989) Effects of hyperthermia on cell survival and patterns of protein synthesis in endothelial cells from different origins. *Cancer Res* 48: 2101-2106.

Keyse, S.M., and Tyrrell, R.M. (1989) Heme oxygenase is the major 32-kDa stress protein induced in human skin fibroblasts by UVA radiation, hydrogen peroxide, and sodium arsenite. *Proc Natl Acad Sci USA* 86: 99-103.

Keyse, S.M., and Tyrrell, R.M. (1990) Induction of the heme oxygenase gene in human skin fibroblasts by hydrogen peroxide and UVA (365nm) radiation: Evidence for the involvement of the hydroxyl radical. *Carcinogenesis (Lond)* 11: 787-791.

Keyse, S.M., Applegate, L.A., Tromvoukis, Y., and Tyrrell, R.M. (1990) Oxidant stress leads to transcriptional activation of the human heme oxygenase gene in cultured skin fibroblasts. *Mol Cell Biol* 10: 4967-4969.

Koizumi, T., Yamauchi, R., Irie, A., Negishi, M., and Ichikawa, A. (1991) Induction of a 31,000-Dalton stress protein by prostaglandins D_2 and J_2 in porcine aortic endothelial cells. *Biochem Pharmacol* 42: 777-785.

Kurata, S-I., and Nakajima, H. (1990) Transcriptional activation of the heme oxygenase gene by TPA in mouse M1 cells during their differentiation to macrophage. *Exp Cell Res* 191: 89-94.

Laemmli, U.K. (1970) Cleavage of structural proteins during the assembly of the head of bacteriophage T4. *Nature* 227: 680-685.

Latchman, D.S. (1991) Heat shock proteins and human disease. *J R Coll Physicians Lond* 25: 295-299.

Levinson, W., Oppermann, H., and Jackson, J. (1980) Transition series metals and sulfhydryl reagents induce the synthesis of four proteins in eukaryotic cells. *Biochim Biophys Acta* 606: 170-180.

Lindquist, S. (1986) The heat-shock response. *Annu Rev Biochem* 55: 1151-1191.

Lunec, J., Brailsford, S., Hewitt, S.D., Morris, C.J., and Blake, D.R. (1987) Free radicals: Are they possible mediators of IgG denaturation and immune complex formation in rheumatoid arthritis. *Int J Immunotherapy* 3: 39-43.

Maines, M.D. (1988) Heme oxygenase: Function, multiplicity, regulatory mechanisms, and clinical applications. *FASEB J* 2: 2557-68.

Mitani, K., Fujita, H., Sassa, S., and Kappas, A. (1990) Activation of heme oxygenase and heat shock protein 70 genes by stress in human hepatoma cells. *Biochem Biophys Res Commun* 166: 1429-1434.

Morimoto, R.I. (1993) Cells in stress: Transcriptional activation of heat shock genes. *Science* 259: 1409-1410.

Nose, K., Shibanuma, M., Kikuchi, K., Kageyama, H., Sakiyama, S., and Kuroki, T. (1991) Transcriptional activation of early-response genes by hydrogen peroxide in a mouse osteoblastic cell line. *Eur J Biochem* 201: 99-106.

Polla, B.S. (1988) A role for heat shock proteins in inflammation? *Immunol Today* 9: 134-137.

Riabowol, K.T., Mizzen, L.A., and Welch, W.J. (1988) Heat shock is lethal to fibroblasts microinjected with antibodies against hsp70. *Science* 242: 433-436.

Stevens, C.R., Williams, R.B., Farrell, A.J., and Blake, D.R. (1991) Hypoxia and inflammatory synovitis: Observations and speculation. *Ann Rheum Dis* 50: 124-132.

Subjeck, J.R., and Shyy, T.T. (1986) Stress protein systems of mammalian cells. *Am J Physiol* 250: C1-C17.

van Eden, W. (1991) Heat-shock proteins as immunogenic bacterial antigens with the potential to induce and regulate autoimmune arthritis. *Immunol Rev* 121: 5-28.

Welch, W.J. (1990) The response of mammalian cells to environmental stress. In: Burdon, R., Rice-Evans C., Blake, D. and Winrow, V. (eds.): *Stress Proteins in Inflammation*, Richelieu Press, London, pp. 13-52.

Welch, W.J. (1992) Mammalian stress response: Cell physiology, structure/ function of stress proteins, and implications for medicine and disease. *Physiol Rev* 72: 1063-1081.

Winrow, V.R., and Blake, D.R. (1991) Stress proteins, hearts, and joints. *Lancet* 337: 614-615.

Winrow V.R., McLean, L., Morris, C.J., and Blake, D.R. (1990) The heat shock response and its role in inflammatory disease. *Ann Rheum Dis* 49: 128-132.

Winrow, V.R., Mojdehi, G., Mapp, P.I., Rampton, D.S., and Blake, D.R. (1990) Immunohistological localisation of stress proteins in inflammatory tissue. In: Burdon, R., Rice-Evans C., Blake, D. and Winrow, V. (eds.): *Stress Proteins in Inflammation*, Richelieu Press, London, pp. 237-51.

Winyard, P.G., Zhang, Z., Chidwick, K., Blake, D.R., Carrell, R.W., Murphy, G. (1991) Proteolytic inactivation of human $_1$ antitrypsin by human stromelysin. *FEBS Lett* 279: 91-94.

Young, D., Lathigra, R., Hendrix, R., Sweetser, D., and Young, R.A. (1988) Stress proteins are immune targets in leprosy and tuberculosis. *Proc Natl Acad Sci USA* 85: 4267-4270.

Oxidative Stress, Cell Activation and Viral Infection
C. Pasquier et al. (eds)
© 1994 Birkhäuser Verlag Basel/Switzerland

The Antioxidant Effects of Glutathione and Ascorbic Acid

A. Meister

Department of Biochemistry, Cornell University Medical College, 1300 York Avenue, New York, N.Y. 10021, USA.

Summary

The functions of glutathione have been explored by use of a model animal system in which the cellular synthesis of glutathione is inhibited. Glutathione deficiency, induced by administration of buthionine sulfoximine (an inhibitor of the first step of glutathione synthesis) leads to mortality in newborn rats and in guinea pigs, animals that are unable to synthesize ascorbic acid, and to tissue damage in these animals and in adult mice, which can synthesize ascorbic acid. Mortality and morbidity are greatly diminished by administration of glutathione esters or of ascorbic acid. These and other findings indicate that cellular glutathione is essential for the physiological function of ascorbic acid, that ascorbic acid can spare glutathione, and that glutathione (supplied as an ester) can spare ascorbic acid. Recent studies showed that administration of glutathione (in an ester form) to guinea pigs fed an ascorbate-deficient diet significantly delays the onset of scurvy, a disease in which oxidative stress leads to inactivation of certain enzymes that catalyze hydroxylation reactions. The present findings demonstrate that ascorbic acid and glutathione function together as an antioxidant couple. Since there is good evidence that glutathione and/or ascorbic acid function in maintaining the reduced forms of other cellular components, such as α-tocopherol, it appears that glutathione is the source of a major portion of cellular antioxidant activity.

Introduction

Glutathione has many functions in cellular metabolism; an important one relates to the protection of cells against the toxic effects of oxygen. Glutathione is of major significance as an antioxidant because it participates directly in the destruction of reactive oxygen compounds and also because it maintains in reduced forms such compounds as ascorbate and -tocopherol, which also exert antioxidant effects. Several reviews of glutathione metabolism and function are available (Larsson *et al.*, 1983; Dolphin *et al.*, 1989; Taniguchi *et al.*, 1989; Meister and Anderson, 1983), and a brief discussion of the antioxidant effects of ascorbate and glutathione has appeared (Meister, 1992a). Early studies indicated that animal tissues have the ability to reduce dehydroascorbate (Szent-Gyorgyi, 1928) and that glutathione is involved in this reaction (Borsook *et al.*, 1937). The reaction has been studied in plants (Hopkins and Morgan, 1936) and in a number of animal tissues (see, for example: Christine *et al.*, 1956; Bigley *et al.*, 1981; Rose *et al.*, 1989; Wells *et al.*, 1992). Recent studies showed that highly purified thiol transferases (glutaredoxin and protein disulfide isomerase) can catalyzed the glutathione-dependent reduction of dehydroascorbate (Wells *et al.*, 1990). It has been reported that mitochondria, which apparently lack thiol transferase activity, can catalyze the pyridine nucleotide-dependent reduction of semi-dehydroascorbate (Stahl *et al.*, 1986; Alcain *et al.*, 1991; Coassin *et al.*, 1991; Iyanagi *et al.*, 1985; Diliberto *et al.*, 1982). In the course of our studies on the effects of glutathione deficiency in adult mice and newborn rats, we have obtained evidence, as summarized here, that glutathione-dependent dehydroascorbate reduction is of major significance *in vivo*.

Various methods have been used for production of experimental glutathione deficiency and of oxidative stress (e.g., application of increased oxygen, oxidizing agents such as diamide and t-butyl hydroperoxide, and thiol reactive agents such as diethylmaleate), but these approaches are non-specific and thus may lead to changes in many cell constituents. In the model of glutathione deficiency developed in our laboratory, the experimental animals (or cell suspensions) are treated with buthionine sulfoximine, a selective transition state inhibitor of -glutamylcysteine synthetase, the enzyme that catalyzes the first and rate-limiting step in the synthesis of glutathione (Meister, 1978, 1983, 1991, 1992a, 1992b). [Certain other S-alkyl homocysteine sulfoximines may also be used, but methionine sulfoximine is not useful because, though it inhibits -glutamylcysteine synthetase, it also inhibits glutamine synthetase, and inhibition of glutamine synthetase in the brain leads to convulsions and death]. Inhibition of -glutamylcysteine synthetase produces markedly decreased levels of cellular glutathione and oxidative tissue damage, which reflects the very significant normal physiological endogenous formation of reactive oxygen intermediates and free radicals; these are normally destroyed by reactions involving glutathione. This model of oxidative stress, which is different from those produced by application of exogenous stress, may be analogous to oxidative stress that occurs in certain degenerative conditions and diseases.

When glutathione deficiency is thus produced in adult mice, certain tissues such as skeletal muscle, lung, jejunum and colon, undergo marked damage associated with mitochondrial degeneration, whereas others (e.g., liver, kidney, heart, stomach) are relatively unaffected. In contrast, glutathione deficiency induced in newborn rats produces severe damage to liver, kidney, brain, lung, cerebral cortex and to formation of cataracts; glutathione deficiency in newborn rats leads to death within a few days. Tissue damage and mortality are markedly decreased by administration of glutathione monoesters, but not by administration of glutathione. Tissue damage and mortality are also greatly decreased by administration of ascorbate. These and related studies, reviewed here, have led to the conclusions that (**a**) glutathione is essential for the *in vivo* functioning of ascorbate, (**b**) glutathione and ascorbate have functions in common and can spare each other, and (**c**) glutathione provides a quantitatively significant and apparently major part of cellular antioxidant activity.

Effects of Glutathione Deficiency in Adult Mice and Newborn Rats.

When adult mice are given L-buthionine-SR-sulfoximine for 6-14 days they develop skeletal muscle degeneration (Martensson and Meister, 1989), damage to lung type 2 cells with destruction of lamellar bodies (Martensson *et al.*, 1989), and degeneration of the mucosa of the jejunum and colon (Martensson *et al.*, 1990a). These effects, which are readily observed by electron microscopy, are associated with mitochondrial degeneration. Mitochondria do not have the synthetases required for glutathione synthesis, but obtain glutathione by transport from the cytosol (Griffith and Meister, 1985; Martensson *et al.*, 1990b). In the course of metabolism, mitochondria produce significant amounts of hydrogen peroxide (Boveris and Chance, 1973; Boveris *et al.*, 1972; Loschen and Flohe, 1971; Forman and Boveris, 1982), which is normally destroyed by the action of glutathione peroxidase. When there is a marked deficiency of glutathione, mitochondrial and other types of cellular damage occur (Meister, 1991).

The effects of glutathione deficiency can be almost entirely protected against by administration of glutathione monoesters, but not by giving glutathione. Glutathione is poorly transported into most cells; its apparent "uptake" by cells, occasionally reported, is usually due to extracellular breakdown, transport of the resulting amino acids or dipeptides, and intracellular synthesis of glutathione (Meister, 1991). Glutathione monoesters are, in contrast to glutathione itself, readily transported into cells and split intracellularly to form glutathione (Puri and Meister, 1983; Wellner *et al.*, 1984; Anderson *et al.*, 1985; Anderson and Meister, 1989).

When glutathione deficiency is produced in newborn rats by giving L-buthionine-SR-sulfoximine, severe tissue damage occurs (Martensson and Meister, 1991). In contrast to the effects seen in adult mice, which do not exhibit liver and kidney damage, the treated newborn rats show focal degeneration of the liver, proximal renal tubular necrosis as well as severe damage to lung (lamellar body degeneration, decrease of surfactant) and cerebral cortex (mitochondrial degeneration). These animals survive for only 4-6 days. These effects of administration of L-buthionine-SR-sulfoximine were not found after giving L-buthionine-R-sulfoximine, a stereoisomer that does not inhibit -glutamylcysteine synthetase. The effects were prevented by giving glutathione esters. These observations support the conclusion that tissue damage and lethality found after giving L-buthionine-SR-sulfoximine are due to glutathione deficiency.

Glutathione-deficient newborn rats were found to have marked depletion of tissue ascorbate (as well as low tissue levels of glutathione) (Martensson and Meister, 1991). The levels of total ascorbate (ascorbate + dehydroascorbate) were decreased in the glutathione deficient tissues; however, the levels of ascorbate in the glutathione deficient tissues (which are normally very close to those for total ascorbate), were lower than those of total ascorbate indicating a significant increase in dehydroascorbate levels in glutathione deficiency. Some representative data on tissue ascorbate levels are given in Table I.

Table I. GSH deficiency decreases tissue ascorbate levels

Tissue	Ascorbate, μmol/g			
	Control	BSO	BSO + ascorbate (2 mmol/kg)	BSO + ascorbate (0.4 mmol/kg)
Kidney	2.58 ± 0.13	0.58 ± 0.16*	2.58 ± 0.24†	2.69 ± 0.21†
	[2.72 ± 0.13]	[1.36 ± 0.27]*	[2.66 ± 0.47]	[2.63 ± 0.30]
Liver	2.51 ± 0.30	0.42 ± 0.15*	3.08 ± 0.50†	2.76 ± 0.28†
	[2.60 ± 0.28]	[1.14 ± 0.23]*	[3.13 ± 0.43]†	[2.80 ± 0.44]†
Lung	2.46 ± 0.30	0.70 ± 0.32*	2.74 ± 0.42†	2.93 ± 0.46†
	[2.56 ± 0.25]	[1.52 ± 0.16]*	[3.12 ± 0.74]†	[2.90 ± 0.90]†
Brain	5.81 ± 0.44	2.71 ± 0.22*	4.86 ± 0.17†	4.77 ± 0.35*†
	[5.98 ± 0.29]	[4.02 ± 0.85]*	[4.75 ± 0.70]	[5.31 ± 0.98]

Tissue levels (values are mean ± SD; $n = 4$) of ascorbate and "total" ascorbate (given in brackets) were determined on samples obtained at 1300 (2 hr after injection of ascorbate or saline) on day 5 (after given seven doses of BSO over 3.5 days). The ascorbate [total ascorbate] values after two doses of BSO (14 hr) for kidney, liver, lung, and cerebral cortex (brain), were, respectively, 0.40 [0.50], 0.13 [0.29], 0.04 [0.46], and 2.00 [2.83] μmol/g. The data for the controls at this age are, respectively, 0.72 [1.01], 0.65 [0.94], 0.60 [0.94], and 1.99 [2.79]. Statistically significantly different data ($P < 0.025$) with respect to controls (*) and BSO (†) groups are indicated.

Mortality of newborn rats treated with L-buthionine-SR-sulfoximine as well as the associated damage to brain, liver, kidney, and lung are largely prevented by giving ascorbate. Interestingly, administration of ascorbate to these animals not only increased the tissue levels of ascorbate to the normal range, but also led to a significant increase in the tissue levels of glutathione. For example, the levels of mitochondrial glutathione increased 2.7 to 6.0 fold in various tissues after administration of ascorbate. These findings demonstrate that ascorbate can spare glutathione under these conditions. Dehydroascorbate was found not to replace ascorbate in this experimental model.

We have also used an experimental protocol in which newborn rats were given only 2 doses of L-buthionine-SR-sulfoximine; these were given on the second and third day of life (Martensson and Meister, 1991). These animals were found to have cataracts when they opened their eyes on days 14-16 of life. Cataract formation in this model was markedly decreased by administration of glutathione monoester (but not of glutathione). When ascorbate was studied in this model cataracts were also almost entirely prevented by administration of ascorbate at doses of 2 mmol/kg/day. Interestingly, administration of dehydroascorbate was also protective, presumably because in this model dehydroascorbate can be reduced to ascorbate. Thus, such animals, which had been given only 2 doses of buthionine sulfoximine, retained appreciable capacity to synthesize glutathione in the liver. The protection observed after administration of dehydroascorbate may be ascribed to its reduction in the liver or other organs and transport of ascorbate via the plasma to the developing lens.

When adult mice are made glutathione deficient they exhibit marked lung type 2 cell damage with degeneration of the lamellar bodies, and analyses reveal substantially decreased levels of phosphatidylcholine (a major constituent of lung surfactant) in the lung and in the bronchoalveolar lining fluid (Jain *et al.*, 1992). Treatment with ascorbate (1-2 mmol/kg/day) prevents the mitochondrial and lamellar body damage as well as the substantial decline in the levels of phosphatidylcholine.

It is notable that glutathione deficiency is lethal to newborn rats and to guinea pigs (see below), animals that are unable to synthesize ascorbate, and that administration of ascorbate prevents lethality. Mortality due to glutathione deficiency has not been seen in adult mice, which are capable of synthesizing ascorbate. Indeed, when adult mice are treated with L-buthionine-SR-sulfoximine, the ascorbate content of the liver increases about 2-fold within a few hours (Martensson and Meister, 1992). Thus, an early effect of glutathione deficiency in adult mice is apparent induction of ascorbate synthesis in the liver. In contrast, no such induction occurs in newborn rats nor would it be expected in animals that are incapable of ascorbate synthesis such as guinea pigs or humans. Induction of ascorbate synthesis in the liver of adult mice treated with buthionine sulfoximine is followed soon by a decline in the level of ascorbate to about control levels, but the level of total ascorbate remains elevated indicating that there is an elevated level of dehydroascorbate. In the kidney, the level of ascorbate declines and that of dehydroascorbate increases. In the lung, both ascorbate and total ascorbate levels decrease, the former more rapidly. The data show that the levels of tissue dehydroascorbate increase markedly when there is glutathione deficiency. The increase in hepatic ascorbate synthesis in adult mice in response to glutathione deficiency may serve to supply some ascorbate to other tissues such as the kidney, but it is apparently insufficient to completely protect the lung or skeletal muscle (Jain *et al.*, 1992).

Effect of Glutathione Ester on Ascorbate Deficiency in Guinea Pigs.

Treatment of adult guinea pigs with buthionine sulfoximine produces results that closely resemble those seen in newborn rats (Griffith *et al.*, 1991). Thus, guinea pigs develop liver damage associated with focal necrosis, proximal renal tubular damage, and damage to the lung characterized by lamellar body degeneration. These changes are accompanied by severe mitochondrial damage and the animals die within a few days. Treatment with ascorbate (1 mmol per kg thrice daily) prevents morbidity and mortality, as found in the studies on newborn rats.

The studies summarized above indicate that administration of ascorbate to glutathione-deficient newborn rats and guinea pigs prevents toxicity and mortality and leads to increased levels of total tissue and mitochondrial glutathione. *Thus, ascorbate spares glutathione*. To answer the converse question - *Does administration of glutathione spare ascorbate?* - we gave glutathione monoethyl ester (which is efficiently transported and split to glutathione intracellularly) to ascorbate-deficient guinea pigs. When guinea pigs are made deficient in ascorbate by feeding a scorbutic diet, they develop signs of scurvy and die within 21-24 days. We found that when ascorbate-deficient guinea pigs were treated with glutathione monoethyl ester, the onset of scurvy (as indicated by weight loss, bone changes, hematomas) was significantly delayed (Martensson *et al.*, 1993). Mitochondrial glutathione levels were decreased in scurvy. The tissue ascorbate levels and glutathione levels of the glutathione ester-treated guinea pigs were higher than those of saline-treated controls. The marked loss of osteoid material (intercellular acidophilic matrix substance) from the long bones typically seen in scurvy was prevented or greatly decreased in the glutathione-ester-treated guinea pigs. The sparing effect of glutathione in scurvy is probably associated with an increase in the reduction of dehydroascorbate (which would otherwise be degraded) and to antioxidant effects of glutathione that are also produced by ascorbate.

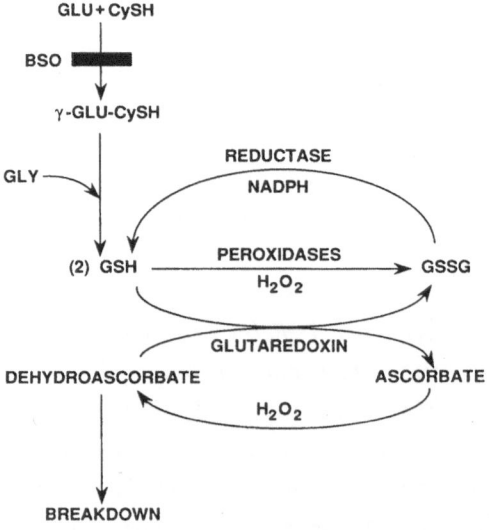

Fig. 1. Ascorbate and GSH act to destroy peroxide (and related active oxygen forms). Glutaredoxin and protein disulfide isomerase catalyze GSH-dependent reduction of dehydroascorbate. GSSG, glutathione disulfide.

Discussion

Scurvy, a disease known for many years, is produced by lack of ascorbate, an essential antioxidant vitamin. In its absence or marked decrease, certain enzymes such as prolyl hydroxylase and probably others that catalyze important hydroxylation reactions are oxidatively inactivated (Englard and Seifter, 1986; Levine, 1986). The consequences of such oxidative stress are severe and lead to pathological phenomena that include decreased collagen and bone formation, and (in guinea pigs) to arterial disease that closely resembles human atherosclerosis (Willis, 1953; Ginter, 1973).

Ascorbate and glutathione can function in the destruction of reactive oxygen compounds (Fig. 1). The observed metabolic redundancy serves to emphasize the physiological importance of this antioxidant function (Meister, 1992a). Glutathione, through the action of the glutathione peroxidases, participates in the reduction of hydrogen peroxide and other peroxides (Mills, 1957; Flohe, 1989; McCay and Powell, 1989; Mannervik, 1985; Burk et al., 1980; Ursini et al., 1980). Ascorbate can also interact with hydrogen peroxide; such reactions may occur nonenzymatically, but are known to be enzyme-catalyzed in chloroplasts, cyanobacteria, and soybean nodules (Dalton et al., 1986), and may also be catalyzed in animal tissues.

The findings indicate that glutathione and ascorbate can "spare" each other, but it seems likely that each is able to participate in certain reactions that the other cannot perform as efficiently. It seems probable that there are critically essential minimum cellular levels of ascorbate and of glutathione. The physiological reduction of dehydroascorbate to ascorbate appears to depend importantly on glutathione (Meister, 1992a). A number of investigations have led to the view that glutathione or ascorbate (or both) are required for maintenance of -tocopherol in the reduced state (Diplock et al., 1989; Packer et al., 1979; Niki et al., 1982; Reddy et al., 1982; Doba et al., 1985; Niki et al., 1987; Leedle and Aust, 1990; Graham et al., 1989; Scholich et al., 1989; Wefers and Sies, 1988). It thus appears that what may be called the "glutathione antioxidant system" is of major importance in the protection of cells against the toxic effects of oxygen. Studies on the experimental model system of glutathione deficiency (Meister, 1978, 1983, 1991, 1992a, 1992b) indicate that normal cellular function requires glutathione, and that ascorbate can serve as an essential antioxidant in glutathione deficiency. Application of this approach to other cellular components and to the testing of potentially useful therapeutic agents should be of interest. The idea that glutathione deficiency may serve as a model or mimic of certain diseases or conditions seems worthy of further consideration.

Glutathione deficiency induced by inhibition of -glutamylcysteine synthetase is currently being tried as a treatment for drug- and radiation-resistant tumors (Meister, 1983, 1988; Vistica et al., 1989; Ozols et al., 1988; Hamilton et al., 1990; O'Dwyer et al., 1992). In this approach it is anticipated that the resistant tumor cells, which often have a higher level of glutathione and a higher requirement for glutathione synthesis than the normal cells, would be more sensitive to anticancer agents after inhibition of glutathione synthesis than the normal cells, which usually have a large excess of glutathione. The possible influence of ascorbate in this type of treatment has been considered (Meister, 1992c) and needs study.

Acknowledgement

Research support from the U.S. Public Health Service, National Institutes of Health, Grant R37-DK-12034, is acknowledged.

References

Alcain, F.J., Buron, M.I., Villalba, J.M., and Navas, P. (1991) Ascorbate is regenerated by HL-60 cells through the transplasmalemma redox system. *Biochim. Biophys. Acta* 1073: 380-385.

Anderson, M.E., and Meister, A. (1989) Glutathione monoesters. *Anal. Biochem.* 183: 16-20.

Anderson, M.E., Powrie, F., Puri, R.N., and Meister, A. (1985) Glutathione monoethyl ester: Preparation, uptake by tissues, and conversion to glutathione. *Arch. Biochem. Biophys.* 239: 538-548.

Bigley, R., Riddle, M., Layman, D., and Stankova, L. (1981) Human cell dehydroascorbate reductase kinetic and functional properties. *Biochim. Biophys. Acta* 659: 15-22.

Borsook, H., Davenport, H.W., Jeffreys, C.E.P., and Warner, R.C. (1937) The oxidation of ascorbic acid and its reduction *in vitro* and *in vivo*. *J. Biol. Chem.* 117: 237-279.

Boveris, A., and Chance B. (1973) The mitochondrial generation of hydrogen peroxide. General properties and effect of hyperbaric oxygen. *Biochem. J.* 134: 707-716.

Boveris, A., Nozomu, O., and Chance, B. (1972) The cellular production of hydrogen peroxide. *Biochem. J.* 128: 617-630.

Burk, R.F., Trumble, M.J., and Lawrence, R.A. (1980) Rat hepatic cytosolic glutathione-dependent enzyme protection against lipid peroxidation in the NADPH-microsomal lipid peroxidation system. *Biochim. Biophys. Acta* 618: 35-41.

Christine (1956) The reduction of dehydroascorbic acid by human erythrocytes. *Clin. Chim. Acta* 1: 557-569.

Coassin, M., Tomasi, A., Vannini, V., and Ursini, F. (1991) Enzymatic recycling of oxidized ascorbate in pig heart: One-electron *vs* two-electron pathway. *Arch. Biochem. Biophys..* 290: 458-462.

Dalton, D.A., Russell, S.A., Hanus, F.J., Pascoe, G.A., and Evans, H.J., (1986) Enzymatic reactions of ascorbate and glutathione that prevent peroxide damage in soybean root nodules. *Proc. Natl. Acad. Sci. USA* 83: 3811-3815.

Diliberto, E.J. Jr., Dean, G., Carter, C., and Allen, P.L. (1982) Tissue, subcellular, and submitochondrial distributions of semidehydroascorbate reductase: Possible role of semidehydroascorbate reductase in cofactor regeneration. *J. Neurochem.* 39: 563-568.

Diplock, A.T., Machlin, L.J., Packer, L., and Pryor, W.A., Eds. (1989) Conference on ascorbic acid. *Ann. N.Y. Acad. Sci.* 570.

Doba, T., Burton, G.W. and Ingold, K.U. (1985) Antioxidant and co-antioxidant activity of vitamin C. The effect of vitamin C, either alone or in the presence of vitamin E or a water-soluble vitamin E analogue, upon the peroxidation of aqueous multilamellar phospholipid liposomes. *Biochim. Biophys. Acta* 835: 298-303.

Dolphin, D., Poulson, R., and Avramovic, O. (Eds.) (1989) *Glutathione: Chemical, Biochemical, and Medical Aspects*, Parts A and B, *Coenzyme and Cofactors Series*, Vol. III. John Wiley, New York.

England, S., and Seifter, S. (1986) The biochemical functions of ascorbic acid. *Annu. Rev. Nutr.* 6: 365-406.

Flohe, L. (1989) The selenoprotein glutathione peroxidase. *In: Glutathione Chemical, Biochemical, and Medical Aspects*, eds. Dolphin, D., Poulson, R. & Avramovic, O., Wiley, New York, Part A., pp. 643-731.

Forman, H.J., and Boveris, A. (1982) Superoxide radical and hydrogen peroxide in mitochondria. *In: Free Radicals in Biology*, (Pryor, W.A., Ed.). Academic Press, New York, Vol. 5, pp. 65-90.

Ginter, E. (1973) Cholesterol: Vitamin C controls its transformation to bile acids. *Science* 179: 702-704.

Graham, K.S., Reddy, C.C., and Scholz, R.W.. (1989) Reduced glutathione effects on -tocopherol concentration of rat liver microsomes undergoing NADPH-dependent lipid peroxidation. *Lipids* 24: 909-914.

Griffith, O.W., Han, J., and Martensson, J. (1991) Vitamin C protects adult guinea pigs against tissue damage and lethality caused by buthionine sulfoximine-mediated glutathione depletion. *FASEB J.* 5: 4708.

Griffith, O.W., and Meister, A. (1985) Origin and turnover of mitochondrial glutathione. *Proc. Natl. Acad. Sci. USA* 82: 4668-4672.

Hamilton, T., O'Dwyer, P., Young, R., Tew, K., Padavic, K., Comis, R., and Ozols, R. (1990) Phase I trial of buthionine sulfoximine (BSO) plus melphalan (L-PAM) in patients with advanced cancer. *Proc. Annu. Meet. Am. Soc. Clin. Oncol.* 9: A281.

Hopkins, F.G., and Morgan, E.J. (1936) Some relations between ascorbic acid and glutathione. *Biochem. J.* 30: 1446-1462.

Iyanagi, T., Yamazaki, I., and Anan, K.F. (1985) One-electron oxidation-reduction properties of ascorbic acid. *Biochim. Biophys. Acta* 806: 255-361.

Jain, A., Martensson, J., Mehta, T., Krauss, A.N., Auld, P.A.M., and Meister, A. (1992) Ascorbic acid prevents oxidative stress in glutathione-deficient mice; Effects on lung type 2 cell lamellar bodies, lung surfactant, and skeletal muscle. *Proc. Natl. Acad. Sci. USA* 89: 5093-5097.

Larsson, A., Orrenius, S., Holmgren, A., and Mannervik, B. (Eds.) (1983) *Functions of Glutathione. Biochemical, Physiological, Toxicological, and Clinical Aspects*. Raven Press, New York.

Leedle, R.A., and Aust, S.D. (1990) The effect of glutathione on the vitamin E requirement for inhibition of liver microsomal lipid peroxidation. *Lipids* 25: 241-245.

Levine, M. (1986) New concepts in the biology and biochemistry of ascorbic acid. *N. Engl. J. Med.* 314: 892-902.

Loschen, G., and Flohe, L. (1971) Respiratory chain linked H_2O_2 production in pigeon heart mitochondria. *FEBS Lett..* 18: 261-264.

Mannervik, B. (1985) Glutathione peroxidase. *Methods Enzymol* 113: 490-4995.

Martensson, J., Han, J., Griffith, O.W., and Meister, A. (1993) Glutathione ester delays the onset of scurvy in ascorbate-deficient guinea pigs. *Proc. Natl. Acad. Sci. USA* 90: 317-321.

Martensson, J., Jain, A., Frayer, W., and Meister, A. (1989) Glutathione metabolism in the lung: Inhibition of its synthesis leads to lamellar body and mitochondrial defects. *Proc. Natl. Acad. Sci. USA* 86: 5296-5300.

Martensson, J., Jain, A., and Meister, A. (1990a) Glutathione is required for intestinal function. *Proc. Natl. Acad. Sci. USA* 87: 1715-1719.

Martensson, J., Jain, A., Stole, E., Frayer, W., Auld, P.A.M., and Meister, A. (1991) Inhibition of glutathione synthesis in the newborn rat: A model for endogenously produced oxidative stress. *Proc. Natl. Acad. Sci. USA* 88: 9360-9364.

110

Martensson, J., Lai, J.C.K., and Meister, A. (1990b) High affinity transport of glutathione is part of a multicomponent system essential for mitochondrial function. *Proc. Natl. Acad. Sci. USA* 87: 7185-7189

Martensson, J., and Meister, A. (1989) Mitochondrial damage in muscle occurs after marked depletion of glutathione and is prevented by giving glutathione monoester. *Proc. Natl. Acad. Sci. USA* 86: 471-475.

Martensson, J., and Meister, A. (1991) Glutathione deficiency decreases tissue ascorbate levels in newborn rats: Ascorbate spares glutathione and protects. *Proc. Natl. Acad. Sci. USA* 88: 4656-4660.

Martensson, J., and Meister, A. (1992) Glutathione deficiency increases hepatic ascorbic acid synthesis in adult mice. *Proc. Natl. Acad. Sci. USA* 89: 11566-11568.

McCay, P.B. and Powell, S.R. (1989) Relationship between glutathione and chemically induced lipid peroxidation. *in: Glutathione Chemical, Biochemical, and Medical Aspects,* Eds. Dolphin, D., Poulson, R., and Avramovic, O., Wiley, New York, Part B, pp. 111-151.

Meister, A. (1978) Inhibition of glutamine synthetase and -glutamylcysteine synthetase by methionine sulfoximine and related compounds. *Enzyme-activated Irreversible Inhibitors* (N. Seiler, N., Jung, M.J., and Koch-Weser, J., Eds.), pp. 187-211. Elsevier-North Holland Biomedical Press, Amsterdam.

Meister, A. (1983) Selective modification of glutathione metabolism. *Science* 220: 471-477.

Meister, A. (1988) Glutathione metabolism and its selective modification. *J. Biol. Chem.* 263: 17205-17208.

Meister, A. (1988) Novel drugs that affect glutathione metabolism. *In: Mechanisms of Drug Resistance in Neoplastic Cells* (Tew, K.D., and Woolley, P.V., Eds.), Chap. 7, pp. 99-126. Bristol Myers Symposium No. 9, Academic Press, New York.

Meister, A. (1991) Glutathione deficiency produced by inhibition of its synthesis and its reversal; Applications in research and therapy. *Pharmacol. Ther.* 51: 155-194.

Meister, A. (1992a) On the antioxidant effects of ascorbic acid and glutathione. *Biochem. Pharmacol.* 44: 1905-1915.

Meister, A. (1992b) A trail of research: From glutamine synthetase to selective inhibition of glutathione synthesis. *ChemTracts Biochemistry & Molecular Biology* 3: 75-106.

Meister, A. (1992c) Depletion of glutathione in normal and malignant human cells *in vivo* by L-buthionine sulfoximine: Possible interaction with ascorbate. *J. Natl. Cancer Inst.* 84: 1601-1602.

Meister, A., and Anderson, M.E. (1983). Glutathione. *Annu. Rev. Biochem.* 52: 711-760.

Mills, G.C. (1957) Hemoglobin catabolism. I. Glutathione peroxidase, an erythrocyte enzyme which protects hemoglobin from oxidative breakdown. *J. Biol. Chem.* 229: 189-197.

Niki, E., Tsuchiya, J., Tanimura, R., and Kamiya, T. (1982) Regeneration of vitamin E from -chromanoxyl radical by glutathione and vitamin C. *Chem. Lett.* 789-792.

O'Dwyer, P.J., Hamilton, T.C., Young, R.C., LaCreta, F.P., Carp, N., Tew, K.D., Padavic, R., Comis, L., and Ozols, R.F. (1992) Depletion of glutathione in normal and malignant human cells *in vivo* by buthionine sulfoximine: Clinical and biochemical results. *J. Natl. Cancer Inst.* 84: 264-267.

Ozols, R.F., Hamilton, T.C., Masuda, H., and Young, R.C. (1988) Manipulation of cellular thiols to influence drug resistance. In: *Mechanisms of Drug Resistance in Neoplastic Cells* (Eds. Tew, K.D., and Woolley, P.V.), Chap. 19, pp. 289-305. Bristol Myers Symposium No. 9, Academic Press, New York.

Packer, J.E., Slater, T.F. and Wilson, R.L. (1979) *Nature (London)* Direct observation of a free radical interaction between vitamin E and vitamin C. 278: 737-738.

Puri, R.N., and Meister, A. (1983) Transport of glutathione as -glutamylcysteinylglycyl ester, into liver and kidney. *Proc. Natl. Acad. Sci. USA* 80: 5258-5260.

Reddy, C.C., Scholz, R.W., Thomas, C.E., and Massaro, E.J., (1982) Vitamin E dependent reduced glutathione inhibition of rat liver microsomal lipid peroxidation. *Life Sci.* 31: 571-576.

Rose (1989) Renal metabolism of the oxidized form of ascorbic acid (dehydro-L-ascorbic acid). *Am. J. Physiol.* 256: F52-F56.

Scholich, H., Murphy, M.E., and Sies, H. (1989) Antioxidant activity of dihydrolipoate against microsomal lipid peroxidation and its dependence on -tocopherol. *Biochim. Biophys.. Acta* 1001: 256-261.

Stahl, R.L., Liebes, L.F., and Silber, R. (1986) Glutathione dehydrogenase (ascorbate). *Methods Enzymol.* 122: 10-12.

Szent-Gyorgyi, A. (1928) CLXXII. Observations on the function of peroxidase systems and the chemistry of the adrenal cortex. Description of a new carbohydrate derivative. *Biochem. J.* 22: 1387-1409.

Taniguchi, N., Higashi, T., Sakamoto, Y., and Meister, A. (Eds.) (1989) *Glutathione Centennial, Molecular Perspectives and Clinical Implications.* Academic Press, New York.

Ursini, F., Maiorino, M., Valente, M., Ferri, L., and Gregolin, C. (1981) Purification from pig liver of a protein which protects liposomes and biomembranes from peroxidative degradation and exhibits glutathione peroxidase activity on phosphatidylcholine hydroperoxides. *Biochim. Biophys. Acta* 710: 197-211.

Vistica, D.T., and Ahmad S. (1989) Acquired resistance of tumors cells to L-phenylalanine mustard: Implications for the design of a clinical trial involving glutathione depletion. *In: Glutathione Centennial: Molecular Perspectives and Clinical Implications* (Taniguchi, N., Higashi, T., Sakamoto, Y., and Meister, A., Eds.), Chap. 21, pp. 301-315. Academic Presss, New York.

Wefers, H., and Sies, H. (1988) The protection by ascorbate and glutathione against microsomal lipid peroxidation is dependent on vitamin E. *Eur. J. Biochem.* 174: 353-357.

Wellner, V.P., Anderson, M.E., Puri, R.N., Jensen, G.L., and Meister, A. (1984) Radioprotection by glutathione ester: Transport of glutathione ester in human lymphoid cells and fibroblasts. *Proc. Natl. Acad. Sci. USA* 81: 4732-4735.

Wells, W.W., Xu, D.P., Yang, Y., and Rocque, P.A. (1990) Mammalian thioltransferase (glutaredoxin) and protein disulfide isomerase have dehydroascorbate reductase activity. *J. Biol. Chem.* 265: 15361-15364.

Wells, W.W., Yang, Y., Deits, T.L., and Gan, Z.-R. (1992) Thioltransferases. *Adv. Enzymol.* 66: 149-201.

Willis, G.C. (1953) An experimental study of the intimal ground substance in atherosclerosis. *Can. Med. Assoc. J.* 69: 17-22.

Oxidative Stress, Cell Activation and Viral Infection
C. Pasquier et al. (eds)
© 1994 Birkhäuser Verlag Basel/Switzerland

Structural consequences of NF-κB inhibition by natural antioxidants: α-lipoic acid and vitamin E

L. Packer and Y.J. Suzuki

Department of Molecular & Cell Biology, University of California, Berkeley, California 94720 USA

Summary

It is conceivable that the use of antioxidants to block NF-κB activation is useful in AIDS therapeutics. Our laboratory concerns on how structures of antioxidants relate to their activities. We show that modifying the anatomy of an antioxidant structure leads to a different behavior in the inhibitory action on TNFα induced NF-κB activation. Understanding such phenomena may lead to accelerated development of effective therapeutic antioxidants. Results from two classes of antioxidants, α-lipoic acid and vitamin E are presented.

Introduction

Acquired immunodeficiency syndrome (AIDS) results from infection with a human immunodeficiency virus (HIV) which eventually destroys a subset (CD4+) of helper T lymphocytes, so that the patient ultimately yields opportunistic infection and neoplasms. The long terminal repeat (LTR) region of HIV-1 proviral DNA contains two binding sites for the transcription factor, nuclear factor kappa B (NF-κB). NF-κB is a mammalian transcriptional activator that is involved in the transmission of signals from the cytoplasm to the nucleus by binding to the sequence 5'-GGGACTTTCC-3' in the κ enhancer where it interacts with transcription apparatus (Baeuerle, 1991). It participates in activation of genes involved in the immune, inflammatory, or acute phase responses, such as various cytokines and surface receptors as well as viruses including HIV-1 (Osborn et al., 1989).

Reactive oxygen species (ROS) have recently been suggested to be involved in the signal transduction pathway which leads to the activation of NF-κB and subsequent HIV-1 activation. This was first discovered by Herzenberg and co-workers who observed that *N*-acetylcysteine (NAC) inhibited HIV LTR-directed expression of β-galactosidase enhanced by tumor necrosis factor-α (TNFα) and phorbol ester in *in vitro* HIV model systems (Roederer et al., 1990), and subsequent findings that NAC inhibited and diamide stimulated NF-κB activation (Staal et al., 1990). NAC is known to increase the intracellular level of reduced glutathione, while diamide increases the amount of oxidized glutathione and depletes reduced glutathione. The finding that NAC blocks TNFα-induced NF-κB responses was also reported by Mihm et al. (1991). Schreck et al. (1991) extended these findings by showing that cell exposure to H_2O_2 leads to NF-κB activation.

Although the exact physiological activator of HIV is not yet determined, TNFα appears to play an important role. A considerable amount of reports point to the generation of ROS in TNFα responses (Matthews et al., 1987; Zimmerman et al., 1989; Yamauchi et al., 1989). Furthermore, TNFα has been found to induce manganous superoxide dismutase in some cells, and its implications as a possible mechanism of cellular resistance to cytotoxicity have been suggested (Wong and Goeddel, 1988; Wong et al., 1989). Thus, these findings establish a signal transduction scheme as follows:

TNFα --> ROS --> NF-κB --> expression of HIV-1.

Given that ROS are involved in the signal transduction mechanisms for the expression of HIV, it is logical to investigate a possibility for the therapeutic use of antioxidants in preventing HIV activation. Indeed, earlier evidence of the involvement of ROS in NF-κB and HIV activation has come from experiments using a pharmacological strategy eliminating ROS by antioxidants (i.e. NAC) as noted above. Thus, the investigation for the use of antioxidants for the treatment of AIDS has already started to emerge.

Understanding the structural consequences of a compound in its action significantly contributes to the accelerated development of therapeutic agents. Elucidation of the structure-activity relationships has led to tremendous success in modern molecular pharmacology. Many specific enzyme inhibitors have been developed and structure-based viral drug design is believed to be the key for a success in AIDS drug development (Erickson et al., 1990; DesJarlais et al., 1990). Assessment of the structure-activity relationships of antioxidants are difficult in that a complex set of information is necessary to cope with the wide range actions of an antioxidant in physiological systems. A compound can exert antioxidant activities in various ways such as by inhibiting ROS generating enzymes, scavenging ROS, preventing reactions of ROS to a specific target, or enhancing activities of other antioxidants (Halliwell and Gutteridge, 1988). Such wide range activities necessitate the considerations of different physicochemical characteristics including hydrophobicity and electrostatic behaviors.

The relationships between structure and activity of antioxidants have been examined in our laboratory using compounds which are structurally related to vitamin E (Kagan et al., 1990c; Suzuki et al., 1993b), ubiquinol (Kagan et al., 1990b) and dihydrolipoic acid (Suzuki et al., 1993a) by assessing the antioxidant properties at chemical and molecular levels. However, the complexity of free radical biology and chemistry has thus far precluded the generalizations as to how structure of an antioxidant relates to its antioxidant activities (Suzuki et al., 1993a). Further investigations examining the structure-activity relationships of various antioxidants in different physicochemical environments are necessary. In this regard, true understandings of the consequences of the structure to antioxidant activities would have to derive from experiments utilizing systems containing various physicochemical parameters. An example of such systems is the cellular system which can satisfy the diverse physicochemical requirements, and also is simple enough to cope with compiling information for theoretical and conceptual generalizations. Thus, we have sought to examine the effects of antioxidants and structurally related compounds on TNFα induced NF-κB activation in order to achieve an understanding of the structure-antioxidant activity relationships at the cellular level. Here, we report such studies on α-lipoic acid and vitamin E.

Materials and Methods

Cell line and cell culture

Jurkat T (human acute leukemia) cells were grown in RPMI 1640 medium supplemented with 10% fetal calf serum (FCS), 1% (w/v) penicillin/streptomycin, 1% sodium pyruvate and 1% glutamine. Cells were plated at a density of 1×10^6 cells/ml. Recombinant $E.$ $coli$-derived human TNFα was kindly provided by Genentech Inc. (South San Francisco, CA). TNF-α (25 ng/ml) was added and cells were incubated for 4 hrs in an atmosphere of 5% CO_2 in air humidified at 37°C, followed by nuclear extraction.

In order to examine the effects of antioxidants, various concentrations of N-acetyl-L-cysteine (Sigma), R,S-α-lipoic acid, R,S-dihydrolipoic acid, R-α-lipoic acid, S-α-lipoic acid, tetranorlipoic acid, bisnorlipoic acid (ASTA Medica), d-α-tocopherol (Henkel Corporation) d-α-tocopheryl succinate (Sigma), vitamin E acetate (Henkel), PMC (gift from the Eisai Company, Tokyo, Japan) or succinate dissolved in ethanol were added before the addition of TNF-α.

Cell viability was determined by the Trypan-Blue exclusion method.

Nuclear extracts and electrophoretic mobility shift assay (EMSA)

Nuclear extracts were prepared from 1×10^6 cells as described previously (Staal et al., 1990; Suzuki et al., 1992). Cells were harvested, centrifuged for 10 min at 1,200 rpm, washed in 1 ml of ice-cold PBS, and centrifuged for 15 sec at 14,000 rpm in an Eppendorf Brinkman-5412 centrifuge at 4°C. Cells were pelleted and washed once in 0.4 ml of buffer A [10 mM Hepes, pH 7.8; 10 mM KCl; 2 mM $MgCl_2$; 1 mM dithiothreitol (DTT); 0.1 mM EDTA; 0.1 mM phenylmethylsulfonyl fluoride (PMSF), 5 μg/ml antipain (Sigma) and 5 μg/ml leupeptin (Sigma)] and incubated on ice for 15 min. Then 25 μl of a 10% Nonidet P-40 solution (Sigma) were added, and cells were vigorously mixed for 15 sec and then centrifuged for 30 sec at 14,000 rpm. Pelleted nuclei were resuspended in 40 μl of buffer C [50 mM Hepes, pH 7.8; 50

mM KCl; 300 mM NaCl; 0.1 mM EDTA; 1 mM DTT; 0.1 mM PMSF; 10% (v/v) glycerol], mixed for 20 min, and centrifuged for 5 min at 14,000 rpm at 4°C. The supernatant containing the nuclear proteins was harvested, protein concentration determined and stored at -80°C.

EMSAs were performed essentially as described earlier (Garner and Revzin, 1981; Fried and Crothers, 1981). Binding reaction mixtures (20 μl) containing 5 μg protein of nuclear extract, 1 μg poly(dI-dC) (Pharmacia), ^{32}P-labeled probe, 50 mM NaCl, 0.2 mM EDTA, 0.5 mM DTT, 2% (v/v) glycerol and 10 mM Tris-HCl (pH 7.5) were incubated for 20 min at 25°C. Proteins were separated by electrophoresis through a native 6% polyacrylamide gel in a running buffer of 12.5 mM Tris borate, 0.25 mM EDTA (pH 8.0), followed by autoradiography. Densitometry scanning was performed on a Personal Densitometer (Molecular Dynamics) to quantify the NF-κB bands. NF-κB and oct-1 oligonucleotide probes (Oncogene Science) were Klenow-labeled with [α-^{32}P]dATP and purified using a NAP-5 column (Pharmacia) in 100 mM NaCl, 1 mM EDTA and 10 mM Tris-HCl (pH 7.5).

HPLC measurements

α-Tocopherol content was determined using HPLC method described by Lang et al. (1987). Cell suspensions was mixed with ethanol and hexane, then subjected to low-speed centrifugation. A portion of the hexane supernatant was removed, dried under nitrogen, and then redissolved in ethanol. The reduced form of α-tocopherol was quantitatively determined by HPLC using a C-18 reverse phase column with electrochemical detection.

Results and Discussion

α-Lipoic acid (see Fig. 1 for structure) exists naturally in physiological systems as a co-factor for decarboxylation reactions. Recently, α-lipoic acid was found to exert antioxidant action *in vivo* and *in vitro*. α-Lipoic acid is thought to be reduced to dihydrolipoic acid (DHLA; see Fig. 1 for structure) intracellularly (Peinado et al., 1989), and this reduced vicinal dithiol compound can act as a potent antioxidant through a variety of mechanisms. It can scavenge superoxide radicals and hydroxyl radicals (Suzuki et al., 1991), prevent lipid peroxidation (Bast and Haenen, 1988; Scholich et al., 1989), and participate in vitamin E recycling (Kagan et al., 1992).

α-Lipoic acid

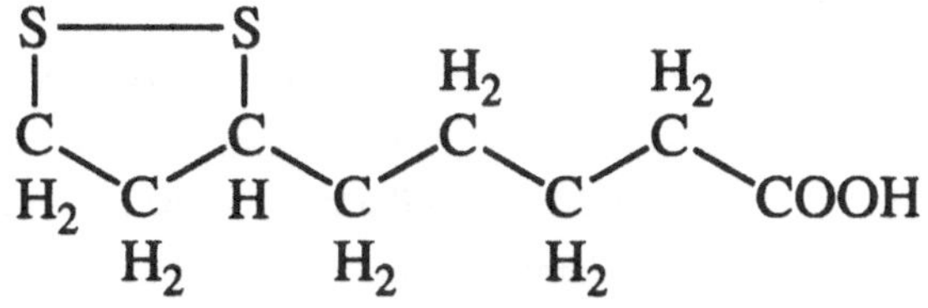

Dihydrolipoic acid

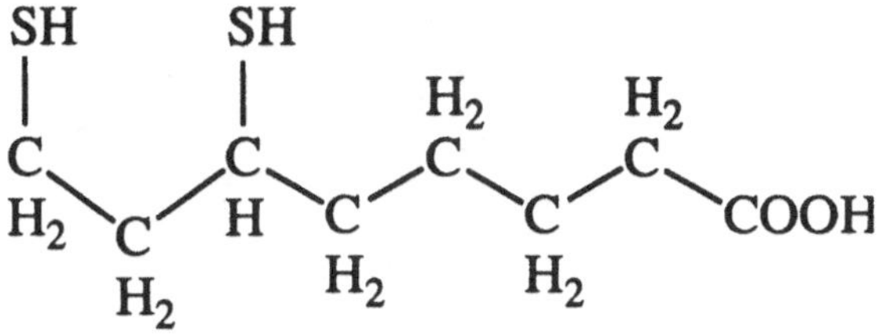

Fig. 1 Structure of a-lipoic acid and dihydrolipoic acid.

α-Lipoic acid

α-Lipoic acid has been shown to inhibit HIV-1 replication in infected cultured cells (Baur et al., 1991). This may be due to the inhibition of NF-κB activation imposed by the antioxidant properties of DHLA generated from α-lipoic acid. Thus, we investigated whether α-lipoic acid inhibits the activation of NF-κB in cultured human T cells (Suzuki et al., 1992). Autoradiograms of EMSAs are shown in Fig. 2. Incubation of Jurkat cells (1×10^6 cells/ml) with 25 ng/ml TNFα resulted in an appearance of a band which can be eliminated by a 1000-fold excess of unlabeled "cold" NF-κB probe. A two-hr pre-incubation of cells with 1 mM α-lipoic acid partially, and 2 mM α-lipoic acid almost completely blocked the activation. Similarly, the activation of NF-κB induced by 50 ng/ml PMA, a protein kinase C activator, was inhibited by α-lipoic acid in a concentration-dependent fashion in which 2 mM exhibited an almost complete inhibition. α-Lipoic acid at these concentrations did not affect either oct-1 DNA binding activity or cell viability. The relative intensity of NF-κB bands obtained by the densitometry scanning is shown in Fig. 3. Both α-lipoic acid and NAC caused inhibition of NF-κB activation in a concentration-dependent fashion. α-Lipoic acid was more potent than NAC as demonstrated by the complete inhibition achieved by 4 mM α-lipoic acid, whereas 20 mM was needed for NAC. A direct supplementation of DHLA in the incubation medium also inhibited the NF-κB activation induced by TNFα as shown in Fig. 4. Although the potency of DHLA on NF-κB inhibition was comparable to that of α-lipoic acid, 3 mM DHLA affected oct-1 DNA binding activity and cell viability (there were no apparent effects at 1 mM). Fig. 5 shows that both *R*-

118

and *S*-enantiomers as well as racemic mixture of α-lipoic acid comparably inhibit NF-κB activation induced by TNFα. Structural homologues of α-lipoic acid, tetranorlipoic acid and bisnorlipoic acid (see Fig. 6 for structures) both inhibited NF-κB activation as well (Fig. 7). Tetranorlipoic acid is extremely potent as 0.5 mM completely inhibited the activation. Bisnorlipoic acid significantly inhibited NF-κB at 0.5 mM, however, it never caused complete inhibition up to 4 mM. Both of these homologues at 4 mM significantly affected oct-1 DNA binding activity and cell viability (there were no apparent effects at 2 mM).

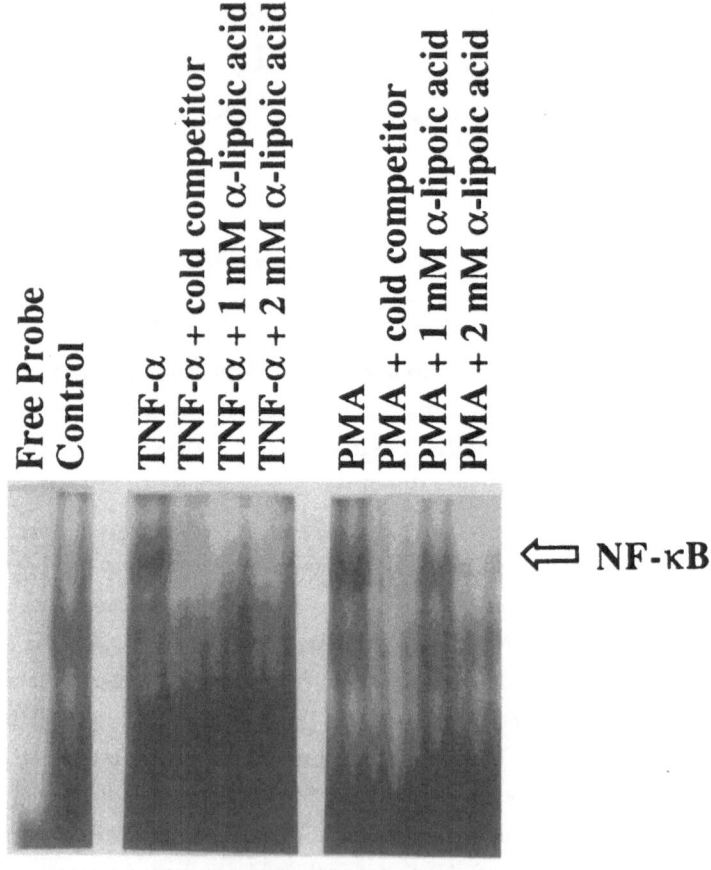

Fig. 2 Effects of α-lipoic acid on NF-κB activation induced by TNF-α or PMA. Jurkat T cells (1 x 10⁶ cells/ml) were incubated for 2 hrs with various concentrations of α-lipoic acid followed by incubation with TNF-α (25 ng/ml) or PMA (50 ng/ml) for 4 hrs. *Lane 1*, free probes; *lane 2*, control; *lane 3*, TNF-α; *lane 4*, TNF-α plus unlabeled competitor (1000-fold excess); *lane 5*, TNF-α plus α-lipoic acid (1 mM); *lane 6*, TNF-α plus α-lipoic acid (2 mM); *lane 7*, PMA; *lane 8*, PMA plus unlabeled competitor (1000-fold excess); *lane 9*, PMA plus α-lipoic acid (1 mM); *lane 10*, PMA plus α-lipoic acid (1 mM). Taken from Suzuki et al. (1992).

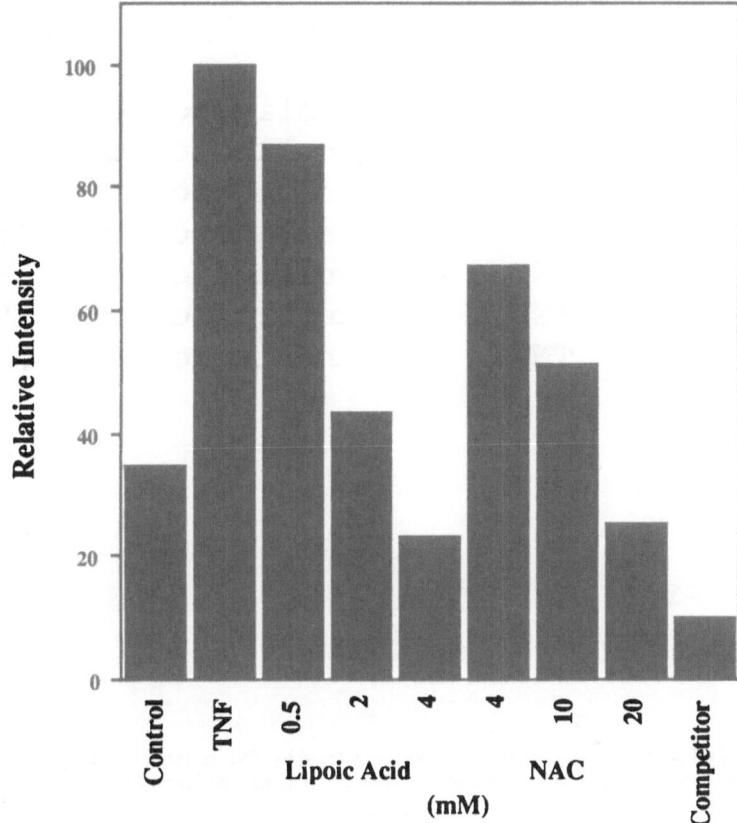

Fig. 3 Effects of α-lipoic acid and *N*-acetylcysteine on NF-κB activation induced by TNF-α. Jurkat T cells (1 x 10^6 cells/ml) were incubated for 2 hrs with various concentrations of α-lipoic acid or *N*-acetylcysteine followed by incubation with TNF-α (25 ng/ml) for 4 hrs. The NF-κB bands of EMSAs were quantified by densitometry. *Bar 1*, control; *bar 2*, TNF-α; *bars 3 - 5*, TNF-α plus α-lipoic acid (0.5, 2 and 4 mM, respectively); *bars 6 - 8*, TNF-α plus *N*-acetylcysteine (4, 10, 20 mM, respectively); *bar 9*, TNF-α plus unlabeled competitor (1000-fold excess). Taken from Suzuki et al. (1992).

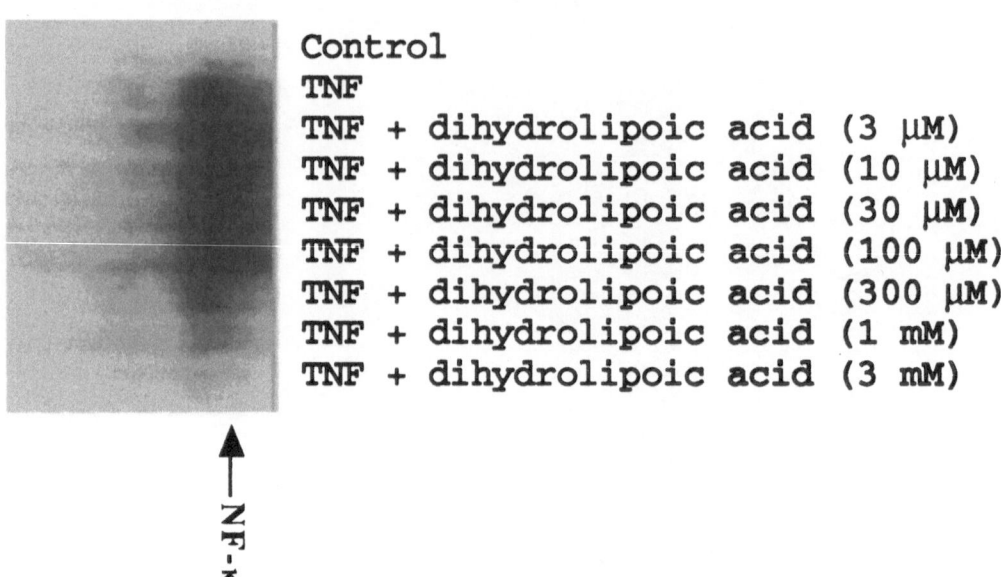

Control
TNF
TNF + dihydrolipoic acid (3 µM)
TNF + dihydrolipoic acid (10 µM)
TNF + dihydrolipoic acid (30 µM)
TNF + dihydrolipoic acid (100 µM)
TNF + dihydrolipoic acid (300 µM)
TNF + dihydrolipoic acid (1 mM)
TNF + dihydrolipoic acid (3 mM)

←— NF-κB

Fig. 4 Effects of dihydrolipoic acid on NF-κB activation induced by TNF-α. Jurkat T cells (1 x 10⁶ cells/ml) were incubated for 30 min with various concentrations of dihydrolipoic acid followed by incubation with TNF-α (25 ng/ml) for 4 hrs.

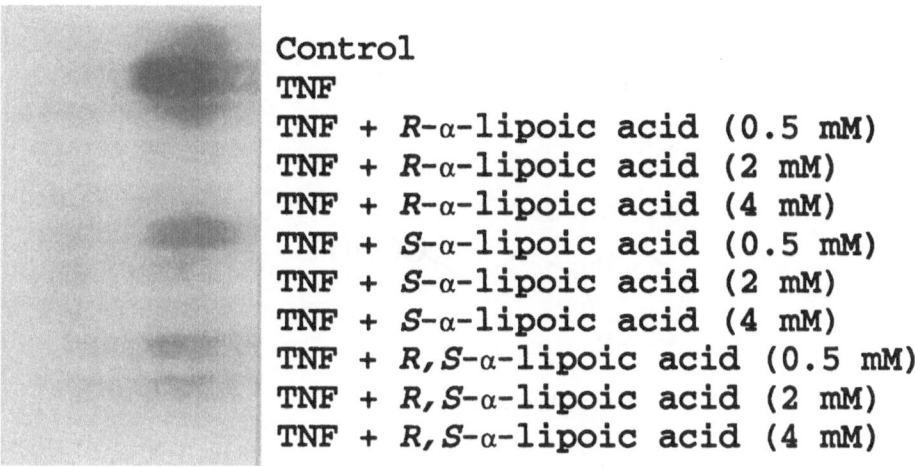

Control
TNF
TNF + R-α-lipoic acid (0.5 mM)
TNF + R-α-lipoic acid (2 mM)
TNF + R-α-lipoic acid (4 mM)
TNF + S-α-lipoic acid (0.5 mM)
TNF + S-α-lipoic acid (2 mM)
TNF + S-α-lipoic acid (4 mM)
TNF + R,S-α-lipoic acid (0.5 mM)
TNF + R,S-α-lipoic acid (2 mM)
TNF + R,S-α-lipoic acid (4 mM)

Fig. 5 Effects of α-lipoic acid enantiomers on NF-κB activation induced by TNF-α. Jurkat T cells (1 x 10^6 cells/ml) were incubated for 2 hrs with various concentrations of R-, S-, R,S-α-lipoic acid followed by incubation with TNF-α (25 ng/ml) for 4 hrs.

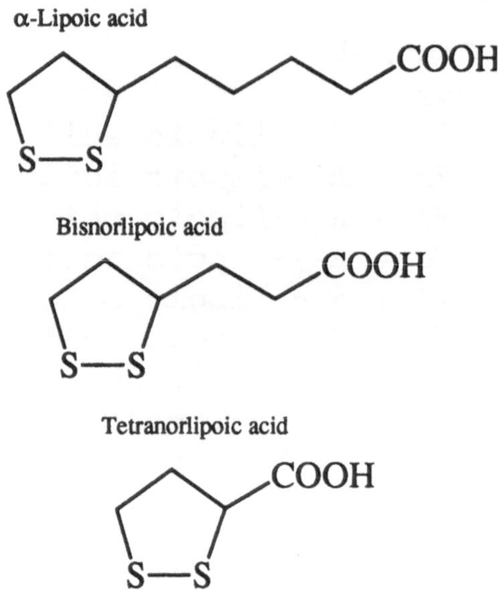

α-Lipoic acid

Bisnorlipoic acid

Tetranorlipoic acid

Fig. 6 Structure of α-lipoic acid, bisnorlipoic acid and tetranorlipoic acid.

Our results demonstrating that both enantiomers of α-lipoic acid as well as reduced form, DHLA exert comparable effects on NF-κB activation suggest that either chirality or redox states may not be particularly important in the mechanism of α-lipoic acid action. The lack of chiral specificity may imply that the enzymatic processes which usually require chiral-specific substrates are not involved in the mechanism. Regarding chirality, our observations are consistent with recent findings by Dr. Leonard Herzenberg and co-workers showing that both L- and D-enantiomers of NAC exert inhibitory actions against NF-κB activation to a similar extent (Staal et al., 1993). Shorter homologues of α-lipoic acid, particularly tetranorlipoic acid exhibited better effects at lower concentrations where cytotoxic influences and non-specific effects were not apparent. These results may be due to facilitated transport of the shorter homologues to reach the molecular target or they may possess different targets from those of longer homologues.

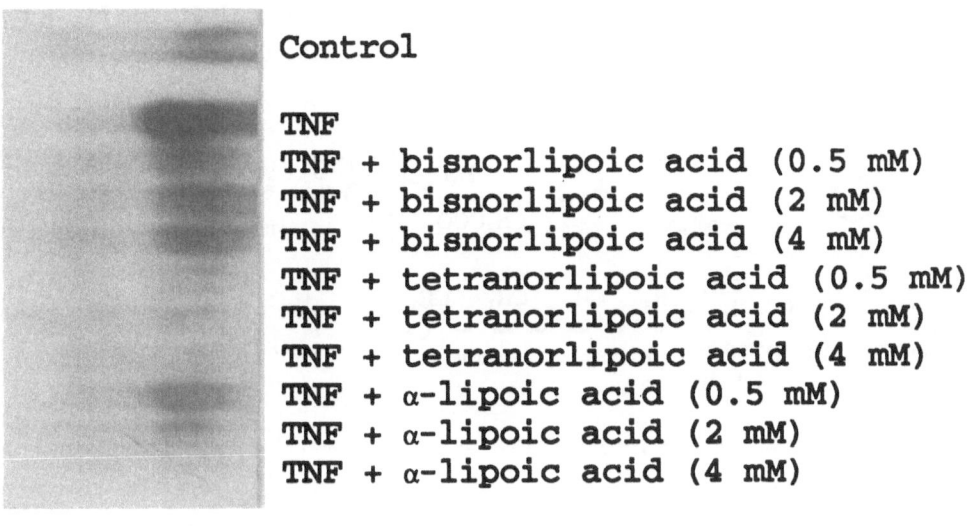

Fig. 7 Effects of structural homologues of α-lipoic acid on NF-κB activation induced by TNF-α. Jurkat T cells (1 x 10^6 cells/ml) were incubated for 2 hrs with various concentrations of tetranor-, bisnor- or α-lipoic acid, followed by incubation with TNF-α (25 ng/ml) for 4 hrs.

Vitamin E

Vitamin E is a well known natural lipophilic antioxidant which protects membranes from lipid peroxidation (Burton and Ingold, 1981; Packer, 1991). Although its preventative role in HIV activation has not yet clearly been determined (Kidd and Huber, 1991), it is one of the most logical natural antioxidants to be examined in terms of its effect on NF-κB activation. In the present study, the effects of vitamin E and its derivatives: α-tocopherol, vitamin E acetate, α-

124

tocopheryl succinate and 2,2,5,7,8-pentamethyl-6-hydroxychromane (PMC; see Fig. 8 for structures) on TNFα induced NF-κB activation were investigated (Suzuki and Packer, 1993).

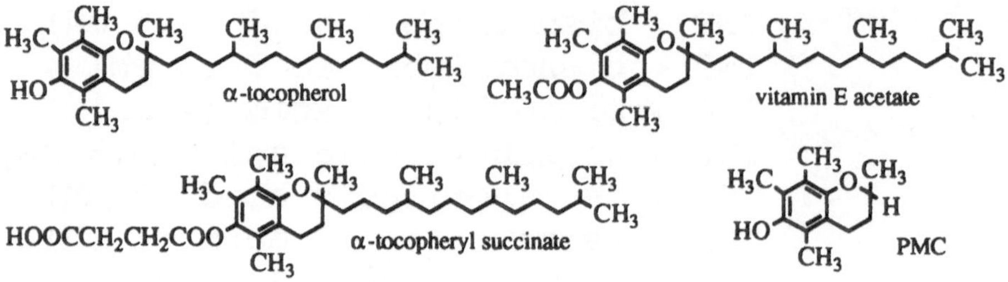

Fig. 8 Structure of α-tocopherol, α-tocopheryl succinate, vitamin E acetate and PMC.

Incubation of Jurkat cells (1×10^6 cells/ml) with 25 ng/ml TNFα resulted in an appearance of the NF-κB band. As shown in Fig. 9, thirty-min pre-incubation of cells with 10 μM - 1 mM α-tocopheryl succinate or vitamin E acetate exhibited a concentration dependent inhibition of NF-κB activation. In contrast, α-tocopherol did not block the activation. PMC was the most potent inhibitor among the vitamin E derivatives examined as 10 μM completely blocked the activation. Each compound was added to the culture medium using ethanol as a vehicle, and the amounts of ethanol used in the present study did not affect the NF-κB activation. None of the vitamin E derivatives at the concentrations examined affected cell viability. Oct-1 DNA binding activity which constitutively exists in these cells was inhibited by α-tocopheryl succinate at concentrations effective in inhibiting NF-κB. In contrast, α-tocopherol, vitamin E acetate and PMC had no effects (Fig. 10). Incubation of the cells with succinate (10 μM - 1 mM) did not inhibit NF-κB activation or oct-1 DNA binding activity (data not shown). α-Tocopheryl succinate, but not any other compounds inhibited the DNA binding activity of the activated NF-κB (data not shown).

HPLC measurements determined that unstimulated Jurkat cells contain 0.45 ± 0.04 pmoles reduced form of α-tocopherol in 10^6 cells. This value was not significantly affected by the incubation of cells with TNF-α (25 ng/ml) for 4 hrs (0.52 ± 0.04 pmol/10^6 cells). These values represent mean $\pm$ S.E. where n=9.

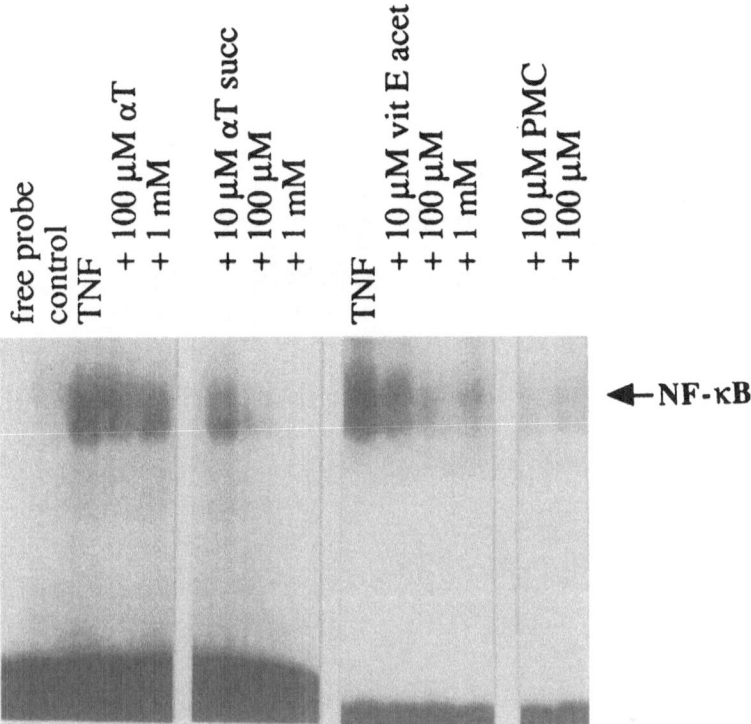

Fig. 9 Effects of vitamin E derivatives on NF-κB activation induced by TNF-α. Jurkat T cells (1 x 10⁶ cells/ml) were incubated for 30 min with various concentrations of α-tocopherol and derivatives followed by incubation with TNF-α (25 ng/ml) for 4 hrs. Taken from Suzuki and Packer (1993).

126

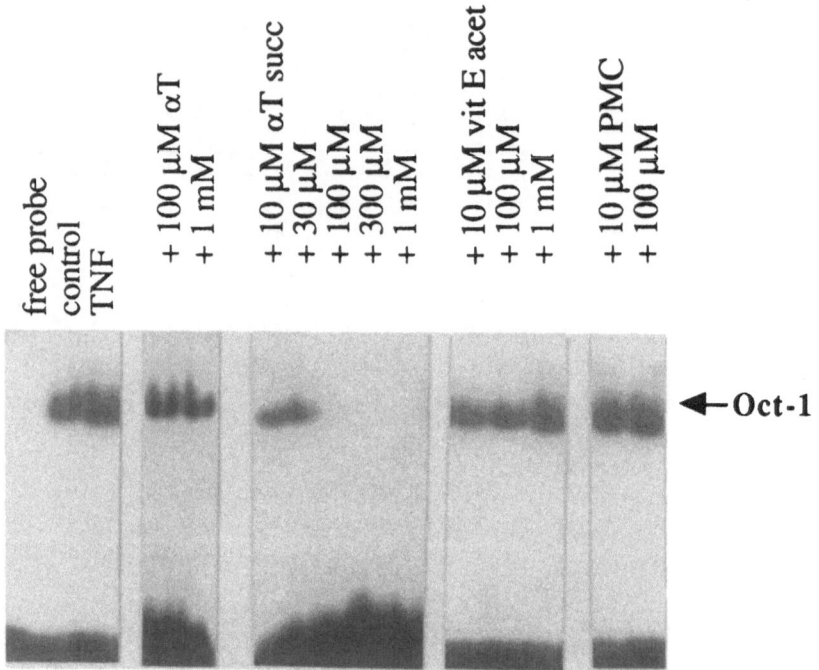

free probe
control
TNF
+ 100 μM αT
+ 1 mM
+ 10 μM αT succ
+ 30 μM
+ 100 μM
+ 300 μM
+ 1 mM
+ 10 μM vit E acet
+ 100 μM
+ 1 mM
+ 10 μM PMC
+ 100 μM

←Oct-1

Fig. 10 Effects of vitamin E derivatives on oct-1 DNA binding activity. Jurkat T cells (1 x 10^6 cells/ml) were incubated for 30 min with various concentrations of α-tocopherol and derivatives followed by incubation with TNF-α (25 ng/ml) for 4 hrs. Taken from Suzuki and Packer (1993).

Results from the present study are intriguing since, for the first time, a lipophilic antioxidant was shown to be effective in inhibiting the TNFα induced NF-κB activation. The natural form of α-tocopherol, vitamin E acetate at 100 μM and 1 mM almost completely blocked the activation, however, never caused complete inhibition. Vitamin E acetate is known to become deesterified to the biologically active antioxidant form, α-tocopherol (Kagan et al., 1990a). This implies that free radical processes involved in the TNFα induced NF-κB activation, at least in part, occurs proximal to the membranes, and membrane oxidation may be an integral step in the signal transduction pathway. The ineffectiveness of direct addition of α-tocopherol in inhibiting the NF-κB activation may also imply that the membrane oxidation processes required for the cell signalling are localized in the internal compartments of cell architecture such as mitochondria where vitamin E acetate, but not α-tocopherol, can be reached before deesterification. Furthermore, observations that vitamin E content does not decrease in response to TNFα treatment suggest that the cell possesses an efficient vitamin E recycling mechanism which has been shown to occur in many cell types (Maguire et al., 1989; Constantinescu et al., 1993). In Jurkat cells, this may be accomplished by utilizing reduced glutathione (GSH) as Staal et al. (1990) observed decreased levels of GSH in response to TNFα exposure. On the other hand, since α-tocopherol has been demonstrated to inhibit protein kinase C (Mahoney and Azzi, 1988; Boscoboinik et al., 1991) and protein kinase C has shown to directly phosphorylate IκB (Gosh and Baltimore, 1990), the inhibition of this enzyme by α-tocopherol may be the mechanism of the NF-κB inactivation. In this case, our HPLC results may imply that the oxidation process involved in NF-κB activation is localized in the cytosolic compartment and membrane oxidation does not occur; α-tocopherol content thus remains unchanged.

α-Tocopheryl succinate appears to exert non-specific effects on DNA binding proteins as demonstrated by the inhibition of NF-κB and oct-1 DNA binding activities. Since neither succinate nor α-tocopherol alone exhibit such actions, a unique behavior by the α-tocopheryl succinate structure must be required.

The ability of PMC to inhibit NF-κB activation should gain a considerable attention. The effectiveness at a low concentration (10 μM) suggests that its free radical scavenging ability may not confer the major mechanism as radical scavenging antioxidants have generally been shown to require higher concentrations for NF-κB inhibition (Schreck et al., 1992). Rather, inactivation of enzymes involved in the TNFα signal transduction cascade may be a more likely event which is responsible for the observation. In fact, we have found that PMC inhibits phospholipase A_2 in human keratinocytes (Pentland et al., 1992) and a role of phospholipase A_2 in TNFα action has been suggested (Krönke et al., 1992). Thus, PMC may inhibit NF-κB activation by inactivating this enzyme. On the other hand, potency of PMC against NF-κB inhibition may be due to the ability of this more hydrophilic compound to reach the critical target where it may engage in the same activity as the longer chain derivatives.

Conclusion

As the AIDS epidemic continues to be destructive (Rosenberg and Fauci, 1991), if not a cure, at least the development of strategies for long-term survival for HIV positive individuals is warranted. Prevention of HIV activation by maintaining integrated proviral DNA in the latently infected stage is, to a degree, a certain success for such a purpose, and thus blocking the HIV transcription at the level of NF-κB using antioxidants may lead to a long-term survival.

128

Furthermore, the elimination of virus by antiviral agents such as reverse transcriptase inhibitors (e.g., AZT, DDC, DDI) exhibits some toxicity against the host and the use of high concentrations is not desirable. This has led to favor an idea of the combination therapy for AIDS (Schinazi et al., 1992). Natural and safe compounds which can affect the HIV lifecycle are attractive in that they may support the actions of aggressive antiviral agents without a risk of toxicity. The effectiveness of NAC was demonstrated earlier (Roederer et al., 1990; Staal et al., 1990). We have added α-lipoic acid (Suzuki et al., 1992) and vitamin E (Suzuki and Packer, 1993) to the list of natural antioxidants which can block the activation of NF-κB.

Further work is needed to elucidate the mechanisms of the inhibitory effects of α-lipoic acid and vitamin E derivatives some of which appear to utilize different processes. This may mean that a small structural alteration of a compound leads to target different components of the signal transduction cascade, and this may mediate both antioxidant and non-antioxidant mechanisms. Understanding such differences in the actions of the structurally related compounds which may be governed by different physicochemical characteristics, resulting in influencing a common biological outcome, may be important in future drug design for AIDS by defining the "structure-activity relationships at cellular level".

Acknowledgments

Supported by the National Institutes of Health (CA47597) and ASTA Medica. This work was done during the tenure of a research Fellowship from the American Heart Association, California Affiliate to YJS. We thank Drs. Bharat Aggarwal, Grace Wong and Helmut Sies for valuable contributions.

References

Baeuerle, P. (1991) The inducible transcription activator NF-κB: regulation by distinct protein subunits. *Biochem. Biophys. Acta* 1072: 63-80.

Bast, A., and Haenen, G.R.M.M. (1988) Interplay between lipoic acid and glutathione in the protection against microsomal lipid peroxidation.*Biochim. Biophys. Acta* 963: 558-561.

Baur, A., Harrer, T., Peukert, M., Jahn, G., Kalden, J.R., and Fleckenstein, B. (1991) Alpha-lipoic acid is an effective inhibitor of human immuno-deficiency virus (HIV-1) replication. *Klin Wochenschr* 69: 722-724.

Boscoboinik, D., Szewczyk, A., Hensey, C., and Azzi, A. (1991) Inhibition of cell proliferation by α-tocopherol. Role of protein kinase C. *J. Biol. Chem.* 266: 6188-6194.

Burton, G.W., and Ingold, K.U. (1981) Autooxidation of biological molecules. 1. The antioxidant activity of vitamin E and related chain-breaking phenolic antioxidants in vitro. *J. Am. Chem. Soc.* 103: 6472-6477.

Constantinescu, A., Han, D., and Packer, L. (1993) Vitamin E recycling in erythrocyte membranes. *J. Biol. Chem.* (in press)

DesJarlais, R.L., Seibel, G.L., Kuntz, I.D., Furth, P.S., Alvarez, J.C., Ortiz de Montellano, P.R., DeCamp, D.L., Babé, L.M., and Craik, C.S. (1990) Structure-based design of nonpeptide inhibitors specific for the human immunodeficiency virus 1 protease. *Proc. Natl. Acad. Sci. USA* 87: 6644-6648.

Erickson, J., Neidhart, D.J., VanDrie, J., Kempf, D.J., Wang, X.C., Norbeck, D.W., Plattner, J.J., Rittenhouse, J.W., Turon, M., Wideburg, N., Kohlbrenner, W. E., Simmer, R., Helfrich, R., Paul, D.A., and Knigge, M. (1990) Design, activity, and 2.8A crystal structure of a C_2 symmetric inhibitor complexed to HIV-1 protease. *Science* 249: 527-533.

Fried, M., and Crothers, D.M. (1981) Equilibria and kinetics of lac repressor-operator interactions by polyacrylamide gel electrophoresis. *Nucleic Acid Res.* 9: 6505-6525.

Garner, M.M., and Revzin, A. (1981) A gel electrophoresis method for quantifying the binding of proteins to specific DNA regions: application to components of the Escherichia coli lactose operon regulatory system. *Nucleic Acid Res.* 9: 3047-3060.

Gosh, S., and Baltimore, D. (1990) Activation in vitro of NF-κB by phosphorylation of its inhibitor IκB. *Nature* 344: 678-682.

Halliwell, B., and Gutteridge, J.M.C. (1988) *Free Radicals in Biology and Medicine.* Clarendon, Oxford.

Kagan, V.E., Bakalova, R.A., Serbinova, E.E., and Stoytchev, T.S. (1990a) Fluorescence measurements of incorporation and hydrolysis of tocopherol and tocopheryl esters in biomembranes. *Meth. Enzymol.* 186: 355-367.

Kagan, V.E., Serbinova, E. A., Koynova, G. M., Kitanova, S. A., Tyurin, V.A., Stoytchev, T. S., Quinn, P. J., and Packer, L. (1990b) Antioxidant action of ubiquinol homologues with different isoprenoid chain length in biomembranes. *Free Rad. Biol. Med.* 9: 117-126.

Kagan, V. E., Serbinova, E.A., and Packer, L. (1990c) Recycling and antioxidant activity of tocopherol homologs of differing hydrcarbon chain lengths in liver microsomes. *Arch. Biochem. Biophys.* 282: 221-225.

Kagan, V. E., Shvedova, A., Serbinova, E., Khan, S., Swanson,C., Powell, R., and Packer, L. (1992) Dihydrolipoic acid--a universal antioxidant both in the membrane and in the aqueous phase. Reduction of peroxyl, ascorbyl and chromanoxyl radicals. *Biochem. Pharmacol.* 44: 1637-1649.

Kidd, P.M., and Huber, W. (1991) *Living with the AIDS Virus.* HK Biomedical, Berkeley.

Krönke, M., Schütze, S., Scheurich, P, and Pfizenmaier, K. (1992) TNF signal transduction and TNF-responsive genes. In: Aggarwal, B.B. and Vilcek, J. (eds.): *Tumor Necrosis Factors. Structure, Function, and Mechanisms of Action,* Marcel Dekker, Inc., New York, pp. 189-216.

Lang, J.K., and Packer, L. (1987) Quantitative determination of vitamin E and oxidized and reduced coenzyme Q by high-performance liquid chromatography with in-line ultraviolet and electrochemical detection. *J. Chromatogr.* 385: 109-117.

Maguire, J.J., Wilson, D.S., and Packer, L. (1989) Mitochondrial electron transport-linked tocopheroxyl radical reduction. *J. Biol. Chem.* 264: 21462-21465.

Mahoney, C.W., and Azzi, A. (1988) Vitamin E inhibits protein kinase C activity. *Biochem. Biophys. Res. Commun.* 154: 694-697.

Matthews, N., Neale, M.L., Jackson, S.K., and Stark, J.M. (1987) Tumour cell killing by tumour necrosis factor: inhibition by anaerobic conditions, free-radical scavengers and inhibitors of arachidonate metabolism. *Immunology* 62: 153-155.

Mihm, S., Ennen, J., Pessara, U., Kurth, R., and Dröge, W. (1991) Inhibition of HIV-1 replication and NF-κB activity by cysteine and cysteine derivatives. *AIDS* 5: 497-503.

Osborn, L., Kunkel, S., and Nabel, G.J. (1989) Tumor necrosis factor α and interleukin 1 stimulate the human immunodeficiency virus enhancer by activation of the nuclear factor κB. *Proc. Natl. Acad. Sci. USA* 86: 2336-2340.

Packer, L. (1991) Protective role of vitamin E in biological systems. *Am. J. Clin. Nutr.* 53: 1050S-1055S.

Peinado, J., Sies, H., and Akerboom, T.P.M. (1989) Hepatic lipoate uptake. *Arch. Biochem. Biophys.* 273: 389-395.

Pentland, A.P., Morrison, A.R., Jacobs, S.C., Hruza, L.L., Hebert, J.S., and Packer, L. (1992) Tocopherol analogs suppress arachidonic acid metabolism via phospholipase inhibition. *J. Biol. Chem.* 267: 15578-15584.

Roederer, M., Staal, F.J.T., Raju, P.A., Ela, S.W., Herzenberg, L.A., and Herzenberg, L.A. (1990) Cytokine-stimulated human immunodeficiency virus replication is inhibited by *N*-acetyl-L-cysteine. *Proc. Natl. Acad. Sci. USA* 87: 4884-4888.

Rosenberg, Z.F. and Fauci, A.S. (1991) Immunopathogenesis of HIV infection. *FASEB J.* 5: 2382-2390.

Schinazi, R.F., Mead, J.R., and Feorino, P.M. (1992) Insights into HIV chemotherapy. *AIDS Res. Human Retroviruses* 8: 963-990.

Scholich, H., Murphy, M.E., and Sies, H. (1989) Antioxidant activity of dihydrolipoate against microsomal lipid peroxidation and its dependence on α-tocopherol. *Biochim. Biophys. Acta* 1001: 256-261.

Schreck, R., Rieber, P., and Baeuerle, P.A. (1991) Reactive oxygen intermediates as apparently widely used messengers in the activation of the NF-κB transcription factor and HIV-1. *EMBO J.* 10: 2247-2258.

Schreck, R., Albermann, K., and Baeuerle, P.A. (1992) Nuclear factor κB: an oxidative stress-responsive transcription factor of eukaryotic cells (a review). *Free Rad. Res. Comms.* 17: 221-237.

Staal, F.J.T., Roederer, M., Herzenberg, L.A., and Herzenberg, L.A. (1990) Intracellular thiols regulate activation of nuclear factor κB and transcription of human immunodeficiency virus. *Proc. Natl. Acad. Sci. USA* 87: 9943-9947.

Staal, F.J.T., Roederer, M., Raju, P.A., Anderson, M.T., Ela, S.W., Herzenberg, L.A., and Herzenberg, L.A. (1993) Antioxidants inhibit stimulation of HIV long terminal repeat-directed transcription. *AIDS Res. Hum. Retroviruses* 9: 299-305.

Suzuki, Y.J., and Packer, L. (1993) Inhibition of NF-κB activation by vitamin E derivatives. *Biochem. Biophys. Res. Commun.* 193: 277-283.

Suzuki, Y.J., Tsuchiya, M., and Packer, L. (1991) Thioctic acid and dihydrolipoic acid are novel antioxidants which interact with reactive oxygen species. *Free Rad. Res. Comms.* 15: 255-263.

Suzuki, Y.J., Aggarwal, B.B., and Packer, L. (1992) α-Lipoic acid is a potent inhibitor of NF-κB activation in human T cells. *Biochem. Biophys. Res. Commun.* 189: 1709-1715.

Suzuki, Y.J., Tsuchiya, M., and Packer, L. (1993a) Antioxidant activities of dihydrolipoic acid and its structural homologues. *Free Rad. Res. Comms.* 18: 115-122.

Suzuki, Y. J., Tsuchiya, M., Wassall, S. R., Choo, Y. M., Govil, G., Kagan, V. E., and Packer, L. (1993b) Structural and dynamic membrane properties of α-tocopherol and α-tocotrienol: implications to the molecular mechanisms of their antioxidant potency. *Biochemistry* (in press)

Wong, G.H.W., and Goeddel, D.V. (1988) Induction of manganous superoixde dismutase by tumor necrosis factor: possible protective mechanism. *Science* 242: 941-944.

Wong, G.H.W., Elwell, J.H., Oberley, L.W., and Goeddel, D.V. (1989) Manganous superoxide dismutase is essential for cellular resistance to cytotoxicity of tumor necrosis factor. *Cell* 58: 923-931.

Yamauchi, N., Kuriyama, H., Watanabe, N., Neda, H., Maeda, M., and Niitsu, Y. (1989) Intracellular hydroxyl radical production induced by recombinant human tumor necrosis factor and its implication in the killing of tumor cells *in vitro*. *Canc. Res.* 49: 1671-1675.

Zimmerman, R.J., Chan, A., and Leadon, S.A. (1989) Oxidative damage in murine tumor cells treated *in vitro* by recombinant human tumor necrosis factor. *Canc. Res.* 49: 1644-1648.

Oxidative Stress, Cell Activation and Viral Infection
C. Pasquier et al. (eds)
© 1994 Birkhäuser Verlag Basel/Switzerland

d-α-Tocopherol and Cell Proliferation

A. Azzi, D. Boscoboinik, E. Chatelain, Nesrin K. Özer and Barbara Stäuble

Institut für Biochemie und Molekularbiologie, Universität Bern, Bühlstrasse 28, 3012 Bern, Switzerland

Summary

Inhibition of cell proliferation by d-α–tocopherol occurs in vascular smooth muscle cells, Balb c/3T3 fibroblasts and neuroblastoma cells. Other cell lines, such as CHO, osteosarcoma and macrophages are not sensitive. The inhibition depends on the signalling path employed by cells to control proliferation. PDGF-BB is the most d-α–tocopherol sensitive mitogenic stimulus while lysophosphatidic acid is the least. d-β–Tocopherol, an analogue of d-α–tocopherol, with similar antioxidant properties, does not inhibit proliferation. Protein kinase C activity is inhibited by d-α–tocopherol but not by d-β–tocopherol, suggesting a central role of this enzyme in the control of cell proliferation by d-α–tocopherol.

Introduction

The regulation of cell proliferation by d-α–tocopherol, has been the subject of numerous investigations in normal and tumor cells. No effect, a stimulation of proliferation and an inhibition of growth have been shown,, suggesting that the type of cell and the conditions of the experiment may play a role in determining the cellular response to d-α–tocopherol. Chen et al (1988) have reviewed the effect of Vitamin E at an epidemiological level and conclude that d-α–tocopherol has been shown to inhibit skin, liver, oral, ear duct, and fore stomach carcinogenesis. Paganelli et al. (1992) also conclude that vitamins A, C, and E have chemopreventive efficacy against colon cancer in animal models. Watson (1986) has also suggested that high intakes of vitamin E can enhance some anticancer, immune defences. It appears thus that in animal models tumor proliferation is moderated by d-α–tocopherol. A larger number of studies have shown, however, at a cellular level a stimulation of proliferation. Morisaki et al. (1982) concluded that vitamin E both increased the cloning potential and the number of population doublings for smooth muscle cells in culture. Kuzuya et al. (1991) described an d-α–tocopherol stimulated endothelial cell proliferation in culture.

Newman (1990) observed that the inhibition of carcinoma and melanoma cell growth by type I transforming growth factor-β was dependent on the presence of polyunsaturated fatty acids and that it was almost completely reversed by the antioxidant vitamin E, suggesting a role for lipid peroxidation in this process.

Burdon et al (1990) observed in immortalised hamster (BHK-21) and rat (208F) cell lines before and after transformation to the malignant state with polyoma virus, or activated H-ras, respectively that d-α–tocopherol addition to serum-free medium was sufficient to restimulate growth. From these and other studies it appears clear that in cells, possibly when their growth is inhibited by lipid peroxidation, an antioxidant (and therefore also d-α–tocopherol) may restimulate proliferation by removing the inhibitory lipid peroxide. The question we have posed two years ago is if d-α–tocopherol had a direct effect as cell regulator and whether this effect was mediated or not by its reduction-oxidation properties, (Boscoboinik et al. 1991a, 1991b). We concluded that d-α–tocopherol had a growth inhibitory effect at a cellular level. This effect was not related to its antioxidant properties since β–tocopherol (an equally potent antioxidant, was not effective (Chatelain et al. 1993). The inhibition by d-α–tocopherol of cell proliferation was shown to be cell type specific and to depend on the mitogen responsible for stimulating growth. These data, together with supplemental new information, are reviewed below.

Materials and Methods

Materials. Tissue culture plastics were purchased from Falcon Labware (Becton Dickinson & Co.) and growth media and serum for cell culturing were obtained from Gibco Laboratories (Grand Island, NY). [Me-^{3}H]Thymidine (25 Ci/mmol) and [γ-^{32}P]ATP (30 Ci/mmol) were from Amersham International plc. d-α–, d-β–tocopherol were generous gifts from Hoffmann La Roche & Co. (Basle, Switzerland) and Henkel Co. (La Grange, IL, USA). The purity of these compounds was checked by thin layer chromatography and HPLC. The peptide (FKKSFKL) used as specific substrate for assaying protein kinase C activity was synthesised from dr. Rolli, Bern. Streptolysin-O (25,000 units) was from Sigma. Anti-protein kinase C polyclonal antibodies were from Gibco and the monoclonal antibodies from UBI, Lake Placid. All the other chemicals were of the purest grade commercially available.

Cell culture. Cell lines used in this work were obtained from the American Type Culture Collection. Human primary smooth muscle cells were a gift from Dr. Resink, Basel. Chinese Hamster Ovary cells were a gift from Dr. P. Gros, Canada. Most of the experiments described in thi article were carried out with the cell line A7r5 (rat aortic smooth muscle). Cells were grown, in Dulbecco's modified Eagle medium (DMEM) containing 25 mM bicarbonate, 60 U/ml penicillin, 60 µg/ml streptomycin, and 10% foetal calf serum (FCS). Cells were usually seeded into 100 mm plastic culture dishes and grown to confluence at 37°C in a humidified atmosphere of 5% CO_2. Culture media were changed every 3 days. In all experiments, media and sera were from the same batch number and source.

Cell synchronisation. Synchronous cultures at the G_1/S boundary were obtained by a combination of serum deprivation and hydroxyurea treatment (Ashihava and Baserga, 1979) Cells were subcultured in 6-well plates or 35-mm culture dishes containing 1.5 ml of DMEM supplemented with 2% FCS. Exponentially growing cultures were made quiescent (G_o) by exposure to serum deficient medium (0.2% FCS) for at least 48 h. Following serum deprivation, cells were restimulated by replacing the serum deficient medium by DMEM containing 2% FCS during 8 h; then, hydroxyurea stock solution was added to each plate (final concentration 1.5 mM). After 14 h of treatment, and when cells were blocked at the G_1/S boundary, medium was removed, cells were washed with PBS and transferred to the fresh complete medium (DMEM-2% FCS) to allow them enter into the S-phase and to progress through the cycle. Onset and duration of the S-phase was determined by pulse labelling with [^{3}H]thymidine for 1 h at different times after stimulation by serum throughout the course of one complete cell cycle. Cell number remained unchanged during 15 h following the removal of hydroxyurea and then cells entered into M-phase. Thus, a cell cycle time of around 24 h was established for A7r5 smooth muscle cells with the following duration of the different phases G_1 : 9h, S: 8h, G2: 6h, M: 1 h.

Measurement of [^{3}H]thymidine incorporation in the S-phase. Cells were pulsed with [^{3}H]thymidine (0.5-1 mCi/well) for 6 h during the S-phase following removal of hydroxyurea. Then, cells were washed twice with PBS supplemented with 10 mg/ml glucose and 1 mg/ml bovine serum albumin, fixed for 30 min with ice-cold 5% trichloroacetic acid, and solubilized in 0.1 M NaOH containing 2% (w/v) Na_2CO_3. The radioactivity incorporated into the acid insoluble material was determined in a liquid scintillation analyser.

Effect of tocopherols on cell proliferation. d-α–Tocopherol and d-β–tocopherol were absorbed to FCS and added in the required amounts to culture media. In all experiments, compounds were added when restimulating the cells with serum (2%) following at least 48 h deprivation. Cells were trypsinized and counted in a hemocytometer in triplicate after the completion of the cell cycle. Viability was assessed by the trypan blue dye exclusion method.

Protein kinase C assay in permeabilized cells. Measurements of protein kinase C activity in permeabilised smooth muscle cells were performed according to the procedure of Alexander et al. (1989) with minor modifications. A7r5 cells in the late G_1 phase of the cycle, preincubated for 8 h in the presence of the indicated tocopherol, were washed twice in PBS, resuspended in intracellular buffer (5.16 mM $MgCl_2$, 94 mM KCl, 12.5 mM Hepes, 12.5 mM EGTA, 8.17 mM $CaCl_2$, pH 7.4) and aliquoted in 220 µl portions (1.5×10^6 cells/ml). Assays were started by adding [γ–^{32}P]ATP (40 cpm/pmol, final concentration 240 µM), peptide substrate (final concentration 250 µM) and Streptolysin-O (0.6 i.u.). The reaction mixtures were incubated at 37°C for 5 min and the reaction was stopped by adding 100 µl of 25% (w/v) trichloroacetic acid in 2 M acetic acid. After being kept on ice for 10 min, samples were centrifuged for 5 min and spotted on P81 ion-exchange chromatography paper (Whatman International) which were then washed several times with 30% (v/v) acetic acid containing 1% (v/v) H_3PO_4 and once with ethanol. The P81 papers were dried, and the bound radioactivity was counted in a liquid scintillation analyser. To estimate the background phosphorylation of the peptide due to a kinase activity other than protein kinase C, assays were performed in cells treated for 24 h with 1 mM PMA (protein kinase C-down regulated cells). The value of ^{32}P incorporated obtained in the latter condition has been subtracted from the experimental data to account for the specific activity.

HPLC determination of d-α– and d-β–tocopherol. Cells were incubated with either d-α- or d-β-tocopherol in groowth medium for 24 h. Control cells were supplemented only with the amount of solvent used for the tocopherol addition. After several washes with BSA containing PBS, the tocopherol content was measured using reverse phase HPLC as described by Hess et al. (1991)

Immunoblot analysis. Cells were harvested at the indicated times and samples (25 µg protein/lane) were resolved by sodium dodecyl sulphate/8% polyacrylamide gel electrophoresis and transferred to PVDF membranes (Immobilon P, Millipore). Blots were saturated with 5% non-fat dry milk in Tris-buffered saline (TBS) for 1 h at room temperature and incubated overnight with rabbit anti-protein kinase C-α peptide antibody (3 µg/ml) in TBS containing 1% non-fat dry milk. Membranes were then washed twice in TBS-containing 0.05% Tween 20, incubated for 3 h with ^{125}I-IgG in TBS-containing 0.05% Tween 20 containing 0.1% non-fat dry milk, washed again in TBS-containing 0.05% Tween 20, dried and exposed to Kodak X-Omat AR film at -70°C. Protein content was determined with the Pierce BCA Protein Assay Reagent kit following the manufacturer's procedure.

Results

d-α– and d-β–tocopherol have different effects on SMC proliferation. The effect of d-α– and d-β–tocopherol on the proliferation 10% FCS is shown in Fig. 1 d-α–Tocopherol (5011M) inhibited cell growth approximately 50%. However, d-β–tocopherol, an analogue of d-α–tocopherol lacking a methyl group in position 7 of the chromanol ring did not show any inhibition of cell proliferation. The amount of α– and d-β–tocopherol present in the cells as measured at 24 hours was not significantly different, indicating that the lack of inhibition was not due to a different uptake of d-β–tocopherol.

Effect of d-α– and β–tocopherol on protein kinase C activity of smooth muscle cells. Itwas previously shown that purified brain protein kinase C could be inhibited, under certain experimental conditions, by d-α–tocopherol (Boscoboinik et al., 1991b). In order to study this reaction at a cellular level, streptolysin-O permeabilized smooth muscle cells and a protein kinase C peptide substrate (Alexander et al., 1989) were employed. As can be seen in Fig. 2 d-α–tocopherol strongly inhibited protein kinase C activity, whereas β–tocopherol was much less effective (Chatelain et. al, 1993).

The effect of α–tocopherol depends on the cell type. Table I shows the effect of d-α-tocopherol on the proliferation of different cell types. All the smooth muscle cell lines tested were inhibited by d-α–tocopherol, including a human primary culture. Balb/3T3 and a neuroblastoma line, were also inhibited. The other tested cell types were not d-α–tocopherol sensitive. The molecular basis of d-α–tocopherol sensitivity is not clear at the present moment. It can be based on a different signalling pathway used for proliferation in the various cell types.It may be possible that d-α–tocopherol transport and metabolism is different, depending on the cell type.

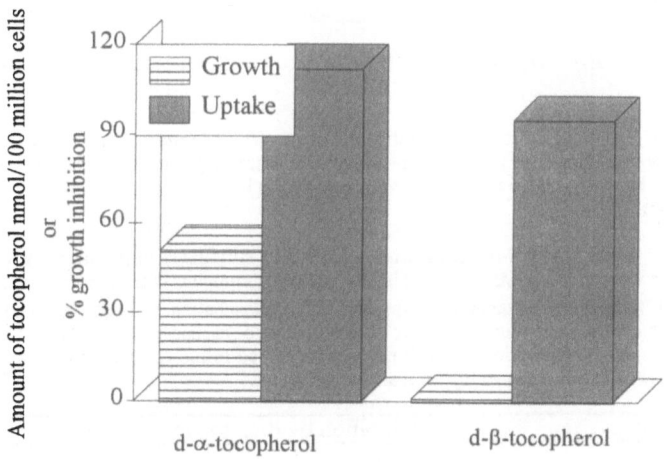

Figure 1. Effect of d-α–tocopherol and d-β–tocopherol on cell growth. Quiescent cells were restimulated to grow and synchronised as described in Methods in the presence of d-α–tocopherol or d-β–tocopherol. After completion of the cell cycle, the cell number was measured and aliquots of the resuspended cells were utilised for the HPLC determination of tocopherols, as described in Methods.

136

Table I. The growth inhibitory effect of d-α-tocopherol on different cell lines

Sensitive lines	Insensitive lines	Tissue and origin
A10		Rat aorta smooth muscle
A7r5		Rat aorta smooth muscle
HAI		Human aorta smooth muscle (primary)
NB2A		Mouse neuroblastoma
Balb/3T3		Mouse fibroblast
	LR73	Chinese hamster ovary
	Saos-2	Human osteosarcoma
	P388 Dl	Mouse monocyte-macrophage

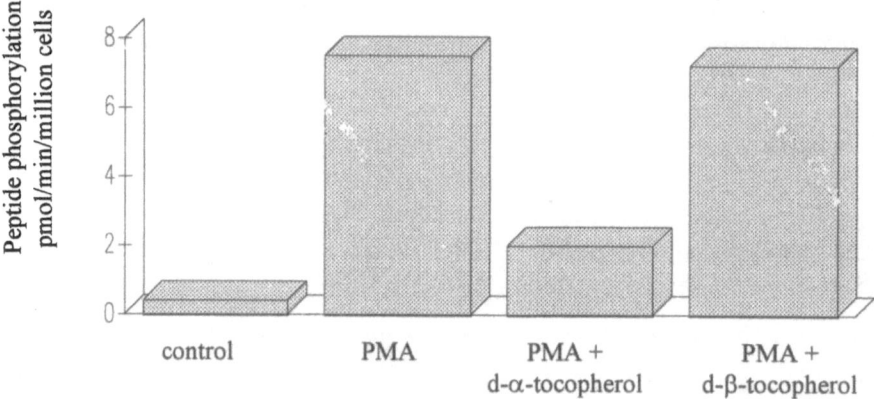

Figure 2. Effect of d-α–tocopherol and d-β–tocopherol on protein kinase C activity. Quiescent cells were restimulated to grow by adding 10% FCS in the presence or absence of d-α-tocopherol or d-β–tocopherol (50 μM). Phorbol myristate acetate was added two hours before the protein kinase C activity assay, which was performed as described in Methods.

Table II Differential inhibition by d-α–tocopherol of A7r5 cells stimulated to proliferate with various mitogens. Quiescent cells were incubated in DMEM containing the indicated mitogens in the presence or absence of 50 mM d-α–tocopherol. [³H]Thymidine incorporation (in the case of lysophosphatidic acid, bombesin and endothelin) or cell number (in the case of the other mitogens) were determined as described in Methods. Results are expressed as percentage of the control incorporation for each mitogen measured in the absence of d-α–tocopherol and are the mean of triplicate determinations from a representative experiment.

Mitogenic stimulus	Inhibition by d-α–tocopherol, %
Lysophosphatidic acid (50 μM)	8±1.5
Bombesin (20 nM)	26±4.5
FCS (2%)	52±2.0
LDL(modified) (5 μg/ml)	78±4.0
PDGF-BB (20 ng/ml)	93±3.0
Endothelin (80 nM)	94±2.5
LDL (5 μg/ml)	99±1.0

Finally it may be conceivable that d-α–tocopherol binding proteins (Nalecz et al., 1992) related to d-α–tocopherol inhibition are present in some cells and not in others. The cell type specificity of d-α–tocopherol inhibition is however important to understand some of the contradictory data in the literature and they indicate also that such inhibition is not caused by some trivial "damaging" effect of d-α–tocopherol, which should affect indiscriminately all cells.

The effect of α–tocopherol depends on the type of mitogen utilised to stimulate cell proliferation. The inhibition of cell proliferation produced by 50 mM d-α–tocopherol is maximal when PDGF, endothelin or native DLL are used as growth stimulant. Other stimulants, such as lysophosphatidic acid, bombesin, FCS and malonaldehyde modified LDL were less effective. It has been shown above that d-α–tocopherol causes inhibition of protein kinase C activity, a signalling element that can regulate cell proliferation in some, but not all cells, depending on the type of mitogen employed to stimulate growth.

The inhibition of protein kinase C activity depends on the time of d-α–tocopherol addition during the cell cycle. The time of addition of d-α-tocopherol during the cell cycle appeared to determine the extent of protein kinase C inhibition observed after several hours of incubation. Fig. 3 shows that addition of d-α–tocopherol at the G_0/G_1 phase transition resulted in an inhibition of the kinase activity (measured towards the end of the G_1 phase) of about 75%. If the addition was made later in the cycle progressively less inhibition of protein kinase C activity was measured. The inhibitory effect disappeared when d-α–tocopherol was added in the late G_1 phase and protein kinase C activity measured in the S phase. In this case a small activation was instead observed. The delayed onset of d-α–tocopherol inhibition was not caused by the time necessary for d-α–tocopherol to reach its target, since this was in both experiments similar. Instead, the time of the cycle when d-α–tocopherol was added appeared to be critical in the onset of d-α–tocopherol inhibition.

The expression of protein kinase C during the cell cycle and the effect of d-α–tocopherol.
In order to establish whether the decrease in protein kinase C activity induced by d-α-tocopherol added in the G_0-phase was due to an decrease in protein kinase C expression, the protein kinase C-α protein level in the G_1 phase was determined by Western blotting analysis. As can be seen in Fig. 4 the protein kinase C-α protein level in the G_1 phase was not affected by d-α-Tocopherol, under condition (cf. Fig. 3) where inhibition of activity was observed. However, in the cycloheximide-treated cells, the expected decrease in the protein kinase C-α protein level was observed.

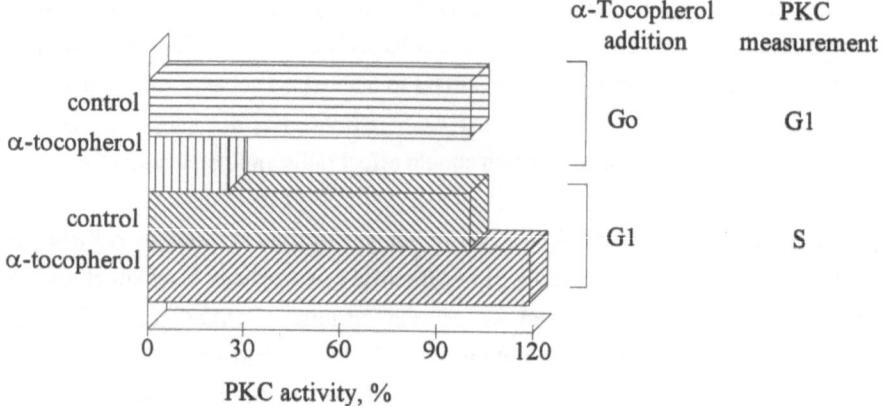

Figure 3. Effect of d-α–tocopherol added at two points in the cycle on the protein kinase C activity in smooth muscle cells. Cells at the G_0/Gl phase were stimulated with 10% FCS in the presence or absence of 50 μM α-tocopherol. 100 nM PMA-activated protein kinase C activity was measured eight hours later (G_1 phase). Alternatively, cells in G_1 phase received fresh media containing 50 μM d-α–tocopherol. and five hours afterwards protein kinase C activity was measured (S phase). Control cells were incubated in the absence of d-α–tocopherol.

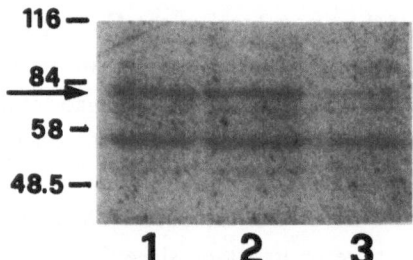

Figure 4. Effect of d-α-tocopherol and cycloheximide on the protein kinase C-α levels in the G_1 phase of the cell cycle. Anti-protein kinase Cα immunoblot of 25 μg of total-cell lysate of A7r5 cells taken 8 h after release from G_0 phase. Lane 1, control cells; lane 2, cells treated with 50 μM d-α-tocopherol; lane 3, cells treated with cycloheximide (2 μg/ml). Molecular weights in kDa are indicated on the left, the arrow corresponds to the position of protein kinase Cα.

Discussion

A number of epidemiological studies and investigations using animal models suggest that d-α–tocopherol is effective in controlling cell proliferation. At a cellular level d-α-tocopherol has been shown to be a stimulant as well as an inhibitor of cell proliferation. The study reported above represent an attempt to rationalise apparently conflicting effects.

Inhibition by d-α–tocopherol of cell proliferation: Radical scavenging versus ligand-interaction models. Under the conditions employed in this study the inhibition by d-α–tocopherol and the lack of inhibition by d-β–tocopherol of cell proliferation and protein kinase C activity indicates that the mechanism involved is not related to the radical scavenging properties of these two molecules, which are essentially equal. On the basis of the analysis of several other tocopherols (Chatelain et al. 1993) and similar compounds, a ligand-interaction type mechanism can be proposed for the action of d-α–tocopherol. The nature of the d-α–tocopherol target and its relationships with protein kinase C are discussed in the subsequent paragraph.

Molecular basis of the inhibition of protein kinase C activity by d-α–tocopherol. The inhibition of protein kinase C activity by d-α–tocopherol has been documented, although the molecular mechanism of inhibition is still elusive. The original hypothesis that protein kinase C may be the direct target of d-α–tocopherol can be now abandoned. The data of Fig. 3 show that d-α–tocopherol is only active if added in the G_0 phase of the cell cycle. Addition of d-α–tocopherol in G_1 phase does not result in inhibition. Since protein kinase C protein levels remain constant during the cell cycle and is not affected by d-α–tocopherol the diminution of activity (associated with a decrease of phorbol ester binding (Boscoboinik et al. 1993) can be reconciled with the following model. d-α–Tocopherol may prevent the activation of the enzyme by hindering post-translational events like phosphorylation (Pears et al., 1992), binding to a receptor protein (Mochley-Rosen et al., 1991a, b) or translocation (Kroft and Anderson, 1983 and for a review see Azzi et al, 1992). Another explanation of the inhibition may be that d-α–tocopherol prevents the synthesis of enzymes necessary for the post-translational events considered in the previous model.

Cell specificity of d-α–tocopherol inhibition. The lack of a molecular description for the inhibitory mechanism of cell proliferation produced by d-α-tocopherol does not diminish the importance of this event, caused by physiological concentrations of d-α-tocopherol, at a cellular level. Since for many cells protein kinase C is an obligatory pathway leading to cell proliferation, inhibition of protein kinase C will result in growth inhibition. Other cells may have alternative, redundant pathway converging on some common, crucial event for cell proliferation. In this case, block of protein kinase C may produce partial inhibition of growth. Proliferation in other cell types may not require protein kinase C activation.

Physiological relevance of d-α–tocopherol inhibition of cell proliferation. The inhibition of smooth muscle cells proliferation by d-α–tocopherol at physiological concentrations may explain the notion that in vivo smooth muscle cells are quiescent and they multiply only under stress condition (Clowes and Schwartz, 1985 and for a review cf. Raines and Ross, 1993). Depletion of d-α–tocopherol may occur locally and generally as a consequence of oxidative stress. Such a condition would be compatible with a stimulation to proliferation. Our proposal can be expressed in terms of the role in the pathogenesis of arteriosclerosis of a dietary or oxidative diminution of d-α–tocopherol as a primary stimulus for the proliferation of smooth muscle cells. It may also be conceivable that high blood cholesterol and LDL levels may result in an increased association of d-α–tocopherol to them with a relative diminution of its availability at the level of arterial wall (Burton and Ingold, 1989: Traber et al..1990). Some tumor cells may also be sensitive in their proliferation to d-α–tocopherol if their growth is controlled by the protein kinase C path. Such a special case may explain why only certain tumours may respond in animal models to vitamin E and why also in vitro some, but not other cells are sensitive.

Conclusion

In all cases in which cell proliferation occurs, stimulated by overexpression of an oncogene, by a virus or other causes, the possibility of a control by d-α–tocopherol cannot be neglected. An elucidation of the molecular target of d-α–tocopherol may lead to a better understanding of the proliferative process and to the possibility, through drug modelling, of producing more potent and selective antiproliferative drugs.

Acknowledgements

The present work has been supported by the Swiss Research Foundation and by Hoffman-La Roche, AG. N.K.Ö. was recipient of a FEBS Fellowship.

References

Alexander, D. R., Hexham, J. M., Lucas, S. C., Graves, J. D., Cantrell, D. A. and Crumpton, M. J (1989) A protein kinase C pseudosubstrate peptide inhibits phosphorylation of the CD3 antigen in streptolysin-O-permeabilized human T- lymphocytes.*Biochem. J*. 260: 893-901.

Ashihara, T. and Baserga, R.(1979) Cell synchronization. *Methods Enzym*. 58, 248-262

Azzi-A, Boscoboinik-D, Hensey-C. 1992 The protein kinase C family. *Eur.J.Biochem*. 208 547-57

Boscoboinik-D, Szewczyk-A, Azzi-A.(1991) Alphα–tocopherol (vitamin E) regulates vascular smooth muscle cell proliferation and protein kinase C activity. *Arch.Biochem.Biophys*., 286, 264-9

Boscoboinik-D, Szewczyk-A, Hensey-C, Azzi-A (1991) Inhibition of cell proliferation by α–tocopherol. Role of protein kinase C. *J.Biol.Chem*., 266, 6188-94

Burdon-R-H, Gill-V, Rice-Evans-C.(1990) Oxidative stress and tumour cell proliferation. *Free Radic.Res.Commun*., 11, 65-76

Burton-G-W, Ingold-K-U. 1989 Vitamin E as an in vitro and in vivo antioxidant. *Ann.N.Y.Acad.Sci* . 570, 7-22,

Chen-L-H, Boissonneault-G-A, Glauert-H-P (1988) Vitamin C, vitamin E and cancer (review). *Anticancer.Res*., 8, 739-48

Clowes, A.W. ans Schwartz, S.M. (1985) Significance of quiescent smooth muscle migration in the injured rat carotid artery. *Circ. Res.* 156, 139-145

Hess-D, Keller-H-E, Oberlin-B, Bonfanti-R, Schuep-W. 1991 Simultaneous determination of retinol, tocopherols, carotenes and lycopene in plasma by means of high-performance liquid chromatography on reversed phase. *Int.J.Vitam.Nutr.Res..*61,232-8

Kraft-A-S, Anderson-W-B. 1983 Phorbol esters increase the amount of Ca2+, phospholipid-dependent protein kinase associated with plasma membrane. *Nature* 301, 621-3

Kuzuya−M, Naito-M, Funaki-C, Hayashi-T, Yamadα−K, Asai-K, Kuzuya−F.(1991) Antioxidants stimulate endothelial cell proliferation in culture. *Artery*, 18, 115-24

Mochly-Rosen-D, Khaner-H, Lopez-J, Smith-B-L 1991a Intracellular receptors for activated protein kinase C. Identification of a binding site for the enzyme. *J.Biol.Chem* . 266 14866-8

Mochly-Rosen-D, Khaner-H, Lopez-J. 1991b Identification of intracellular receptor proteins for activated protein kinase C. *Proc.Natl.Acad.Sci.U.S;A* .88 3997-4000

Mochly-Rosen-D, Miller-K-G, Scheller-R-H, Khaner-H, Lopez-J, Smith-B- L. 1992 p65 fragments, homologous to the C2 region of protein kinase C, bind to the intracellular receptors for protein kinase C. *Biochemistry* 31 8120-4

Morisaki-N, Stitts-J-M, Bartels-Tomei-L, Milo-G-E, Panganamalα−R-V, Cornwell-D-G. (1982) Dipyridamole: an antioxidant that promotes the proliferation of aorta smooth muscle cells. *Artery* 11, 88-107

Newman-M-J.(1990) Inhibition of carcinoma and melanoma cell growth by type 1 transforming growth factor beta is dependent on the presence of polyunsaturated fatty acids. *Proc.Natl. Acad;Sci;U.S.A* 87, 5543-7

Paganelli-G-M, Biasco-G, Brandi-G, Santucci-R, Gizzi-G, Villani-V, Cianci-M, Miglioli-M, Barbarα−L.(1992) Effect of vitamin A, C, and E supplementation on rectal cell proliferation in patients with colorectal adenomas. *J.Natl.Cancer.Inst* 84, 47-51, .

Raines, E.W. and Ross, R. (1993). Smooth muscle cells and the pathogenesis of the lesion of atherosclerosis. Br. *Heart J.*, 69. 30S-37S

Traber-M-G, Burton-G-W, Ingold-K-U, Kayden-H-J.1990 RRR- and SRR-alpha-tocopherols are secreted without discrimination in human chylomicrons, but RRR-alpha-tocopherol is preferentially secreted in very low density lipoproteins. *J.Lipid.Res* 31 675-85

Traber-M-G, Sokol-R-J, Burton-G-W, Ingold-K-U, Papas-A-M, Huffaker-J- E, Kayden-H-J.1990 Impaired ability of patients with familial isolated vitamin E deficiency to incorporate alpha-tocopherol into lipoproteins secreted by the liver.*J.Clin.Invest* 85 397-407

Watson-R-R.(1986) Immunological enhancement by fat-soluble vitamins, minerals, and trace metals: a factor in cancer prevention. *Cancer.Detect.Prev* 9, 67-77

Oxidative Stress, Cell Activation and Viral Infection
C. Pasquier et al. (eds)
© 1994 Birkhäuser Verlag Basel/Switzerland

Effects of Intracellular Redox Status on Cellular Regulation and Viral Infection

G. Rotilio, L. Knoepfel, C. Steinkuhler, *A. T. Palamara, M. R. Ciriolo and *E. Garaci

Depts. of Biology and *Experimental Medicine Biochemical Sciences, University of Rome "Tor Vergata", Rome, Italy.

Summary

Cell activation by changes of cell redox state is in many instances related to oxidative modifications of the transcription factors. In particular, we investigated the effects of redox state on the DNA binding capacity of the zinc-finger transcription factor Sp1. H_2O_2, diamide or dialysis under nonreducing conditions completely abolished the ability of Sp1 to bind to its consensus DNA sequence. This inhibition was quantitatively reversed by reduced glutathione (GSH), dithiotreitol (DTT) or β-mercaptoethanol (β-ME), but not by ascorbate. Using the thiol alkylating agents N-ethylmaleimide (NEM) and iodacetamide we showed that Sp1 contains SH groups that are necessary for DNA binding. We suggest that the NEM-reactive SH groups confer redox-susceptibility to the Sp1 molecule and are most likely located within the zinc finger domain.

On the other hand, virus infection may be adversely affected by thiol reagents, since formation and stability of S-S bridges is often essential to proteins involved in virus binding and infectivity. In particular, we have investigated the effects of exogenous glutathione on Sendai virus replication, a virus-host cell system in which virus replication is associated with a marked decrease in total glutathione content. Addition of reduced glutathione inhibited the production of Sendai virus in AGMK cells. The effect was dose-dependent, associated with an increase of the intracellular glutathione level, while it did not affect cellular metabolism. The antiviral effect of glutathione was accompanied by decrease and inactivation of the hemagglutinin-neuraminidase (HN) Sendai virus glycoprotein, which normally assembles into oligomers via formation of disulfide bonds. Reductive cleavage of the S-S bridges by GSH may, at least in part, be responsible for the lower amount of HN detected in virus produced by AGMK cells.

Introduction

Evidence is growing for an important role played by redox mechanisms in the regulation of different cellular functions and in viral infections of mammalian cells. Among the cellular processes found to be regulated by a redox mechanism there are hormone-receptor interactions (Grippo et al., 1985), translational regulation (Hentze et al., 1989; Klausner et al., 1989), bacteriophage DNA replication (Holmgren et al., 1989), protein binding to mRNA (Clerch et al., 1992) and binding of several transcription factors to cognate DNA.

Regulation of trascription by reducing conditions was found in the case of the Fos-Jun heterodimer. Abate (Abate et al., 1990) showed that treatment with the sulfhydryl oxidizing agent diamide abolishes DNA binding of the Fos-Jun heterodimer, while it does not alter DNA-binding capacity of Fos-Jun proteins point mutated at a critical cysteine to serine.

Hutchinson et al. showed that DNA-binding of the glucocorticoid receptor is abolished by H_2O_2 treatment and that it can be restored by addition of both zinc and a thiol disulfide exchange reagent such as dithiotreitol (Hutchinson *et al.*, 1991). Moreover, they proved that the inhibition by peroxide is due to disulfide-bridging within the zinc fingers containing DNA-binding domain of the glucocorticoid receptor.

On the other hand, Schreck *et al.* showed that the activation of the multisubunit transcription factor NF-κB is mediated by a redox mechanism (Schreck *et al.*, 1991). Dissociation of the inhibitory subunit IκB, resulting in activation of NF-κB to its DNA-binding state, is induced by many different agents (reviewed in Bauerle and Baltimore, 1991), including mitogens (phorbol esters, lectins), cytokines (interleukin-1, tumor necrosis factor), viruses (HIV-1, cytomegalovirus), double stranded RNA and agents provoking oxidative stress. Schreck *et al.* found that the antioxidant and radical scavenger N-acetyl-L-cysteine prevents activation of NF-κB not only by oxidants like peroxide, but also by many of the other activating agents. Based on these results, the authors suggested that the activating pathways triggered by different agents may converge in a unifying mechanism to induce DNA-binding of NF-κB: transient increase of the intracellular concentration of reactive oxygen intermediates.

Also infection by several viruses seems to be associated to oxidative stress. Hennet *et al.* found a marked decrease of the reduced forms of pulmonary antioxidants in the lungs of mice after infection with influenza A PR8 virus (Hennet *et al.*, 1992). In individuals seropositive for the human immunodeficiency virus (HIV), decreased levels of total acid-soluble thiols and of reduced glutathione (GSH) have been found in plasma, peripheral blood monocytes and lung epithelial lining fluid (Bull *et al.*, 1989; Eck *et al.*, 1989; Staal *et al.*, 1992a). Kalebic *et al.* (1991) have demonstrated that GSH and GSH esters can block cytokine-stimulated HIV-transcription. Moreover N-acetyl-L-cysteine, which enhances intracellular glutathione, inhibits cytokine-stimulated HIV-replication in many systems *in vitro* (Staal *et al.*, 1992b). Thiol-mediated regulation of HIV activation has been related to the influence exerted by GSH on NF-κB, the nuclear factor that increases HIV transcription and replication (Staal *et al.*, 1992b). These results are consistent with the ones obtained by Schreck *et al.*, described above.

In this study we investigated the effects of redox state on the DNA-binding capacity of the zinc finger transcription factor Sp1. Sp1 was originally identified as a protein from Hela cells that binds to the GC-boxes in the 21 bp repeat elements of SV40, activating in vitro transcription from the SV40 early promoter (Dynan and Tjian, 1983; Gidoni *et al.*, 1984, 1985). Following studies revealed that Sp1, which contains in its DNA-binding domain three contiguous zinc finger structures of the Cys_2His_2 type, activates transcription of a variety of cellular and viral promoters. We found that specific DNA-binding of Sp1 requires the reduced state of the transcription factor.

We have also investigated redox regulation of Sendai virus (SV) replication in African Green Monkey Kidney (AGMK) cells. SV is an enveloped virus with six major structural proteins. Nucleoprotein (NP), large (L) and P proteins are associated with the nucleocapside, while hemagglutinin-neuraminidase (HN), fusion (F) and matrix (M) proteins constitute the envelope. Results presented in this study indicate that an increase in total intracellular glutathione content has a remarkable inhibitory effect on viral replication associated with a defective function of proteins of the viral envelope.

Materials and Methods

Preparation of Nuclear Extracts

K562 nuclear extracts were prepared following the standard procedure described by Dignam (Dignam *et al.*, 1983) with the only modification that DTT in all buffers was 0.1 mM instead of 0.5 mM. DTT and PMSF were added to the buffers just prior use.

Synthesis and Labelling of Oligonucleotides

Oligo-GC is a Sp1 binding 32-mer that contains two tandem copies of a 16 bp segment from the Herpes Simplex virus immediate-early 3 gene promoter. The 32-mer used for competition experiments, termed oligo-SOD, is a 32-mer segment of the cytoplasmatic superoxide dismutase gene of zea mais. The sequences of the oligonucleotides are: oligo-GC 5'-GGATGGGCGGGGCCGGGGATGGGCGGGGCCGG-3'; oligo-SOD 5'-CGTGTTGCCTGT GGTATCATTGGGCTCCAAGG-3'. Oligonucleotides with sequences complementary to oligo-GC and oligo-SOD were also synthesized and annealed to form doublestranded oligonucleotides.

Electrophoretic Mobility Shift Assay (EMSA)

In the gel retardation experiments reactions were carried out by incubating 8 μg nuclear proteins in a buffer consisting of 20 mM Tris, pH 7.6, 8% Ficoll, 50 mM KCl, 0.06 μg/μl poly(dI-dC) with different reagents. When treating with oxidants, nuclear extracts were preincubated with the oxidizing reagents H_2O_2 or diamide (see figure legends for concentrations) for 1 h at 4°C. The reaction with H_2O_2 was stopped by addition of 100 U catalase. Where specified, oxidized nuclear extracts were subsequently subjected to reduction by incubation for 30 min at room temperature (RT) with GSH, DTT, β-ME or ascorbate. Incubation with NEM or iodacetamide was for 45 min at RT. Dialysis of nuclear extracts under nonreducing conditions was performed for 5 hrs against buffer D according to Dignam, except that DTT was omitted (Dignam *et al.*,

1983). For competition experiments, 8 µg nuclear proteins were incubated with 5 pmoles unlabelled oligo-GC or oligo-SOD. Binding assays were subsequently performed by incubating pretreated nuclear extracts with50 fmoles ^{32}P-endlabelled Oligo-GC for 20 min at RT. The reaction mixtures were loaded on a 4% polyacrylamide gel in 0.25xTBE buffer and electrophoresed at 0.75 mA/cm at RT. Gels were subsequently autoradiographed.

AGMK-Cell Culture and Viral Infection

AGMK cells were grown in RPMI 1640 medium supplemented with 2% fetal calf serum in humidified atmosphere containing 5% CO_2 at 37° C .

Confluent monolayers were washed with phosphate-buffered saline (PBS) and infected with egg grown sendai virus (SV) (10 hemagglutinating units (HAU) x 10^5 cells). 1 hr after infection the virus inocula were removed, monolayers were washed three times with PBS and incubated in medium with or without GSH. Sendai virus titration was performed by measuring HAU present in the culture medium. Intracellular glutathione content was assayed according to Anderson by a DTNB- glutathione reductase recycling assay (Anderson, 1985).

DNA, RNA and protein synthesis were measured in confluent monolayers of cells treated with 20 mM GSH after the infection period. Cells were labelled for 18 hrs, starting 5 hrs after infection, with 5 µCi/ml/2 x10^5 cells of ^{3}H-thymidine, ^{3}H--uridine and ^{35}S- methionine. Cells were then washed three times with PBS; 0.4 ml of 5% TCA was added to each culture, and radioactivity in acid-soluble material was counted. Acid- insoluble radioactivity was measured after washing for three times the TCA precipitates with ethanol, drying under an infrared lamp and dissolving the samples in 0.4 ml of a solution containing 0.1 M NaOH, 0.5% SDS.

For study of GSH treatment on SV production by AGMK cells, confluent monolayers of cells infected with SV were treated with 20 mM GSH after the infection period and labelled with ^{35}S-methionine (5 µCi/ml). Culture supernatants were collected 24 hrs post infection and clarified at 60000 g for 20 min. Supernatants were diluted in PBS and centrifuged at 12000 g for 2hrs (Hsu et al., 1979). After centrifugation the viral particles were resuspended in 50 µl of polyacrylamide gel electrophoresis sample buffer and, after determining radioactivity incorporated, electrophoresed on a 10 % gel. Proteins were revealed by autoradiography of the dried gel.

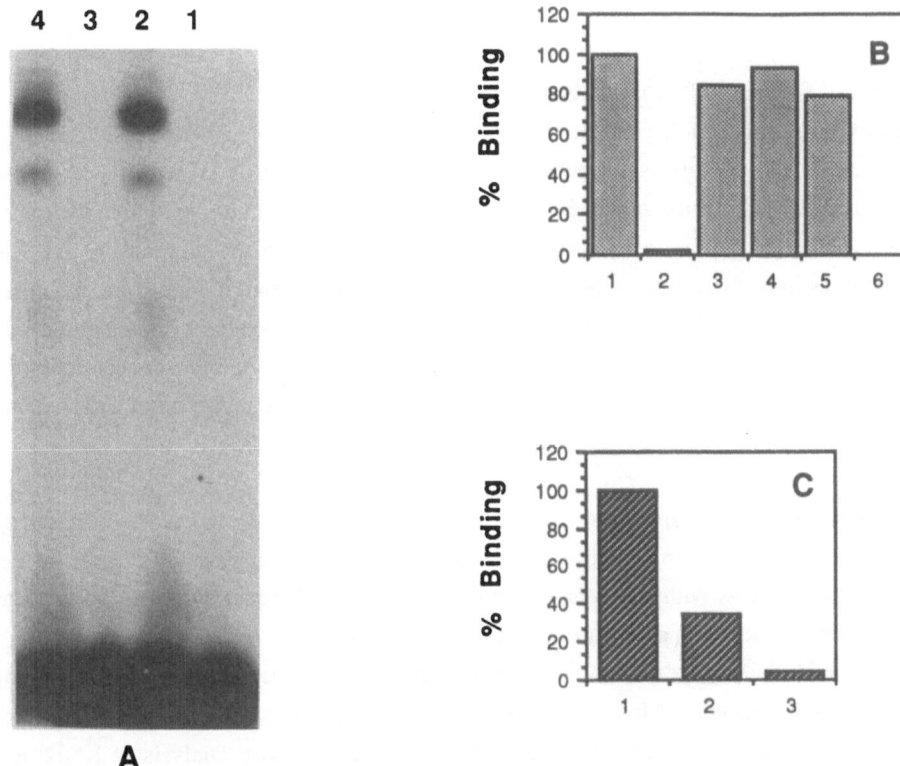

Figure 1.(A) Detection of Sp1 in K562 cells nuclear extracts. *Lane 1*, end-labelled oligo-GC; *lane 2*, addition of K562 cells nuclear extract; *lane 3*, competition with 5 pmoles unlabelled oligo-GC; *lane 4*, competition with 5 pmoles (unlabelled) oligo-SOD.

(B) Effect of oxidation and reduction of nuclear extract on the DNA-binding capacity of Sp1. Autoradiographs of the gels were subjected to densitometric analysis to quantify DNA-binding of Sp1. Percent binding is expressed. *Column 1*: untreated nuclear extract (=100%); *columns 2-6*: incubation with 3 mM diamide for 1 h at 4°C; *columns 3-6*: subsequent incubation with 20 mM GSH, 20 mM DTT, 20 mM βME and 40 mM ascorbate respectively, for 30 min at RT.

(C) Effect of treatment with the thiol alkylating reagents NEM and iodacetamide on DNA-binding capacity of Sp1. Autoradiographs of the gels were subjected to densitometric analysis to quantify DNA-binding of Sp1. Percent binding is expressed. *Column 1*: untreated nuclear extract (=100%); *column 2*: incubation with 25 mM iodacetamide for 45 min at RT; *column 3*: incubation with 3 mM NEM for 45 min at RT.

Results

Detection of Sp1 in K562 Cells Nuclear Extracts

To detect the presence of Sp1 in K562 cells nuclear extracts we used an electrophoretic mobility shift assay (emsa). In this assay, an end-labelled DNA fragment is incubated with a nuclear extract and is then electrophoresed through a native polyacrylamide gel. Oligo-GC, a 32 bp oligonucleotide containing a strong Sp1 binding site present in the promoter of the Herpes simplex virus immediate-early 3 gene, was used as probe. As shown in Figure 1A, the observed band shift was specific for Sp1, since complex formation was prevented by competition with an excess of unlabelled oligo-GC oligonucleotide and unaffected by the same concentration of the unrelated 32-mer oligo-SOD. These results indicate that selective detection of Sp1 is achieved with the bandshift assay using crude nuclear extracts.

Oxidants Inhibit DNA-binding by Sp1 in a Reversible Manner, Acting on the SH-groups of Sp1

To investigate if changes in the redox-state affect the DNA-binding capacity of Sp1, we treated K562 nuclear extracts with different oxidizing agents and subsequently quantified changes in binding with densitometric analysis of the shifted bands (Fig.1B).

Treatment with the sulfhydryl oxidizing agent diamide completely abolished DNA-binding by Sp1 at a concentration of 3 mM (Fig. 1B). Similar results were obtained using H_2O_2, which abolished binding at a concentration of 10 mM (data not shown). Dialysis of K562 nuclear extracts for 5 hrs under nonreducing conditions (i.e. omitting dithiotreitol in the dialysis buffer) also inhibited DNA-binding of Sp1, whereas dialysis under reducing conditions had no measurable effects. The abolition of DNA-binding by the thiol oxidizing reagent diamide indicates that SH-groups confer oxidant susceptibility to Sp1.

In each of the above cases, binding was quantitatively restored by subsequent treatment of the oxidized nuclear extract with millimolar amounts of the reducing agents DTT, GSH and β-ME. No restoration of DNA-binding activity of Sp1 could be observed by treatment with the reductant ascorbate at concentrations between 4 and 40 mM, suggesting that only thiol disulfide exchange reagents may be able to reverse the effects of oxidants on Sp1.

These findings indicate that Sp1 transcription factor is able to undergo reversible oxido-reductions and that changes in the redox state of its thiol groups modulate DNA-binding capacity.

To further investigate the role of SH-groups in DNA-binding by Sp1, we studied the effects of the SH alkylating agents NEM and iodacetamide on Sp1 (Fig. 1C).

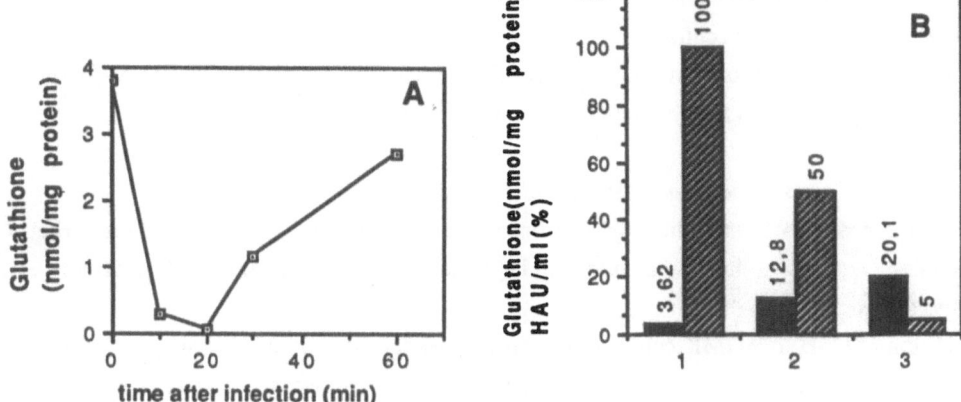

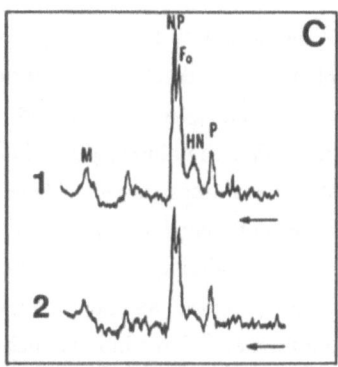

Figure 2. (A) Intracellular glutathione content of AGMK cells infected with Sendai virus. After infection, the cells were washed twice with PBS, detached by gentle scraping and used for total glutathione assay. The data are from a single experiment representative of three.

(B) Effect of addition of exogenous GSH on intracellular glutathione levels and on viral infection. 1) untreated; 2) cells were treated with GSH for 24 hrs before infection; 3) GSH was added 1 hr after infection and maintained for 24 hrs. In the case of 1) and 2) intracellular glutathione content was measured befor infection, while in the case of 3) it was measured 24 hrs after infection. ■ : intracellular glutathione (nmol/mg protein); ▨ : HAU/ml (%) 24 hrs after infection.

(C) Effect of GSH treatment on Sendai virus production. Densitometric analysis of SDS-PAGE-autoradiography of ^{35}S-methionine-labelled proteins from purified virus. 1) infected 2) infected GSH-treated. See text for details.

Binding of Sp1 to its cognate DNA was almost completely abolished by incubation with 3 mM NEM, while higher concentrations of iodacetamide (25 mM) sensibly diminished the binding without completely abolishing it. These results further support the hypothesis of SH groups playing a central role in the binding of Sp1 to its cognate DNA.

Inhibition of Sendai Virus Replication by GSH

The time profile of intracellular glutathione content in AGMK cells was studied during SV infection. Figure 2A shows that intracellular glutathione levels decline rapidly at very early times (20 min).

Addition of GSH to AGMK cells led to a dose dependent inhibition of virus production up to 20 mM GSH (data not shown). In particular, when GSH was added to the cells 24 hrs before the infection a 50% inhibition in the virus titer was observed. On the other hand, treatment of monolayers after the infection period resulted in a 93-97% inhibition of the hemagglutinating titers (Figure 2 B). Furthermore, GSH treatment strongly increased total internal glutathione levels in both cases (Figure 2B). At the dose used in these experiments (20 mM) GSH did not produce any toxic effect on uninfected cells, as determined either by microscopic examination or by vital dye exclusion. Furthermore, DNA, RNA synthesis and uptake of precursors were not significantly affected by GSH treatment in both uninfected and infected cells, while the rate of protein synthesis was slightly inhibited (data not shown). These data suggest that the GSH inhibitory effect on virus replication was not related to an aspecific cytopathic effect.

To further investigate the role of GSH in antiviral action we purified the virus from supernatants of confluent monolayers treated with 20 mM GSH and labelled with ^{35}S-methionine for 24 hrs after the infection period. A 69% inhibition of ^{35}S-methionine incorporation in the viral proteins after virus purification was observed. Furthermore, autoradiography of the SDS-PAGE of the virus particles indicates that GSH strongly decreased the amount of virus proteins. This result was confirmed by densitometric analysis of the autoradiography, which shows a marked decrease for HN glycoprotein (50%), and a less pronounced decrease for NP plus F and M proteins (28 and 26% respectively) (Figure 2C).

Discussion

Viral proliferation is a multistage process occurring in different cellular and extracellular compartments. Many single steps are affected by changes in the redox potential of the milieu in which they take place. We have shown that interventions aimed at modulating redox state are likely to have pleiotropic effects on viral infection, the net result depending on a complex balance of different factors. For example, transcription of viral genes may be both stimulated or repressed under oxidant conditions, depending on the exact nature of the involved transcription factors. As a matter of fact, oxidants stimulate DNA-binding of NF-κB (Schreck *et al.* 1991) but inhibit nucleic acid recognition of transcription factors whose functionality critically depends on the reduced state of single SH-groups (Abate *et al.*, 1990; Hutchinson *et al.*, 1991). In this context, transcription factors such as Sp1, which contain the zinc finger DNA-binding motif, consisting of either 2 Cys/2 His, 3 Cys/1 His or 4 Cys residues as ligands for the zinc cofactor, may be particularly prone to oxidation processes, resulting in impaired or abolished DNA-binding activities. Expression of viral genes under the control of promoters containing binding sites for this class of transcription factors may therefore be reduced under oxidant conditions.

On the other hand, both productive folding and assembly of viral proteins as well as recognition of cell surface receptors by the virus may depend on the formation or the presence of disulfide bridges. Reducing conditions may impede correct disulfide bridge formation of folding polypeptides resulting in malfolded proteins. Furthermore, external addition of thiol agents such as glutathione may cause reduction of S-S bridges on the viral envelope, thereby causing structural changes that abolish critical recognition steps.

This mechanism is a likely explanation for our findings of the effects of GSH on Sendai virus replication. In particular, the HN glycoprotein, which is involved in attachment of the virus to receptors, agglutination of erythrocytes, neuraminidase activity (Scheid and Choppin, 1974),and the F protein, which is responsible for virus penetration of host cell membranes, virus-induced hemolysis and cell fusion, (Scheid and Choppin, 1974) normally assemble into oligomers. This oligomerisation occurrs via formation of disulfide bonds. Reductive cleavage of these S-S bonds exerted by the higher intracellular GSH levels, may be responsible for both the loss of hemagglutinating activity and, at least in part, for the lower amount of HN detected in virus produced by AGMK cells in the presence of added GSH. Furthermore, the fact that the F protein is influenced to a lesser extent with respect to HN, is in line with a higher resistence of this protein to reducing agents, due to the presence of unreactive S-S bridges (Vidal et al., 1989).

Acknowledgements

C.S. was recipient of a fellowship awarded by the Deutsche Forschungsgemeinschaft. This work was supported by the CNR Target Project "Chimica fine 2" and "Biotecnologie e Biostrumentazione".

References

Abate, C., Patel, L., Rauscher III, F. J., and Curran, T. (1990) Redox regulation of Fos and Jun DNA-binding activity in vitro. *Science* 249: 1157-1161.

Anderson, M. E. (1985) Determination of glutathione and glutathione disulfide in biological samples. *Methods Enzymol.* 113: 548-555.

Bauerle, P. A., and Baltimore, D. (1991) In Cohen, P. and Foulkes, J. G. (eds), *Molecular aspects of cellular regulation, hormonal control regulation of gene transcription.* Elsevier/North Holland Biomedical Press, Amsterdam pp. 409-432.

Buhl, R., Jaffe, H.A., Holroyd, K.J., Wells, F.B.,Mastrangeli, A., Saltini, C., Cantin, A.M., and Crystal, R.G. (1989) Systematic glutathione deficiency in symptom-free HIV-seropositive individuals . *Lancet* ii: 1294-1297.

Clerch, L. B., and Massaro, D. (1992) Oxidation-reduction-sensitive binding of lung protein to rat catalase mRNA. *J. Biol. Chem.* 267: 2853-2855.

Dignam, J.D., Lebovitz, R. M., and Roeder, R.G. (1983) Accurate transcription initiation by RNA polymerase II in a soluble extract from isolated mammalian nuclei. *Nuc. Acids Res.* 11: 1475-1489.

Dynan, W. S., and Tjian, R. (1983) Isolation of transcription factors that discriminate between different promoters recognized by RNA-polymerase II. *Cell* 32: 669-680.

Eck, H-P., Gmunder, H., Hartmann, M., Petzoldt, D., Daniel, V., and Drodge, W. (1989) Low concentrations of acid-soluble thiol (cysteine) in the blood plasma of HIV-1 infected patients. *Biol. Chem. Hoppe-Seyler* 370: 101-108.

Gidoni, D., Dynan, W. S., and Tjian, R. (1984). Multiple specific contacts between a mammalian transcription factor and its cognate promoters. *Nature* 312: 409-413.

Gidoni, D., Kadonaga, J. T., Barrera-Saldana, H., Takahashi, K., Chambon, P., and Tjian, R. (1985). Bidirectional SV40 transcription mediated by tandem Sp1 binding interactions. *Science* 230: 511-517.

Grippo, J.F., Holmgren, A., and Pratt, W. B. (1985) Proof that endougenous, heat stable glucocorticoid-activating factor is thioredoxin. *J. Biol. Chem.* 260: 93-97.

Hennet, T., Peterhans, E., and Stocker, R. (1992) Alterations in antioxidant defences in lung and liver of mice infected with influenza A virus. *J.Gen. Virol.* 73: 39-46.

Hentze, M. W., Rouault, T. A., Harford, J. B., and Klausner, R. D. (1989) Oxidation-reduction and the molecular mechanism of a regulatory RNA-protein interaction. *Science* 244: 357-359.

Holmgren, A. (1989) Thioredoxin and glutaredoxin systems. *J. Biol. Chem.* 264: 13963-13966.

Hsu, M., Scheid, A., and Choppin, P.W. (1979) Reconstitution of membranes with individual paramyxovirus glycoproteins and phospholipid in cholate solution. *Virology* 95: 476-491.

Hutchinson, K. A., Matic, G., Meshinchi, S., Bresnick, E. H., and Pratt, W. B. (1991) Redox manipulation of DNA-binding activity and BuGR epitope reactivity of the glucocorticoid receptor. *J. Biol. Chem.* 266: 10505-10509.

Kalebic, T., Kinter, A., Poli, G., Anderson, M.E., Meister, A., and Fauci, A.S. (1991) Suppression of human immunodeficiency virus expression in chronically infected monocytic cells by glutathione, glutathione ester and N-acetylcysteine. *Proc. Natl. Acad. Sci.* 88: 986-990.

Klausner, R. D., and Harford, J. B. (1989) Cis-trans models for post-translational gene regulation. *Science* 246: 870-871.

Scheid, A., and Choppin, P.W. (1974) Identification of biological activities of paramyxovirus glycoproteins: activition of cell fusion, hemolysis and infectivity by proteolytic cleavage of an inactive precursor protein of sendai virus. Virology 57: 475-490.

Schreck, R., Rieber,P., and Bauerle, P. A. (1991) Reactive oxygen intermediates as apparently widely used messengers in the activation of the NF-kB transcription factor and HIV-1. *EMBO J.* 10: 2247-2258.

Staal, F.J.T., Roeder, M., Israelski, D.M., Bub, P.J., Mole, L.A., McShane, D., Deresinski, S.C., Ross, W., Sussman, H., Raju, P.A., Anderson, M.T., Moore, W., Ela, S.W., Herzenberg, L.A., and Herzenberg, L.A. (1992)a Intracellular glutathione levels in T cell subsets. Decrease in HIV infected individuals. *AIDS Res. Hum. Retroviruses.* 8: 305-311.

Staal, F.J.T., Ela, S.W., Roederer, H., Anderson, M.T., Herzenberg, L.A., and Herzenberg, L.A. (1992)b Glutathione deficiency and human immunodeficiency virus infection. *Lancet* 339:909-912.

Vidal, S., Mottet, G., Kolakofsky, D.,   and Roux, L. (1989) Addition of high-mannose sugars must precede disulfide bond formation for proper folding of sendai virus glycoproteins. *J. Virol.* 63: 892-900.

Oxidative Stress, Cell Activation and Viral Infection
C. Pasquier et al. (eds)
© 1994 Birkhäuser Verlag Basel/Switzerland

Mechanisms of Regulation of Cell Growth by Cytokines of the Immune System

B. B. Aggarwal and K. Totpal

Cytokine Research Laboratory, Department of Clinical Immunology and Biological Therapy, University of Texas M.D. Anderson Cancer Center, 1515 Holcombe Boulevard, Houston, Texas 77030 U.S.A.

Summary

The growth of cells in culture is regulated in part by a balance between growth stimulatory and growth inhibitory polypeptide molecules. Unrestrained proliferation of cancer cells may result from abnormal production and release of certain polypeptide growth factors or from their failure to express or respond to growth inhibitory polypeptides. Tumor necrosis factor (TNF) and lymphotoxin (LT), secreted products of activated monocytes and lymphocytes, respectively, inhibit the growth of wide variety of tumor cells while stimulating the growth of certain normal cells. The mechanism by which these cytokines transduce the signal is not understood. Depending on the cell type, early signal transduction mechanisms involve the role of transcriptional factors, kinases, phosphatases, phospholipases, sphingomyelinase, superoxide radical, nitric oxide, and proteases. However, at later stages the antiproliferative effects of these cytokines appear to involve damage to subcellular organelles such as the plasma membrane, mitochondria, and lysosomes.

Introduction

The immune system plays an important role in the regulation of normal and abnormal cell growth. Different cells of the immune system, including lymphocytes, monocytes, neutrophils, and mast cells, once activated by viruses, bacteria, parasites, foreign antigen, or tumor cells, secrete discrete factors, commonly called cytokines. During the last ten years over 35 novel cytokines have been identified that either stimulate or inhibit cell proliferation (Aggarwal and Gutterman 1992). In vivo, these cytokines have been implicated in both the pathogenesis and treatment of different diseases. Our group has purified, sequenced, and cloned the cDNA for two cytokines, tumor necrosis factor (TNF) and lymphotoxin (LT), which are important for regulation of growth of normal and tumor cells. In this report, we describe some of their growth regulatory properties and their possible mechanism of action.

Materials and Methods

Cells

All cell lines were obtained from the American Type Cell Culture Collection (Rockville, MD). TNF-resistant BT-20 (TNFR) and NIH-3T3 (LTR 1000) are variants of parent cell lines generated in our laboratory as previously described (Totpal et al. 1991). KG-1 and KG-1a were kindly provided by Dr. Takis Pantazis of the Stehlin Foundation in Houston, and normal human foreskin fibroblasts were provided by Dr. Olivia Smith of the Baylor College of Medicine in Houston. Cells were tested for mycoplasma contamination by the DNA-based assay kit purchased from Gen-Probe (San Diego, CA). RPMI 1640 medium and FBS were obtained from Grand Island Biological Company (Grand Island, NY). Recombinant TNF was obtained from Genentech, Inc. (South San Francisco, CA). All cell cultures were maintained in continuous exponential growth by weekly passage of cells. Cells were routinely grown in RPMI 1640 medium supplemented with glutamine (2 mM), penicillin (100 units/ml), streptomycin (100 μg/ml), and FBS (10%) in a humidified incubator containing 5% CO_2 in air.

Antiproliferative Assays

5×10^3 cells in 0.1 ml of the medium were plated in 96-well Costar plates. After overnight incubation, the medium was removed and a serial dilution of the test sample was layered in a 0.1ml volume. After 72 h of incubation at 37°C, the medium was removed and viable cells were monitored by crystal violet staining according to procedures described elsewhere (Sugarman et al. 1985). For some experiments, cell proliferation was examined by thymidine incorporation assays as described earlier (Totpal et al. 1992).

Arachidonic Acid Release

Cells (10,000/1ml) were labeled with 0.1μCi of tritiated arachidonic acid (specific activity 60-100 Ci/mmole) for 16 h at 37°C and then incubated with TNF (0.2 μg/ml) for 18 h. The release of arachidonic acid from the cells into the supernatants was monitored. All determinations were made in triplicate.

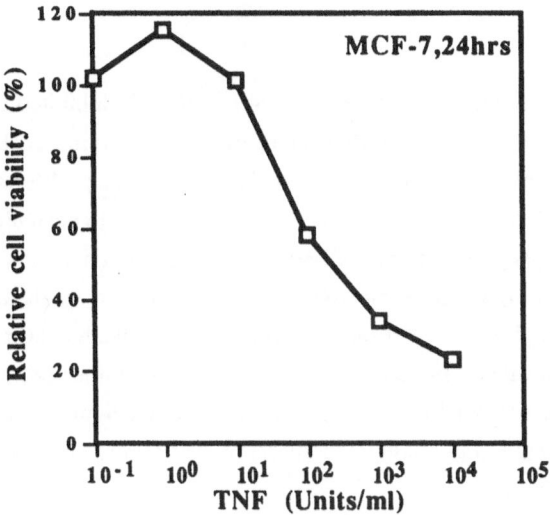

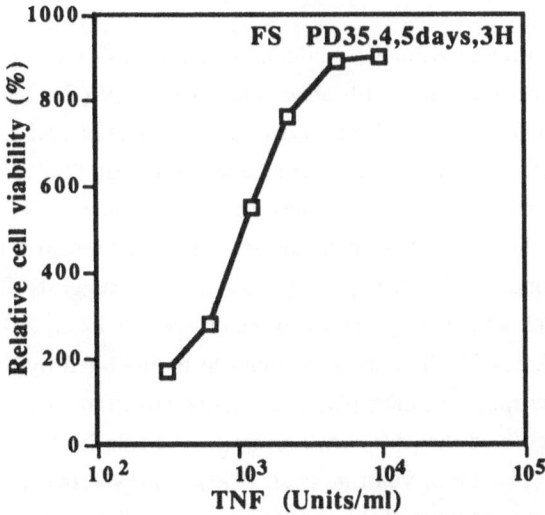

Figure 1. Effect of different concentrations of TNF on the proliferation of human breast tumor cell line MCF-7 and human foreskin fibroblast cells. 5×10^3 or 8×10^3 (fibroblast) cells were incubated for 1 and 5 days, respectively, with the cytokine and then relative cell viability monitored by thymidine incorporation method.

Results and Discussion

Figure 1 (upper panel) shows that treatment of cells for 24 h with TNF inhibited the growth of human breast tumor cells (MCF-7) in a dose dependent manner as monitored by thymidine incorporation. Low concentrations of TNF slightly (10-20%) enhanced the proliferation of tumor cells. Approximately 10-20% of the population of these cells are resistant to even higher doses of the cytokine. TNF reduced the growth not only of breast tumor cells but also of a wide variety of other tumor cell types, including lymphocytic and monocytic leukemia, melanoma, cervical carcinoma, retinoblastoma, and glioblastoma (Table I). Some tumor cells, however, appeared to be more sensitive to TNF than others. In contrast to its inhibition of tumor cell growth, TNF enhanced the proliferation of normal human foreskin fibroblasts (Figure 1, lower panel). Approximately a nine-fold increase in thymidine incorporation was observed within 5 days after cytokine treatment at the highest dose.

How cell growth is modulated by TNF is not fully understood but evidences for several mechanisms have been found. Our results in Figure 2 suggest some of the mechanism involved in action of TNF.

In particular, the presence of a protein synthesis inhibitor such as cycloheximide enhanced the cytotoxic effects of TNF against the murine fibroblast cell line L-929 (Figure 2; upper left panel), indicating first that protein synthesis is not needed for this effect of TNF. Secondly, the enhancement of the cytokine effect by cycloheximide also suggests that TNF may induce the synthesis of certain cellular proteins that counteract the cytotoxic effects of TNF. Furthermore, we also found that TNF causes the release of arachidonic acid from cells sensitive to the growth inhibitory effects of the cytokine (Figure 2; upper right panel). This suggests that TNF may induce the activation of phospholipase A2, which may be critical for its anticellular effects. Since the antiproliferative effects of TNF were also found to be inhibited by orthovanadate (Figure 2; lower, left panel); phosphatases must play a role in the growth modulatory effects of TNF. Several groups have suggested that different kinases play a role in the action of TNF (Guesdon *et al*. 1993, Kalthoff *et al*. 1993, Van Lint *et al*. 1992). However, our results shown in Figure 2 (lower, right panel) demonstrate that staurosporin, an inhibitor of protein kinase, enhanced the antiproliferative effect of TNF. These results are consistent with a recent report (Beyaert *et al*. 1992). Our results both with staurosporine and orthovanadate suggest that dephosphorylation of certain proteins may be a necessary event in the mechanism of action for the anticellular effects of this cytokines.

Table I. Effect of human tumor necrosis factor on different human and murine tumor and normal cell lines

Cell Lines	Relative Cell Viability (%)
Erythroblastoid cell line (K562)	94
Histiocytic Lymphoma (U-937)	54
Histiocytic Lymphoma (U-937-CF-1)	13
Promyelocytic Lymphoma (HL-60)	100
Burkitt Lymphoma (Raji)	76
T cell Lymphoma (Jurkat)	74
Myelogenous Leukemia (KG-1)	43
Myelogenous Leukemia (KG-1a)	60
Myelogenous Leukemia (ML-1a)	59
Myelogenous Leukemia (ML-1b)	35
Monocytic Leukemia (THP-1)	81
Breast Carcinoma (BT20)	24
Breast Carcinoma (BT20TNFR)	60
Breast Carcinoma (MCF7)	1
Breast Carcinoma (SK-BR3)	52
Breast Carcinoma (ZR-75-1)	6
Melanoma (RPMI 7951)	73
Melanoma (A375)	76
Epidermoid Carcinoma (A-431)	82
Cervical Carcinoma (ME-180)	20
Ovarian Carcinoma (OVCAR-3)	27
Cervical Carcinoma (HeLa)	55
Hepatoma (HepG-2)	83
Retinoblastoma (Weri-Rb-1)	60
Retinoblastoma (Y-79)	92
Glioblastoma (LG)	84
Murine Fibroblasts (NIH 3T3)	11
Murine Fibroblasts (LTR1000)	85
Murine Fibroblasts (L-929)	0
Normal Human Foreskin Fibroblasts	311

Cells (5000/well/0.1 ml) were incubated with human TNF (0.2 µg/ml) for 72 h at 37°C and then relative cell viability (%) was determined by thymidine incorporation.

160

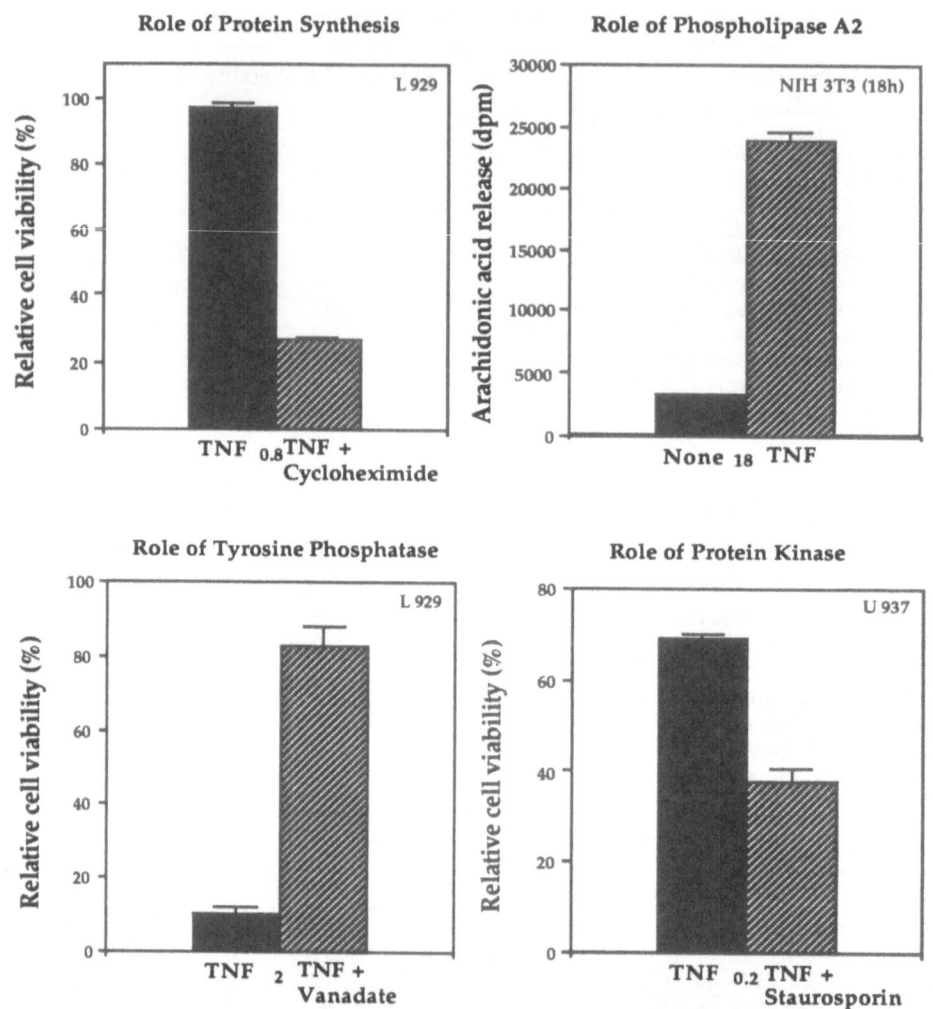

Figure 2.Effect of cycloheximide (1 ʋg/ml) (upper left), orthovanadate (1 µg/ml) (lower left) and staurosporin (5 ng/ml) (lower right) on the cytotoxic effects of TNF (0.8-200 ng/ml) against different cell types (5,000-20,000 cells/0.1 ml). Effect of TNF (0.2 µg/ml) on arachidonic acid release from murine fibroblasts (10,000 cells/1ml exposed for 18 h) is shown in upper right panel.

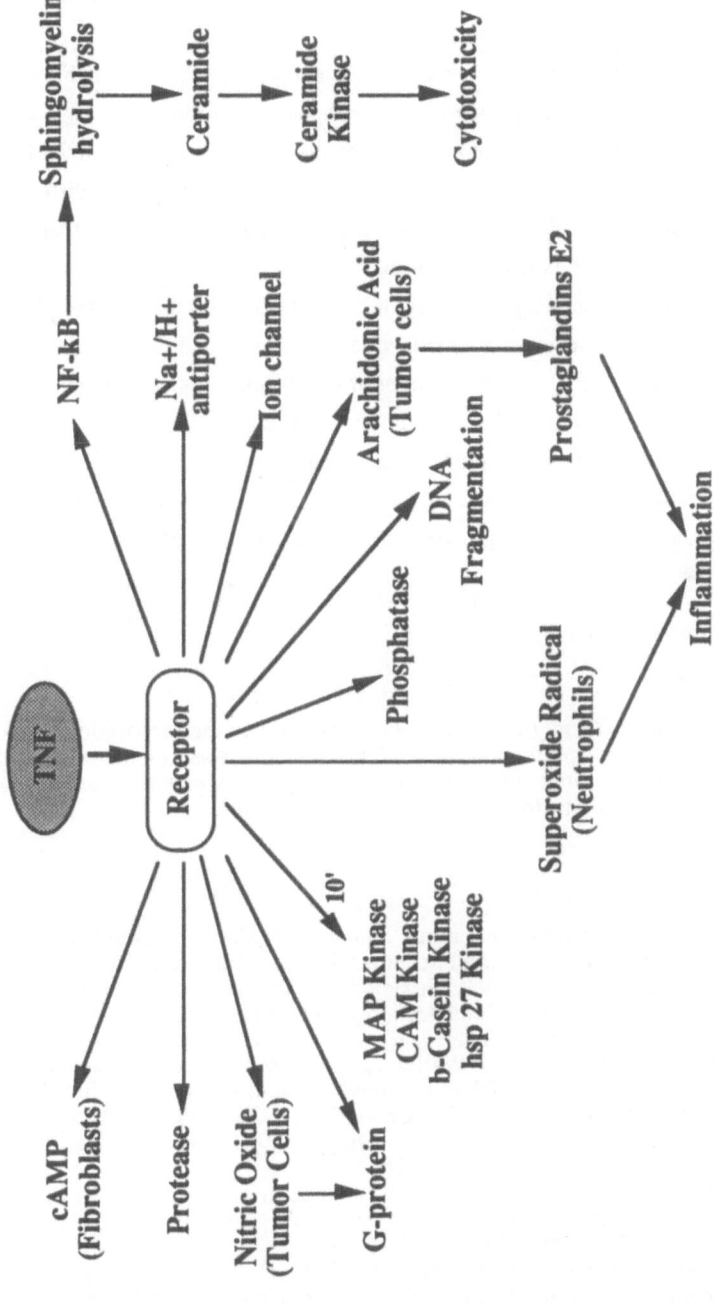

Figure 3. Various modes of signal transductions of TNF in different cell types. For reference see Kronke, et al.

Besides the mechanisms demonstrated in this report, several other intermediates in the signal transduction pathway of TNF have been described (Figure 3) (Kronke *et al.* 1992). Since TNF is a highly pleiotropic cytokine, whether different signals are responsible for different effects of TNF is not known. It is also not clear whether a signal identified for a given effect of TNF is universal for all cell types. Some of our results suggests that the same effect on different cells may be due to different signals (Totpal *et al.* 1992). Recently, two different receptors for TNF have been identified. Whether these two receptors transduce different signals is not clear. Some investigators indicate that the antiproliferative signals of TNF are mediated through the p60 receptor, whereas the proliferative signals are transduced through the p80 form (Tartaglia and Goeddel 1992). However, subsequent reports demonstrated the transduction of antiproliferative signals through the p80 receptor (Heller *et al.* 1992). Thus the relationship of the p60 receptor and p80 receptors for signal transduction is not clear. In a recent report we showed a trans-downmodulation of these receptors (Higuchi and Aggarwal 1992), suggesting a direct or indirect communication between the two receptors. Future studies will be directed to further elaboration of the mechanism by which growth-inhibitory cytokines such as TNF transduce the signal for their action.

Acknowledgments

We wish to thank Wendy Chaffin for her assistance in the preparation and Mr.Walter Pagel for his careful editing of this manuscript. This research was conducted, in part, by The Clayton Foundation for Research and was supported, in part, by new program development funds from The University of Texas M. D. Anderson Cancer Center.

References

Aggarwal, B. B. and J. U. Gutterman. *Human Cytokines: A Handbook for Basic and Clinical Researchers* 1-464 (Blackwell Scientific Publications, New York, 1992).

Beyaert, R., B. Vanhaesebroeck, K. Heyninck, D. De Valck, E. Boone, G. Haegeman, F. Van Roy and W. Fiers (1992) Sensitization of tumor cells to tumor necrosis factor by the protein kinase inhibitor staurosporine. *Eur Cyt Network* 2: 178.

Guesdon, F., N. Freshney, R. J. Waller, L. Rawlinson and J. Saklatvala (1993) Interleukin 1 and tumor necrosis factor stimulate two novel protein kinases that phosphorylate the heat shock protein hsp27 and b-casein. *J Biol Chem* 268: 4236-4243.

Heller, R., K. Song, N. Fan and D. J. Chang (1992) The p70 tumor necrosis factor receptor mediates cytotoxicity. *Cell* 70: 47-56.

Higuchi, M. and B. B. Aggarwal (1992) Modulation of two forms of tumor necrosis factor receptors and their cellular response by soluble receptors and their monoclonal antibodies. *J Biol Chem* 267: 20892-20899.

Kalthoff, H., C. Roeder, M. Brockhaus, H.-G. Thiele and W. Schmiegel (1993) Tumor necrosis factor up-regulates the expression of p75 but not p55 TNF receptors, and both receptors mediate, independently of each other, upregulation of transforming growth factor-a and epidermal growth factor receptor mRNA. *J Biol Chem* 268: 2762-2766.

Kronke, M., S. Schutz, P. Scheurich and K. Pfizenmaier. TNF signal transduction and TNF responsive genes. In *Tumor Necrosis Factors: Structure, Function and Mechanism of Action* (eds. Aggarwal, B.B. & Vilcek, J.) 189-216 (Marcel Dekker, New York, 1992).

Sugarman, B. J., B. B. Aggarwal, P. E. Hass, I. S. Figari, M. A. Palladino and H. M. Shepard (1985) Recombinant human tumor necrosis factor-a: Effects on proliferation of normal and transformed cells in vitro. *Science* 230: 943-945.

Tartaglia, L. A. and D. V. Goeddel (1992) Two TNF receptors. *Immunol Today* 13: 151-153.

Totpal, K., S. Agarwal and B. B. Aggarwal (1992) Phosphatase inhibitors modulate the growth-regulatory effects of human tumor necrosis factor on tumor and normal cells. *Cancer Res* 52: 2557-2562.

Totpal, K., R. LaPushin, H. N. Ananthaswamy and B. B. Aggarwal (1991) In vitro selection of NIH-3T3 cells for resistance to lymphotoxin induces resistance to activated macrophages and enhances tumorigenicity in vivo. *Lymphokine Cytokine Res* 10: 359-367.

Van Lint, J., P. Agostinis, V. Vandevoorde, G. Haegeman, W. Fiers, W. Merlevede and J. R. Vandenheede (1992) Tumor necrosis factor stimulates multiple serine/thronine kinase protein kinases in swiss 3T3 and L-929 Cells: Implication of casein kinase-2 and extracellular signal-regulated kinases in the tumor necrosis factor signal transduction pathway. *J Biol Chem* 267: 25916-25921.

Oxidative Stress, Cell Activation and Viral Infection
C. Pasquier et al. (eds)
© 1994 Birkhäuser Verlag Basel/Switzerland

Oxidization of human low density lipoproteins measured by Laser Doppler electrophoresis

D. Bonnefont-Rousselot[1], B. Arrio[2], J. Catudioc[3] and L. Packer[3]

1- *Laboratoire de Biochimie, Hôpital de la Salpétrière, 47 blvd de l'Hôpital, 75651 Paris Cedex 13, France*
2- *Laboratoire de Bioénergétique Membranaire, Bâtiment 432, Université de Paris-Sud, 91405 Orsay, France*
3-*Department of Molecular and Cell Biology, University of California, Berkeley, California 94720, U.SA*

Summary

Oxydative modification of human low density lipoproteins (LDL) is generally believed to initiate macrophage uptake by scavenger receptors which eventually leads to the formation of fatty streaks and atherosclerotic plaques that precede cardiovascular disease. Hence, sensitive methods to assess the existence of LDL subpopulations and knowledge about antioxidant prevention of LDL oxidative modifications are important. In this study, laser Doppler electrophoresis (LDE) was used to detect changes in electrophoretic mobility of LDL populations before and after exposure to oxidative stress. Before oxidative stress, LDL suspensions exhibited homogenous populations of net negative charge but after exposure to hydroxyl and superoxide radicals (simultaneously generated by steady state 60Cobalt gamma radiolysis), peroxyl radicals produced by the water-soluble azo-initiator 2,2'-azobis(2-amidinopropane)HCl (AAPH), LDL populations exhibited mobility changes. The major population of LDL became more negatively charged in agreement with agarose gel electrophoresis (AGE). However, greater heterogeneity of LDL in some instances can be discerned by liquid as compared to solid phase electrophoresis. Hence, LDE is a sensitive method for detecting the appearance of subpopulations of LDL of differing surface charge density during oxidative modification. The effect of lipophilic antioxidants in protection against these changes in physicochemical properties of LDL was investigated. LDL obtained from a vitamin E deficient patient pretreated with β-carotene afforded protection. After exposure to peroxyl radicals generated by AAPH, a more narrow range of subpopulations was observed. Thus, β-carotene can protect LDL against oxidative modification even when endogenous vitamin E levels are low. In other experiments, with native LDL pretreated with betatene (an almost equivalent mixture of all-trans- and cis-ß-carotene), a protection was also observed.

Introduction

Oxidative modifications of human low density lipoproteins (LDL) is likely to be involved in the constitution of atherosclerotic plaques (Goldstein et al., 1979; Esterbauer et al., 1991; Haberland et al., 1984; Jessup et al., 1990; Jialal et al., 1991) that precede cardiovascular disease. Hence, sensitive methods to assess the existence of LDL subpopulations and knowledge about antioxidant prevention of LDL oxidation are important. In this study, laser Doppler electrophoresis (LDE) was used to detect changes in surface charge of LDL populations before and after exposure to oxidative stress.

Material and Methods

Laser Doppler electrophoresis

Electrophoretic light scattering is a technique which has been recently introduced to study the surface charge properties of macromolecules, micelles, vesicle membranes and colloidal particles.

Detailed presentation of quasi elastic laser light scattering principles, instrumentation and applications have been already described by Bloomfield and Lim (1978) and Rivière et al. (1988), therefore an intuitive background and the basic formula are given. Laser light scattering can be used to measure diffusion coefficients by measuring the Doppler broadening of the frequency of the light scattered due to the Brownian motion of the scattering particles.

Moreover, when an electric field is placed perpendicular to the incident beam, the broadening of the laser line is accompagnied by a line shift proportional to the velocity of the particles. Doppler frequency shifts of laser light are caused by scatterers (macromolecules, vesicles etc. driven by an electric field strength E at constant velocity v :

$$v = \mu E \quad (1)$$

where μ is the electrophoretic mobility depending on the surface charge density s of the moving particles, according to the relation :

$$\mu = \frac{\sigma}{\kappa \eta} = \frac{2\pi \Delta v}{KE \cos \dfrac{\theta}{2}} \quad (2)$$

where η is the viscosity of the medium, θ the observation angle of the scattered lightand Δv the measured Doppler shift, $K = \dfrac{4\pi n}{\lambda} \sin \dfrac{\theta}{2}$ is the wave vector of the laser light propagating in a medium of refractive index v

κ is the Debye-Huckel parameter *i.e*, the attenuation factor of the electrostatic porential created by the charge density σ. This parameter has the dimension of a reciprocal length. In water and at 25°C, it may be calculated from the relation :

$$\kappa = 3.288\sqrt{I} \quad (nm^{-1}) \text{ in water at } 25°C$$

An increase of the ionic strength I, causes the the potential of the particle to fall off and consequently the electrophoretic mobility to decrease.

The electrophoretic velocity is superimposed on the random thermal motion, therefore, the result of a measurement is a frequency spectrum converted in mobility according to the previous relations (1) and (2). The shape of the frequency dependent signal is represented by the relation :

$$P_{(v)} \approx \frac{\dfrac{K^2 D}{2\pi}}{\left(\dfrac{K^2 D}{2\pi}\right)^2 + (v \pm Kv)^2}$$

in which appears the Brownian contribution :

$$D = \frac{kT}{6\pi\eta r}$$

where k is the Boltzman constant, T, the absolute temperature, h, the medium viscosity and r the hydrodynamic radius of the moving particles. An example of spectrum is given in Fig 1 showing the different informations available.

Isolation of LDL

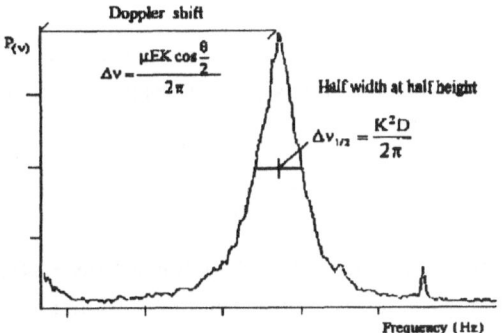

Fig. 1 Typical LDE spectrum

Human LDL (1.019<d<1.063) were isolated by sequential ultracentrifugation from sera of normolipidemic donors in the presence of EDTA (0.04%). LDL were dialyzed before use against a 10^{-2} M sodium phosphate buffer pH 7 and were adjusted to a defined concentration (5 g.l^{-1}) by dilution with this buffer. The apolipoprotein B concentration in LDL was measured by the means of an immunonephelometric technique described by Avogaro et al. (1988).

Genetic abnormalities in vitamin E absorption have been described in humans. One such subject who could not accumulate vitamin E into lipoproteins or tissues has been particularly studied by Kohlschütter et al. (1988). Relevant LDE measurements were performed on LDL isolated from this patient, which were taken several days after withdrawal from dietary vitamin E supplementation. These LDL had barely detectable amounts of vitamin E (1.35 mM α-tocopherol).

Gamma radiolysis

Gamma irradiations were performed with a ^{60}Co irradiator (activity ~ 80 Ci), on 5 ml of LDL solutions in 10^{-2} M phosphate buffer at pH 7 (dose rate = 3.4 x 10^{-2} Gy.s^{-1}) according to the procedure already described by Bonnefont et al. (1992). Dosimetry was determined by the method designed by Fricke and Morse (1927). Before irradiation, the LDL solutions were saturated with oxygen for 40 min at 25°C. Under these conditions, hydroxyl and superoxyde radicals were simultaneously produced withrespective yields of 2.8×10^{-7} and 3.4×10^{-7} mol.J^{-1}. For each experimental set, 5 ml of non-irradiated LDL solution was taken as control.

Oxidation by AAPH

Oxidation was induced by the thermolytic decomposition of the water-soluble azo initiator of peroxyl radicals, 2,2'-azobis(2-amidinopropane)HCl (AAPH) at a concentration of 5 mM in 10 mM HEPES buffer pH 7.3, for 4 hours at 40°C.

Oxidation was arrested by the addition of 40 µM butylated hydroxytoluene (BHT) and refrigeration.

To study protective effects of carotenoids, normal LDL were preincubated (8h) in the presence of 1.25µM Betatene from Henkel Corp.. Vitamin E deficient LDL were preincubated (24h) in the presence of 0.47mM β-carotene.

Control measurements were carried out with LDL solutions incubated in the same conditions without AAPH and in the presence of ethanol, when the effects of carotenoids were studied.

Results and Discussion

Gamma radiolysis

The mobility of the irradiated samples were dose dependent. The highest dose of 620 Gy delivered resulted in a very negative average mobility and had the most complex mobility spectrum with positive components which were absent from an intermediate dose of 310 Gy (data not shown). The oxygenated and irradiated sample exhibited peaks with mobilities indicating the formation of a new subpopulation: One with a low negative charge and other subpopulations with a net negative charge. The hydrodynamic radius of the irradiated LDL, measured by quasi elastic laser light scattering, was not affected by this treatment.

Aggregation occurred at a little extent and was eliminated by centrifugation before electrophoretic mobility measurements.

If the reaction between free radicals, gamma radiolysis generated (Bonnefont-Rousselot et al., 1991-1993), and the protein produced a cleavage leading to isolated fragments, the hydrodynamic radius would be affected. In absence of significant variation, the overall structure of LDL was maintained, even in the presence of an intramolecular cleavage.

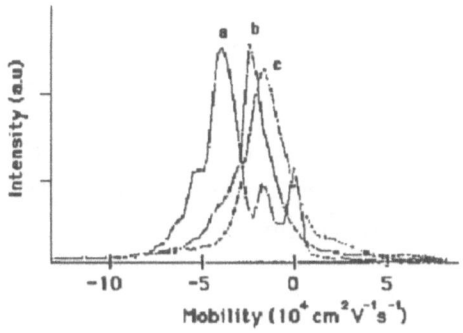

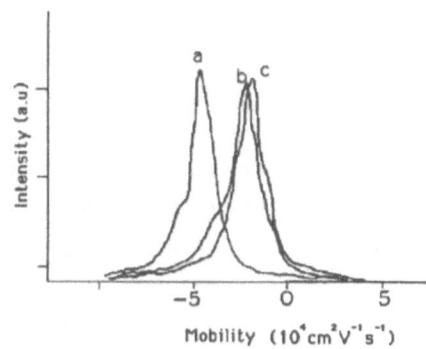

Fig.2 Effect of OH ˙ / O₂⁻ free radicals produced
by gamma radiolysis on human LDL electrophoretic
mobility.
a -rradiated LDL (irradiation dose : 620 Gy)
b-oxygenated non irradiated LDL
c-control LDL(non oxygenated, non irradiated)

Fig.3 Protective effect of betatene pretreatment
on AAPH oxidized LDL
a-AAPH oxidized LDL
b-betatene preincubated LDL
c-LDL preincubated without AAPH
and betatene

At least one of these negative populations had an almost identical mobility to that observed in the oxygenated non-irradiated control (Fig.2). Betatene protective effects were demonstrated by the less net negative mobility of the oxidized and irradiated LDL pretreated with betatene as compared to oxidized LDL that were not pretreated.

AAPH oxidation

AAPH oxidation resulted in an overall increase in net negative charge, but exhibited a broad mobility distribution. Betatene exhibited a protective effect on LDL treated with AAPH i.e less net negative mobility after Betatene preatreatment (Fig.3).

The vitamin E deficient LDL showed a great hetereogeneity with a broad distribution. Moreover, the vitamin E deficient sample exhibited a slightly greater average negative charge than a control sample of normal LDL.

After oxidation induced by AAPH, the LDL isolated from the vitamin E deficient patient showed a more net negative charge than a control sample of normal LDL, indicating its increased susceptibility to oxidation (Fig.4).

LDE spectrum were analyzed from vitamin E deficient LDL preparations with and without exogenous addition of 0.47 mM ß-carotene (in ethanol solution). Incorporation was monitored at 470 nm after dialysis for 4 hours.

170

Subsequent AAPH treatment of these preparations showed a more homogenous mobility spectrum in the preparation with β-carotene whereas the preparation lacking β-carotene exhibited a more negative mobility (Fig.5).

These LDE experiments indicate that in this individual LDL preparation, deficient in vitamin E by natural means (due to lower absorption by the body), β-carotene can protect against surface charge modifications induced by exposure to peroxyl radicals produced in the aqueous phase by AAPH.

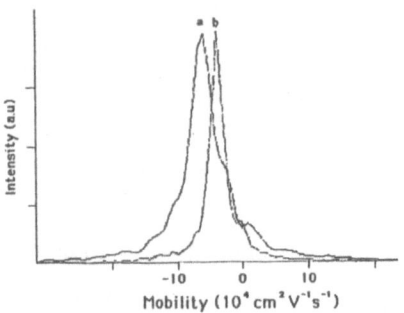

Fig.4 Susceptibility of vitamin E deficient LDL to AAPH oxidation.
a - vitamin E deficient LDL oxidized by AAPH
b - normal LDL oxidized by AAPH

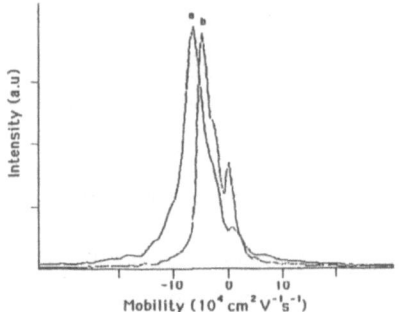

Fig.5 Protective effect of β-carotene on the electrophoretic mobility changes of vitamin E deficient LDL oxidized by AAPH.
a - vitamin E deficient LDL oxidized by AAPH
b - vitamin E deficient LDL preincubated with β-carotene and oxidized by AAPH

Conclusion

LDE is more sensitive than agarose gel electrophoresis to the real charges at the surface (i.e., is not affected by the porosity of the gel) and can clearly discriminate the heterogeneity of the population of charged particles. Thus, analysis of average peak values for mobility, peak width, changes in negatively and positively charged particles as a sample ages or is exposed to various oxidation systems in the absence or presence of antioxidants, and difference spectra between populations that arise, can be readily determined. Moreover, the same sample can be recovered and used for other measurements, particularly biochemical separations. In the present investigation when liquid phase electrophoresis was performed using LDE, it was clearly observed that homogeneous populations exist after isolation of LDL from human serum, but after exposure to oxidative stress a complex profile of LDE subpopulations arise (i.e. populations of different surface charge densities). The majority of particles are more negatively charged and fall into several subclasses depending on the degree and type of oxidant exposure.

References

Avogaro, P., Bittolo-Bon, G. and Cazzolato, G. (1988). Isolation and partial characterization of an oxidized LDL in humans. In: *Oxygen Radicals in Biology and Medicine* (Simik, M.G., Taylor, K.A., Ward, J.F. and von Sonntag, C., Eds., Plenum Press, New York, pp. 391-396.

Bloomfield, V.A. and Lim, T.K. (1978). Quasi elastic light scattering. In: *Methods in Enzymology*. 48: 415-494.

Bonnefont-Rousselot, D., Gardés-Albert, M., Lepage, S., Delattre, J. and Ferradini, C. (1992). Effect of pH on low-density lipoprotein oxidation by $O_2^{\bullet-}/HO_2^{\bullet}$ free radicals produced by gamma radiolysis. *Radiat. Res.* 132: 228-236.

Bonnefont-Rousselot, D., Delattre, J., Galli, A., Gardés-Albert, M. and Ferradini, C. (1991). Peroxydation des lipoprotéines de basse densité (LDL) par divers radicaux libres oxygénés produits par radiolyse gamma. *J. Chim. Phys.* 88: 913-925.

Bonnefont-Rousselot, D., Gardés-Albert, M., Delattre,J. and Ferradini, C. (1993). Oxidation of low density lipoproteins by OH $\cdot$ and OH$\cdot$ / O_2 $^-\cdot$ free radicals produced by gamma radiolysis. *Radiat. Res.* 134 : 271-282.

Esterbauer, H., M. Dieber-Rotheneder, G. Striegl, and G. Waeg. (1991). Role of vitamin E in preventing the oxidation of low-density lipoprotein. *Am.. J. Clin. Nutr.* 53: S314-S321.

Fricke, H. and Morse, S. (1927). The chemical action of Roëntgen rays on dilute ferrosulfate solutions as a measure of dose. *Am. J. Roentgenol. and Rad. Therapy.* 18: 430-432.

Goldstein, J. L., Y. K. Ho, S. K. Basu, and . Brown, M. S. (1979). Binding site on macrophages that mediates the uptake and degradation of acetylated low density lipoprotein, producing massive cholesterol deposition. *Proc. Natl. Acad. Sci. USA.* 76: 333-33

Haberland, M.E., Olch, C.L., and Fogelman, A.M. (1984). Role of lysines in mediating interaction of modified low-density lipoproteins with the scavenger receptor of human monocyte macrophages. *J. Biol. Chem.* 259: 11305-11311.

Jessup, W., Dean, R. T., . de Whalley, C. V., Rankin, S. M. and Leake. D. S. (1990). The role of oxidative modification and antioxidants in LDL metabolism and atherosclerosis. *Adv. Exp. Med. Biol.* 264: 139-142.

Jessup, W., Rankin, S. M., de Whalley, C. V., Hoult, R. S., Scott, J. and Leake, D. S. (1990). α-tocopherol consumption during low density-lipoprotein oxidation. *Biochem. J.* 265: 399-405.

Jialal, I., Norkus, E.P., Cristol, L., Grundy, S.M. (1991). ß-Carotene inhibits the oxidative modification of low-density lipoprotein. *Biochim. Biophys. Acta.* 1086: 134-138.

Kohlschütter, A., Hubner, C., Jansen, W., and Lindner, S.G. (1988). A treatable familial neuromyopathy with vitamin E deficiency, normal absorption, and evidence of increased consumption of vitamin E. *J. Inherited Met. Dis.,* 11 Suppl. 2: 149-152.

Riviere, M.-E., Johannin, G., Gamet, D., Molitor, V., Peschek, G.A., and Arrio, B. (1988). Laser Light Scattering techniques for determining size and surface charges of membrane vesicles from Cyanobacteria. In: *Methods in Enzymology*. 167: 691-700.

Oxidative Stress, Cell Activation and Viral Infection
C. Pasquier et al. (eds)
© 1994 Birkhäuser Verlag Basel/Switzerland

Selective modulation of brain antioxidant defense capacity by genetic or metabolic manipulations

M. Merad-Boudia, L. Fideler, A. Nicole, B. Aral, P.M. Sinet , P. Briand* and I. Ceballos-Picot

*Laboratoire de Biochimie Génétique, CNRS URA 1335, Hôpital Necker-Enfants Malades, 149 rue de Sèvres, 75015 Paris-France and * ICGM, 22 rue Méchain, 75014 Paris-France.*

Summary

Oxidative stress may play an important role in the pathogenesis of brain injuries associated with aging and neurodegenerative disorders such as Down's syndrome, Alzheimer's disease, Parkinson's disease and amyotrophic lateral sclerosis. Brain contain several antioxidant enzymes, most importantly, superoxide dismutases (MnSOD and CuZnSOD), glutathione peroxidase (GSH-Px), glutathione reductase (GSSG-R) and catalase. The tripeptide reduced glutathione (GSH) serves as one of the major endogenous antioxidants in protecting mammalian cells against oxidative stress. GSH has a role not only as a substrate for GSH-Px to break down hydrogen peroxide and lipid peroxides but also as free radical scavenger. The individual contribution of these antioxidants in neuronal protection remains unknown and requires elucidation. A first approach to this question was to establish a comprehensive profile of the brain antioxidant defense potential during aging in control mice. A second approach was to use genetic manipulations to construct transgenic mice that overexpress CuZnSOD. These constructed transgenic mice which exhibit increased CuZnSOD activity in the brain would provide a suitable model to further investigate the role of an increased CuZnSOD activity in neuronal aging and Alzheimer-like neuropathology in Down syndrome. A third strategy in assessing the role of a particular antioxidant in brain fonction was to use metabolic manipulations to deplete brain GSH levels with electrophiles that conjugate with GSH in the presence of glutathione S-transferases, such as diethylmaleate (DEM), and to correlate this depletion with brain injury or susceptibility to oxidative stress. Obtention of such a variation of a particular antioxidant in the brain and neurobiological consequences will be evaluated and discussed.

Introduction

The brain is considered to be sensitive to free radical attack, owing to its high rate of oxidative metabolism, its high concentration of oxidizable substances such as polyunsaturated fatty acids and catecholamines, its relatively low levels of antioxidants, consists of post-mitotic cells and had regions that are particularly enriched in iron (e. g. substantia nigra). Most of the oxygen consumed by the brain is reduced to H_2O in the mitochondrial respiratory chain, but under normal conditions, a small fraction of the oxygen forms free radical species such as the superoxide anions ($O_2°-$), hydrogen peroxyde (H_2O_2) and the hydroxyl radicals ($OH°$) via stepwise one-electron reduction (Chance et al., 1979). H_2O_2 is also produced in nerve terminals in connection with the catabolism of catecholamines and indoleamines by monoamine oxidase (MAO). These reactive oxygen species are powerful oxidants and oxidants that escape the body's numerous antioxidant defenses can produce an array of oxidative-damage in the cell such as peroxidative damage to lipids within mitochondrial and cell membranes, DNA damage lesions, impaired DNA replication, single-strand DNA breakage, mutagenesis, formation of aldehyde breakdown products, crosslinkage of macromolecules (Freeman and Crapo, 1982), inactivation by metal-catalyzed oxidation reactions of metabolic enzymes such as glutamine synthetase and glucose-6-phosphate dehydrogenase, and oxidative damage to proteins resulting in protein carbonyl derivatives (Smith et al., 1991). A reversal of the age-related increase in oxidative damage to brain proteins and improvement in memory after administration of a free radical-trapping compound have also been reported (Carney et al., 1991). In view of the wide range of damage resulting from free radical-mediated oxidants, it has been suggested that they play a signifant role in the processes of aging and can promote early aging, mutations and carcinogenesis ("Free radical theory of aging", Harman, 1981; Sohal and Allen; 1990b).

In order for cells not to succumb to damage induced by oxidants resulting from cellular oxygen metabolism, enzymatic antioxidant defenses exist to cope with oxygen free radicals, e.g, superoxide dismutases (SODs; EC 1.15.1.1), which catalyze the dismutation of superoxide anions (O_2°-) to hydrogen peroxide (H_2O_2). Two different SODs have been identified, namely the cytosolic CuZn (CuZnSOD) and the mitochondrial Mn (MnSOD). Catalase (EC 1.11.1.6) and seleno-dependent glutathione peroxidase (GSH-Px; EC 1.11.1.9) catalyze the degradation of H_2O_2. GSH-Px catalyzes the reduction of H_2O_2 at the expense of reduced glutathione (GSH). In addition, numerous other hydroperoxides, including those originating from unsaturated fatty acids, also serve as substrates for GSH-Px. The regeneration of GSH as a substrate for GSH-Px is catalyzed by glutathione reductase (GSSG-R; EC 1.6.4.2). These enzymes make up an important defense system that ensures the degradation of cytotoxic oxygen species to less harmful compounds. Under physiological conditions a balance may exist between the rate of H_2O_2 formation via dismutation of superoxide by SODs and its removal by GSH-Px. If SODs are up regulated out of proportion to the activity of GSH-Px, a built up of H_2O_2 could ensue (Yim et al., 1990). Then, hydroxyl radicals are formed from H_2O_2 in the presence of iron or copper by the Fenton reaction leading to oxidative stress. Recent interest has focused on the role of oxidative stress in nervous system injury and its possible involvement in the etiology of neurodegenerative disorders such as Alzheimer's disease (Blass and Gibson, 1991; Ceballos et al., 1989; 1990a,b; 1991a; Delacourte et al., 1988; Evans et al., 1989; Sinet and Ceballos-Picot, 1992); Down's syndrome (Ceballos et al., 1988; 1991b,c 1992a,b; ; Sinet, 1982; Sinet and Ceballos-Picot, 1992), Parkinson's disease (Ceballos et al., 1990a,b; Di Monte et al., 1992) and more recently in amyotrophic lateral sclerosis (Rosen et al., 1993).

Oxidative stress in the aged brain

The brain is believed to be particularly vulnerable to oxidative damage. Thus oxidative damage could conceivably have a significant role in the process of brain aging (Carney et al., 1991). Protein oxidation products are increased in both aged rodent and human brain, and the activity of enzymes vulnerable to oxidation, such as glutamine synthetase, is decreased (Carney et al., 1991; Fucci et al., 1983; Smith et al., 1991). In the human brain there is an age-related alteration from oligodendrocytes to astrocytes in the cellular distribution of ferritin, an iron-binding protein (Connor et al., 1990) and there is an overall reduction in the content of human brain membrane lipids with increasing age (Svennerholm et al., 1991). A clear indication of age-related alterations in brain lipid metabolism and possible increased lipid peroxidation is the accumulation of intraneuronal lipofuscin. The accumulation of this "aging pigment" has essentially a linear correlation with the age of the organism (Dowson et al., 1982). Any factor that weakens the brain antioxidant defense systems would render the brain more susceptible to extensive oxidative damage. The activity of antioxidant enzymes in normal dividing cells has been reported to increase or to decrease, depending on the tissue and age of the animals studied (Cao Danh et al., 1983; Ceballos-Picot et al., 1992b). Furthermore, aging may affect each individual antioxidant enzyme in a particular way (Ceballos-Picot et al., 1992b,c; Vanella et al., 1989).

The individual contribution of the antioxidants in neuronal protection during aging remains unknown. A first approach to this question was to establish a comprehensive profile of the brain antioxidant defense potential during aging in control mice (Ceballos-Picot et al., 1992b). We have first determined the age-related profiles of CuZnSOD, MnSOD and glutathione-related enzyme (GSH-Px and GSSG-R) activities from 2 months and continuing to senescence (28 months) in the brain of 40 control mice. To elucidate the physiological significance of the two SODs in the aging process, we developed sensitive assays for cuZnSOD (cyanide sensitive) and MnSOD (cyanide-insensitive) activities in the mouse brain. The results are reported in Table I.

Table I (from I. Ceballos-Picot et al., 1992b)

AGE-RELATED CHANGES IN ANTIOXIDANT ENZYME ACTIVITIES IN THE BRAINS OF CONTROL AND TRANS-GENIC MICE

	Age (months)			
	2	12	28	
CuZnSOD				
Control	43.2 ± 5.6	50.3 ± 5.1 [a]***	50.3 ± 5.4 [a]**	A, $P < 0.01$
	(n = 13)	(n = 12)	(n = 15)	
Transgenic	81.0 ± 8.3 [c]****	81.8 ± 9.3 [c]****	84.7 ± 7.0 [c]****	A, NS
	(n = 11)	(n = 11)	(n = 10)	
MnSOD				
Control	35.6 ± 6.6	35.8 ± 2.6	32.4 ± 2.7	A, NS
	(n = 13)	(n = 12)	(n = 15)	
Transgenic	44.2 ± 7.9 [c]***	40.3 ± 6.1 [c]**	40.0 ± 2.9 [c]****	A, NS
	(n = 11)	(n = 13)	(n = 10)	
GSH-PX				
Control	10.4 ± 1.9	12.6 ± 1.8 [a]*	14.6 ± 1.4 [a]****,[b]**	A, $P < 0.0001$
	(n = 13)	(n = 12)	(n = 15)	
Transgenic	11.8 ± 2.4	12.4 ± 1.9	14.4 ± 1.5 [a]***	A, $P < 0.01$
	(n = 11)	(n = 13)	(n = 10)	
GSSG-R				
Control	41.0 ± 6.5	40.2 ± 7.0	41.7 ± 7.1	A, NS
	(n = 13)	(n = 12)	(n = 15)	
Transgenic	38.9 ± 7.3	39.8 ± 4.5	37.9 ± 4.5	A, NS
	(n = 11)	(n = 14)	(n = 10)	

Results are mean ± SD; n, number of mice; CuZnSOD and MnSOD: U/mg of protein; GSH-PX and GSSG-R: nmole of NADPH oxidized/min/mg of protein.
A, aging effect (correlation between age and activity); NS, not significant; [a,b] Significant differences among age groups (Student's *t*-test or Wilcoxon test); [a] significantly different from 2 months. [b] significantly different from 12 months; [c] significant difference between transgenic and control mice: * $P < 0.05$; ** $P < 0.01$; *** $P < 0.001$; **** $P < 0.0001$.

We found a positive correlation between age and CuZnSOD activity (r=0.47; P<0.01) and age and GSH-Px activity (r=0.72; P<0.0001). Another positive correlation was found between CuZnSOD and GSSG-R activities when the 40 mice of the 3 groups were considered (r=0.34; P<0.05). Morever, when this correlation was studied for each age group, a strong positive correlation was only found at 28 months (r=0.82; P<0.001). No modification with age in MnSOD and GSSG-R activities was observed. Evidence of free radicals production during aging is provided by increased SOD and GSH-Px activities in the brain. Thus, it is clear that the cellular antioxydant system is increased rather than weakened in senescent brain. Our data showing increased CuZnSOD activity with age are compatible with the other results in rats (Bracco et al., 1986; Cand et al., 1989). However, these findings contradict other authors who report no age-related increase in rats (Barja de Quiroga et al., 1990, Cao Danh et al., 1983; Kellog and Fridovich, 1976) or in mice (Reiss et al., 1976) or even a decrease in rats (Benzi et al., 1989b; Bracco et al., 1986; Geremia et al., 1990; Gupta et al. 1991; Massie et al., 1979, Mizuno and Ohta, 1986, Semsei et al., 1991) or mice (Massie et al., 1979). It was demonstrated that the rate of mitochondrial $O_2°$- generation is correlated with aging (Sawada et al., 1987; Sohal et al., 1990a). Thus, the observed increase in CuZnSOD acitivity in the central nervous system of mice during aging may constitute a brain protection against superoxide anion elevation. Moreover, cytochrome oxidase activity increases in the brains of aging rats (Vanella et al., 1989), this suggests a close relationship between CuZnSOD activity and the rate of oxygen reduction by brain mitochondria. We found no changes in MnSOD activity with age. A possible explanation for the different CuZn and MnSOD activity profile during aging is that the two enzymes are differently regulated in the brain. Few data are available for MnSOD activity in brain: an increase of MnSOD activity in the brain of old rats and mice has been reported (Bracco et al., 1986; Cao Danh et al., 1983, Scarpa et al.,1987) although another group showed a decrease (Ledig et al., 1982).

The pattern of changes in glutathione-related enzyme activities in the mouse brain are of particular interest: GSH-Px activity showed a significant positive correlation with increasing age, whereas GSSG-R activity did not vary with age. Vitoria et al., (1984) also found increasing GSH-Px activity with age in the soluble fraction of mitochondria. Other studies were unable to detect any variation in GSH-Px activity in the aging rat brain (Cand and Verdetti, 1989) or mouse brain (Hazelton and Lang, 1985), even a decrease has been reported (Sohal et al., 1990a). For GSSG-R, a small but significant increase with age in rat brain has been reported (Hothersall et al., 1981; Sohal et al., 1990a) and confined to some brain areas such as the hippocampus and the cerebellum (Mizuno and Ohta; 1986). A decrease has also been reported (Benzi et al., 1989b). In brain, the GSH-Px/GSSG-R-catalyzed GSH-GSSG cycle plays a key role in metabolizing H_2O_2 and organic peroxides. The observed increase of GSH-Px activity with age suggest an increase in H_2O_2 formation in the cytoplasm and/or mitochondria, an adaptive phenomenon (Ceballos et al., 1988), and could underlie the age-related increase in the rate of prooxidant generation such as H_2O_2, as a candidate for being a biomarker of aging (Sohal, 1991). Both CuZnSOD and GSH-Px activities increase with age. However, there was no direct correlation between the two activities. The strong positive correlation we have observed between CuZnSOD and GSSG-R in the 28 months control group has no evident biological explanation but was also observed in human erythrocytes of subjects aged 65 years and over (Berr et al., 1993).

The tripeptide glutathione (GSH: L-γ-glutamyl-L-cysteinyl-glycine), the most abundant non-protein thiol in the cell, has been shown to participate in several protective mechanisms including the enzymatic reduction of intracellular peroxides via GSH-Px, maintenance of reduced disulfide bounds in cellular proteins via GSH-S-transhydrogenase, facilitation of the non-enzymatic reduction of free radicals, and in the direct conjugaison of toxicants via GSH S-transferases. In this manner, GSH plays an important role in the prevention of oxidative damage that is induced by either exogenous toxicants or by endogenous products of oxidative metabolism, such as lipid peroxides and the products of monoamine degradation (Meister and Anderson, 1983). While the role of GSH in oxidation-reduction and conjugation reactions as well as in amino-acid transport has been extensively studied in other tissues (Meister and Anderson, 1983), little is known regarding the brain. The measurement of reduced glutathione (GSH) and glutathione disulfide (GSSG) (Anderson, 1985; Griffith, 1980) and of the GSSG/GSH ratio are sensitive index of redox status of the cell and reflect the equilibrium between the generation of prooxidant challenge and the antioxidants to counteract such a challenge. Thus, their determination complements other antioxidants for the study of oxidative stress during aging.

The purposes of our study was twofold: first- to measure the levels of GSH and GSSG in the brain, cerebellum, liver and lung of mice. The GSH level was higher in liver (6,50 µmol/g wt) than in the brain (1,18), lung (1,07) and cerebellum (0,70) (Fig. 1A). The GSSG level was higher in lung (0,340 µmol/g wt) and liver (0,152) than in brain (0,022) and cerebellum (0,022). (Fig. 1B). The role of increased level of GSH within liver as compared to other tissues analyzed remains to be determined. It is reasonable, however, to predict that a larger intracellular pool of GSH reflects a greater potential to supply GSH to those reactions involved in detoxification and cellular protection from xenobiotic and oxidative damage. Second- we determined whether or not brain aging is associated with variations of GSH and GSSG levels. We examined these paramaters in young (3 months) and old (24 months) mice. No variation in GSH level was observed according to age in the different tissues analyzed (Fig. 1A). However, aging was associated with a significant increase in GSSG levels in brain (0,030 µmol/g wt; $P<0.05$), cerebellum (0,060; $P<0.001$) and lung (0,589; $P<0.001$) but not in liver (0,145) (Fig 1B). The high level of GSH within liver is consistent with a role in protecting liver from oxidative damage associated with age. Previous reports have shown a decrease in GSH levels in several peripheral organs with age (Abraham et al., 1978; Hazelton and Lang, 1980; Naryshkin et al., 1981; Schneider et al., 1982) and in some brain regions of aged rats like cerebellum, cortex and olfactory bulb (Ravindranath et al., 1989) and hippocampus (Chen et al., 1989; Ravindranath et al., 1989). In contrast but in agreement with our results, Farooqui et al. (1987) found low levels of GSH in whole brain homogenates at one month, which increase 6-fold at 2.5 months and return to the one month levels in 3-year-old rats. Benzi et al. (1989a) reported similar changes in GSH levels in rat forebrain. The use of homogenates provides a general idea of overall concentration, but such techniques cannot resolve the relative contribution of distinct cell populations and subtypes to the total GSH content within the tissue. Since the brain exhibit important regional and cellular heterogeneity, it will be necessary to quantify GSH in different regions areas of the young and aged mice brain and will allow the GSH distribution to be examined at the cellular level. Differences in regional and cellular distribution of GSH (Philbert et al., 1991) in heterogeneous tissue such as the nervous system are likely critical determinants of differential sensitivity to oxidative stress. GSH appears primarily localized in the neuropil and white matter tracts in the rat brain; the neuronal somata do not appear to contain appreciable amount of GSH (Philbert et al., 1991). Similarly, Slivka et al (1987) suggested that GSH may be primarily localized to glia and/or axons and nerve terminals and Raps et al. (1989) have shown high concentrations of GSH in cultured astrocytes but not in cultured neurons.

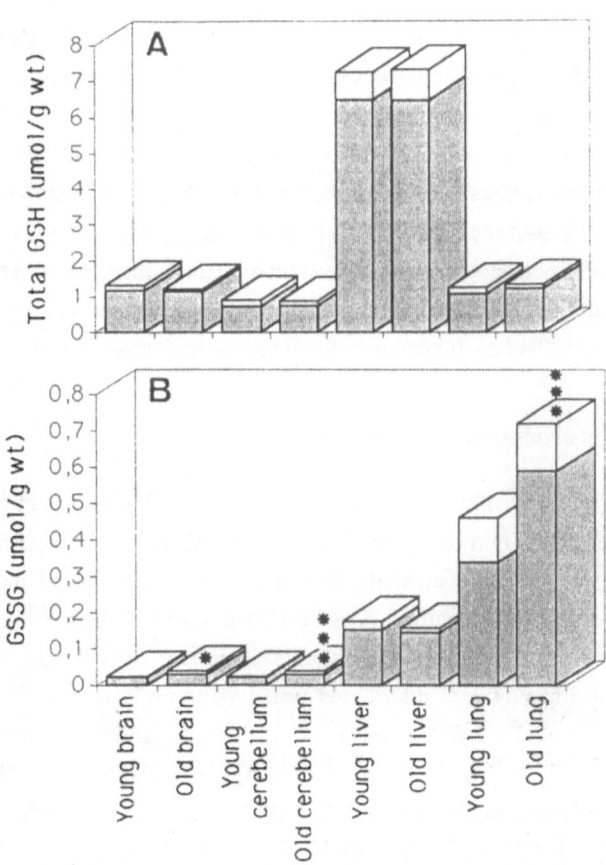

Figure 1. Total (GSH) and oxidized (GSSG) glutathione levels in different tissues of mice and variations with aging. A: Total GSH levels (mean ±sd; n=10) ; B: GSSG levels (mean ± sd, n=10). For each tissue, variation with age was compared using t-test. * P<0.05; ** P<0.01; *** P<0.001.

Enzymes associated with GSH functions might be expected to colocalize with the tripeptide. Although data are limited, the distribution of these proteins appears to follow the general pattern of GSH distribution. GSH-Px and GSH-S-transferases have both been observed in glial cells in rat brain (Cammer et al., 1989; Ushijima et al., 1986). These data suggest that glia may play a dominant role in the detoxification of oxidants and/or excretion of xenobiotics and that neurons may play a more passive role. However, CuZnSOD protein and mRNA was preferentially expressed in neurons in mice (Ceballos-Picot et al., 1991b) and humans, in areas particularly vulnerable during aging, i.e., the substantia nigra (Ceballos et al., 1990a) and hippocampus (Ceballos et al., 1989; Delacourte et al., 1988; Somerville et al., 1991). The heterogeneous distribution of GSH and enzymes involved in the neutralisation of oxygen free radicals and/or excretion of xenobiotics in the nervous sytem may form a basis for selective cellular and/or regional expression of neuronal vulnerability. Concentrations of GSH are partly regulated by GSH-Px and GSSG-R. GSSG-R reduces GSSG to GSH. The increased activity of GSH-Px together with the unchanged GSSG-R activity in the brain during aging (Table I) could explain the age-associated increase in GSSG concentration reported here (Fig. 1B). This finding is in agreement with those of previous authors for the brain but not for the liver (Vina et al., 1992) but is different from other (Chen et al., 1989; Hazelton and Lang, 1980). This Change in glutathione redox status denote an imbalance between proxidant and antioxidant in favor of oxidation. GSH plays a salient role in the regulation of intracellular redox status, and the higher level of GSSG in the aging brain may reflect an increased susceptibility to oxidative injury

Oxidative stress in neurodegenerative disorders

Down's syndrome (DS) or trisomy 21 is the most frequent birth defect in humans, afflicting about one newborn in 1000. In most of the cases, DS results from the presence of an extra copy of the entire chromosome 21 in all cells. Dysfunctions of the central nervous system are manifested by two pathological features: mental retardation and Alzheimer-like neuropathology. Neuropathology is encountered in potentially all DS patients over the age of 35, with senile plaques and neurofibrillary tangles distributed in the same brain territories as in Alzheimer's disease (AD). Assessment of the AD lesions in DS patients at various ages compared to the age-related occurence of AD lesions in the general population leads to the conclusion that the neuropathological processes might start in a DS brain some 50 years earlier than in a normal brain (Rumble et al., 1989). Thus, DS can be regarded as a disease model for identifying biological mechanisms leading to accelerated aging and AD-like neuropathology. We can infer from this model that excess of one or several genes on chromosome 21 is responsible for the pathogenesis of these features. Human CuZnSOD is encoded by a gene residing on the 21q22.1 band of chromosome 21 (Sinet et al., 1976). Overexpression of the CuZnSOD gene, due to gene dosage, has been recorded in all tissues of trisomic 21 patients and in a variety of trisomy 21 cells in culture (Sinet, 1982) as well as in the brains of DS patients (Brooksbank and Balazs; 1984). In the human brain, its level of expression is particularly high in neurons (Ceballos et al., 1989; 1990a; 1991a). Immunohistochemistry (Ceballos et al., 1991a; Delacourte et al., 1988) and cDNA-mRNA in situ hybridization (Ceballos et al., 1989; Ceballos et al., 1991a; Somerville et al., 1991) studies on human hippocampus have shown that the amounts of CuZnSOD protein and mRNA are high in large pyramidal neurons and granular cells. An intense labeling is observed with the CuZnSOD cDNA probe over the pyramidal cell layers of Ammon's horn (CA), the CA1-CA4 fields, the subiculum and in the granule cell layer of the dentate gyrus (Fig. 2A). The specificity of the hybridization pattern is demonstrated by the absence of labeling with a radioactive cDNA unrelated to the hippocampus such as the tyrosine hydroxylase cDNA probe (Fig. 2B). These results suggest that the production of superoxide radicals within these cells might be particularly elevated.

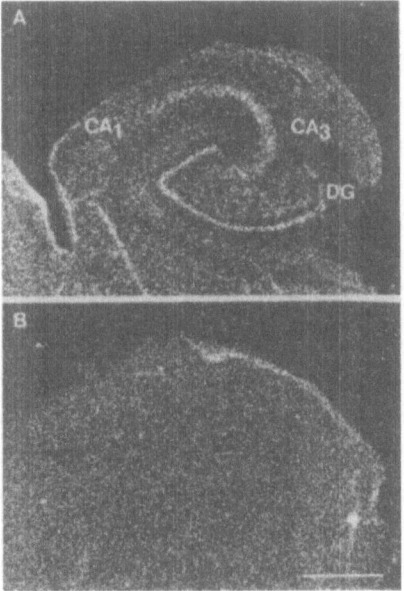

Figure 2.(from Ceballos et al., 1989)
Flms autoradiograms of adjacent sagittal sections of a human hippocampus hybridized with a human ^{35}S-labeled CuZnSOD cDNA probe (A) or tyrosine hydroxylase cDNA probe (B). Bar=2mm.

Moreover, elevated CuZnSOD activity has been reported in two other neurodegenerative disorders. First, in brains of patients with AD, increased CuZnSOD levels were found in certain regions (Marklund et al., 1985) and interestingly, we have found that the protein and mRNA are localized in neurons that are selectively vulnerable in AD (Ceballos et al., 1991a; Delacourte et al., 1988). Thus, high levels of copper-zinc SOD in neurons which are preferentially vulnerable in neurodegenerative processes suggest that oxidative stress may be involved in the pathogenesis of these lesions and in the mechanism of nerve cell death in aging. Second, increased SOD activities have also been found in the substantia nigra of patients who suffer from Parkinson's disease (Marttila et al., 1988) and the expression of CuZnSOD mRNA is particularly high in the age-vulnerable dopaminergic melanized neurons (Ceballos et al., 1990a). Neuromelanin is formed through autoxidation of dopamine and there is a concomitant production of reactive oxygen species (Graham; 1978). Furthermore, H_2O_2 is formed in connection with the catabolism of catecholamines by monoamine oxidase and elevated concentrations of iron have been found in the substantia nigra of Parkinsonian patients . Taken together, these findings suggest that reactive oxygen species are involved in neurodegeneration.

Since CuZnSOD catalyzes the dismutation of superoxide radicals $O_2°-$ to oxygen and hydrogen peroxide, it has been hypothesized that increased CuZnSOD might induce overproduction of H_2O_2 within cells and accelerate oxidative damage to cell components (Sinet et al., 1982). Recent reports have shown that CuZnSOD is able to catalyze hydroxyl radicals $(OH°)$ production (Yim et al., 1990) and that elevated level of SOD is cytotoxic (Norris and Hornsby, 1990) and enhances the cytotoxicity of active oxygen species (Scott et al., 1987; 1989). Such a mechanism may be in part responsible for certain clinical symptoms of DS patients and for the accelerated aging and AD pathology associated with the Down's phenotype (Sinet et al., 1982). It could also contribute to the neuropathology associated with AD and Parkinson's disease (Ceballos et al., 1990b). Recently, Rosen et al. (1993) report tight genetic linkage between Familial Amyotrophic Lateral Sclerosis (FALS) and CuZnSOD gene. Given this linkage and the potential role of free radical toxicity in other neurodegenerative disorders, the authors have investigated CuZnSOD gene as a candidate gene in FALS and identified 11 different CuZnSOD missense mutations in 13 different FALS families. Rather than decreasing CuZnSOD activity, it is plausible that FALS mutations, with amino-acids changes at these residues, are of increased stability and hence of increased activity of CuZnSOD protein.

Whether the increased CuZnSOD activity play a role in neurodegenerative disorders is now strongly suggested but is not totally established. While exploring this possibility cellular clones and transgenic mice overexpressing CuZnSOD have been constructed (for review, Ceballos-Picot et al., 1992a)

Specific increase of copper-zinc superoxide dismutase by genetic manipulations

To investigate the possible involvement of CuZnSOD overproduction in the etiology of these neurodegenerative diseases, the cloned gene and cDNA could be introduced into cells as part of recombinant plasmids. Work with expression vectors containing CuZnSOD indicates that an oversupply of this enzyme is not necessarily beneficial to a mammalian cell (Ceballos-Picot et al., 1992a) and can even cause increased lipid peroxidation (Ceballos et al., 1991b; Elroy-Stein et al., 1986). Transfecting an expression vector containing a CuZnSOD gene into cultured adrenocortical cell provoked cell fusion, nuclear fragmentation and subsequent propagation of cell death (Norris and Hornsby; 1990). PC12 cell lines expressing the transfected human CuZnSOD gene had impaired uptake of neurotransmitters such as dopamine and norepinephrine, resulting from modification of membrane properties of chromaffin granules, probably secondary to lipid peroxidation (Elroy-Stein and Groner, 1988).

In order to further explore the deleterious effect of excess CuZnSOD, transgenic mice for the human CuZnSOD were obtained by us (Ceballos-Picot et al., 1991b; 1991c) and other (Epstein et al., 1987). In our transgenic strain, one copy was integrated in the mouse genome and CuZnSOD activity was higher than in control mice in almost all tissues (Ceballos et al., 1991b). In the brain, CuZnSOD activity was increased by a factor of 2 and was similar in each age group studied (Table I).

Despite the increased CuZnSOD activity in the brains of transgenic mice, GSH-Px activity was not modified in the 3 age groups studied when compared to the corresponding age group control. This contrasts with what has been observed in DS erythrocytes (Sinet et al., 1982), in lymphoïd cells and fibroblasts (Anneren et al., 1987) and in transfected cells expressing the human CuZnSOD gene (Ceballos et al., 1988), but corresponds to findings in DS fetal brain (Brooksbank et Balazs; 1984). The reason for the discrepancy between the results in the brains of transgenic mice and in the transfected cells, for the same human CuZnSOD transgene, could be a different cellular localization for CuZnSOD and GSH-Px enzymes in the brain. We have demonstrated in a previous report a specific neuronal localization of endogenous CuZnSOD and human CuZnSOD in the brain of transgenic mice particularly in the pyramidal neurons of the CA1-CA4 fieds of Ammon's horn (Ceballos-Picot et al., 1991a). GSH-Px and glutathione S-transferases proteins have both been observed in astrocytes and oligodendrocytes in rats (Cammer et al., 1989; Ushikima et al., 1986). These data underline the heterogeneous cellular distribution of the enzymes implicated in oxygen free radical detoxification and may form a basis for selective and/or regional expression of free radical toxicity in the brain.

The MnSOD activity was significantly higher in transgenic mice compared to control mice when the 35 mice of the 3 groups were considered as well as in a separate analysis for each age group (Table I). Thus, the expression of the human CuZnSOD gene in the brain of transgenic mice resulted in increase MnSOD activity (Przedborski et al., 1992; Ceballos-Picot et al., 1992b). MnSOD enzyme was localized in mitochondria, the increased CuZnSOD activity in transgenic mice presumably leads to an augmentation in the steady-state superoxide concentration inside the mitochondria, which may in turn trigger MnSOD expression. Increasing activity of MnSOD has been suggested to be due to an enhancement in the respiration state in aged rats (Chen et al., 1972). In mitochondria, $O_2°-$ is generated mainly from the membrane-borne respiratory chain, possibly at the NADH-ubiquinone reductase and ubiquinone-cytochrome c reductase steps (Chance et al., 1979). The increase MnSOD activity in transgenic mice might be a defense response to protect mitochondria from oxidative damage. Moreover, MnSOD gene expression and enzyme activity have been shown to be increased by tumor necrosis factor α and β and by interleukin 1α and 1β, both in vivo and in different cell lines (Whong and Goeddel; 1988) and the possibily that neurotrophic factors (such as Nerve Growth Factor) are involved in the regulation of antioxidant enzymes in the brain has a potential interest. Alternatively the increased MnSOD activity could lead to enhanced production of H_2O_2 inside the michondria and, as a consequence, could lead to increased lipid peroxidation of mitochondrial membranes and oxidative damage to mitochondrial genome (Bandy and Davison, 1990; Wallace; 1992), since we have demonstrated the absence of adaptative increase in GSH-Px activity in brains of transgenic mice. Other interesting features have been observed in these transgenic mice, such as ultrastructural changes in the neuromuscular junction of the tongue muscle similar to those seen in aging rats and mice as well as in the tongues muscles of Down's syndrome patients (Avraham et al., 1988). These observations suggest that CuZnSOD excess could have deleterious effects on the nervous system.

Modulation of brain glutathione concentration following metabolic manipulations

Although GSH is widely distributed in the brain, its biological function in the central nervous system is not fully understood. However, its large concentration and rapid synthesis suggest an important role in this organ (Griffith and Meister, 1979). Concentrations of GSH in tissues of the mouse and rat decline with age for some authors (Hazelton and Lang, 1980; Ravindranath et al., 1989; Teramoto et al., 1992) as well as in peripheral tissues in humans (Naryshkin et al., 1981). Moreover, a decrease in GSH levels has been reported in the brains of parkinsonian patients as compared to control subjects (Perry et al., 1982) which appeared to be selective for the substantia nigra. These changes have been interpreted as the result of increased oxygen radical production in the nigrostriatal pathway of Parkinsonian patients, and have led to the speculation that Parkinson's disease may be attributable to nigral GSH deficiency (Perry et al., 1982; Di Monte et al., 1992).

A number of links exist between oxidative stress and mitochondrial damage in the brain and GSH may represent one such link: first- depletion of GSH in the brain of newborn rats by administration of buthionine sulfoximine (L-BSO: an inhibitor of GSH synthesis) caused mitochondrial damage assessed by both electron microscopy and measurements of the activity of citrate synthase, a mitochondrial matrix marker enzyme (Jain et al., 1991). Mitochondrial degeneration may be ascribed to accumulation of H_2O_2 and consequent oxidative damage.

Second- the level of a specific mitochondrial DNA deletion (MtDNA[4977]) increases with age, particularly in the human substantia nigra (Wei Soong et al., 1992). This brain area is characterized by a high dopamine metabolism. The breakdown of dopamine by mitochondrial MAO produces H_2O_2 which can lead to free radicals formation. Like the levels of MtDNA deletion, mitochondrial oxygen radical production appears to increase with age (Bandy and Davison, 1990). Wei Soong et al. (1992) suggest that mtDNA[4977] may be the "tip of the iceberg" of the spectrum of somatic mutations produced by oxidative damage. Thus, excessive production of H_2O_2 within mitochondria may lead to depletion of mitochondrial GSH and impairment of mitochondrial function, providing an example of the relationship between the glutathione status, oxidative stress and mitochondrial damage. If the loss of GSH may cause mitochondrial damage, it is also conceivable that impairment of mitochondrial function can lead to a decrease in cytosolic GSH. GSH synthesis requires ATP, and a deficiency of energy supplies by mitochondria is likely to affect the cellular turnover of glutathione.

Thus, the long-term effects of GSH depletion on mitochondrial function and mutagenesis in the brain need to be studied. Selective modulation of brain GSH and observation of the effects of such modulation on neuronal aging might help in understanding the significance of GSH in the brain protection during aging and neurodegenerative diseases.

Various compounds have been used to modulate the levels of tissue GSH, namely, Diethylmaleate (DEM) which is known to deplete GSH in many organs by forming GSH conjugates (Gerard-Monnier et al., 1992). Other methods used for depleting tissue GSH involve the inhibition of GSH synthesis in the cell using L-BSO, a selective inhibitor of γ-glutamylcysteine synthetase, has been used extensively to deplete GSH levels in tissues (Meister and Anderson, 1983). But L-BSO does not cross the blood-brain barrier and is effective in depleting brain GSH only when the blood brain barrier is non-functional, for example in preweanling animals (Slivska et al., 1988). Depletion of GSH by L-BSO in adult animals requires intracerebroventricular administration using stereotaxic techniques (Pileblad and Magnusson, 1989). This is difficult to perform when using a large number of animals to study the effects of a chronic depletion of GSH on the central nervous sytem. Inhibition of GSH synthesis by L-BSO could be a model for endogenously produced oxidative stress, since in the newborn rat, proximal renal tubular, hepatic and brain damages occur. Cellular damages were associated with major destruction of mitochondria (Martensson et al., 1991).

Our work was undertaken to develop simple methods for a selective depletion of GSH in the brain. We examined the dose-dependent GSH depletion in brain, cerebellum and liver of mice following intraperitoneal administration of DEM (3 and 6 mmol/kg). The time course of GSH depletion was examined at earlier time points (3 and 6 hours after DEM injection) (Fig. 3).

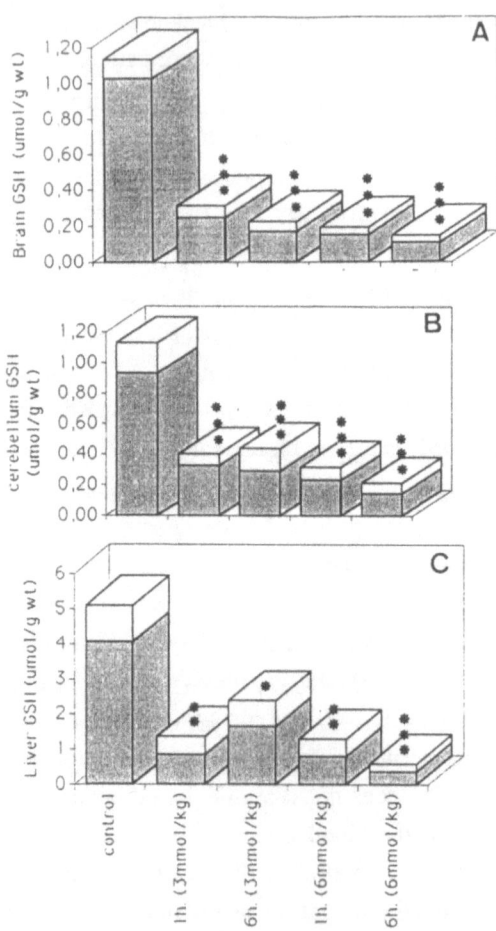

Figure 3. Effect of Diethylmaleate injection on glutathione levels in various tissues of mice
A: Brain GSH concentration; B: Cerebellum GSH concentration and C: liver GSH concentration. DEM was administered by intraperitoneal injection (3 or 6 mmol/kg) and was dissolved in corn oil. Animals treated (n=12 for 3 mmol/kg and n=12 for 6 mmol/kg)). control mice (n=6) received only corn oil. Animals were killed 1 and 6 hours following injection. * P<0.05; ** P<0.01; *** P<0.001; t-test.

The fall of GSH concentration reached an upper value of 90% in liver (Fig. 3C) and brain (Fig. 3A) and 84% in cerebellum (Fig. 3B) 6 hours afer acute intoxication with 6 mmol DEM/kg. This study shows that DEM may be used to deplete brain GSH by 90% in vivo in mice. Neurobiological consequences of such a depletion will be evaluated. Moreover, these studies provide an animal model for the human diseases involving GSH deficiency. Since the brain exhibits considerable regional heterogeneity, the effect of GSH modulation by DEM on various regions of the brain must be examined. Moreover, brain exhibits considerable cellular heterogeneity, it is possible that the depletion of GSH within a particular cell type is more or less than the overall value determined. The physiological relevance of such depletion could be understood by examining the GSH levels in individual cell types following DEM treatment.

Conclusion

There is an increasing body of evidence that indicates that major neurodegenerative diseases occurs in the human brain if the endogenous antioxidant mechanisms are not able to limit free radicals destruction. Specifically, we are actively testing the hypothesis that neuronal death and progressive neurodegeneration associated with aging and neurodegenerative disorders: DS, ALS and both PD and AD are caused by increased oxidative stress and the inabilility of the brain to deal with radical oxygen species attack.

The use of genetic manipulations to construct transgenic cells or transgenic mice that specifically lack or overexpress a single antioxidant as well as metabolic manipulations to deplete selectively a single antioxidant represent a mean for understanding the role of individual antioxidant in cellular protection and/or neurodegeneration. Extended biochemical and ultrastructural studies of the brains and biochemical and genetic studies of the mitochondria of the transgenic mice and of GSH-depleted mice described here should explain whether CuZnSOD excess or GSH depletion participate in neurodegenerative processes and aging.

References

Abraham, E.C., Taylor, J.F., Lang, C.A. (1978) Influence of mouse age and erythrocyte age on glutathione metabolism. *J. Biochem.* 174: 819-825

Anderson, M.E. (1985) Determination of glutathione and glutathione disulfide in biological samples. *Methods Enzymol.* 113: 548-555.

Anneren, G. and Epstein , C.J. (1987) Lipid peroxidation and superoxide dismutase-1 and glutathione peroxidase activities in trisomy 16 fetal mice and human trisomy 21 fibroblasts. *Pediat. Res.* 21: 88-92.

Avraham, K.B., Schickler, M., Sapoznikov, D, Yarom, R., Groner, Y. (1988) Down's syndrome: abnormal neuromuscular junction in tongue of transgenic mice with elevated levels of human CuZn superoxide dismutase. *Cell* 54: 823-829.

Bandy, B. and Davison, A.J. (1990) Mitochondrial mutations may increase oxidative stress: implications for carcinogenesis and aging? *Free Rad. Biol. Med.* 8: 523-539

Barja de Quiroga, G., Perez-Campo, R., Lopez-Torrez, M. (1990) Anti-oxidant defenses and peroxidation in liver and brain of aged rats. *Biochem J.* 272: 247-250.

Benzi, G., Pastoris, O., Marzatico, F. and Villa, R.F. (1989a) Age-related effect induced by oxidative stress on the cerebral glutathione system. *Neurochem. Res.* 14: 473-481.

Benzi, G., Marzatico, F., Pastoris, O., Villa, R.F.(1989b) Relationship between aging, drug treatment and the cerebral enzymatic antioxidant system. *Exp. Gerontol.* 24: 137-148.

Berr, C., Nicole, A., Godin, J., Ceballos-Picot, I., Thevenin, M., Dartigues, JF., Alperovitch, A. (1993) Relationships between plasma selenium, erythrocyte selenium, oxygen metabolizing enzyme concentrations and age in elderly community residents: a pilot epidemiologic study. *J. Am. Geriatr. Soc.* 41: 143-148.

Blass, J.P. and Gibson, G.E. (1991) The role of oxidative abnormalities in the pathophysiology of Alzheimer's disease. *Rev. Neurol.* 147: 513-525.

Bracco, F., Burlina, A.P., Malesani, R., Rigo, A., Battistin, L. (1986) Free-radical related enzymes in the aging brain. In: A. Bès et al. (Eds), *Senile dementia, early detection,* John Libbey Eurotext, pp. 293-297

Brooksbank, B.W.L., Balazs, R. (1984) Superoxide dismutase, glutathione peroxidase and lipoperoxidation in Down syndrome fetal brain. *Dev. Brain Res.*16: 37-44.

Cammer, W., Tansey, F., Abramovitz, M., Ishigaki, S., Listowsky, I. (1989) Differential localization of glutathione-S-transferase Yp and Yb subunits in oligodendrocytes of rat brain. *J. Neurochem.* 52: 876-883.

Cand, F., Verdetti, J.(1989) Superoxide dismutase, glutathione peroxidase, catalase and lipid peroxidation in the major organs of the aging rats. *Free Rad. Biol. Med.* 7: 59-63.

Cao Danh, H., Strolin Benedetti, M., Dostert, P.(1983) Differential changes in superoxide dismutase activity in brain and liver of old rats and mice *J. Neurochem.*40: 1003-1007.

Carney, J.M., Starke-Reed P.E., Oliver, C.N., Landum, R.W., Cheng, M.S., Wu, J.F., Floyd, R.A. (1991) Reversal of age-related increase in brain protein oxidation, decrease in enzyme activity, and loss in temporal and spatial memory by chronic administration of the spin trapping compound N-tert-butyl-a-phenylnitrone. *Proc. Natl. Acad. Sci.* 88: 3633-3636.

Ceballos, I., Delabar, J.M., Nicole, A., Lynch, R.E., Hallewell, R.A., Kamoun, P., Sinet P.M.(1988) Expression of transfected human CuZnSOD in mouse L cells and NS2OY neurobastoma cells induces enhancement of glutathione peroxidase activity. *Biochim. Biophys. Acta* 949: 58-64.

Ceballos, I., Javoy-Agid, F., Hirsch, E.C., Dumas, S., Kamoun, P., Sinet, P.M., Agid, Y.(1989) Localization of copper-zinc superoxide dismutase mRNA in human hippocampus by in situ hybridization. *Neurosci. Lett.* 105: 41-46.

Ceballos, I., Lafon, M., Javoy-Agid, F., Hirsch, E., Nicole, A., Sinet, P.M., Agid, Y. (1990a) Superoxide dismutase and Parkinson's disease. *The Lancet* 335: 1035-1036.

Ceballos, I., Javoy-Agid, F., Delacourte, A., Defossez, A., Nicole, A., Sinet, P.M. (1990b) Parkinson's disease and Alzheimer's disease: neurodegenerative disorders due to brain antioxidant system deficiency? In: Emerit et al (Eds): *Antioxidants in Therapy and Preventive Medicine,* Plenum Press, New York, pp. 493-498.

Ceballos, I., Javoy-Agid, F., Delacourte, A., Defossez, A., Lafon, M., Hirsch, E., Nicole, A., Sinet, P.M., Agid, Y. (1991a) Neuronal localization of copper-zinc superoxide dismutase protein and mRNA within the human hippocampus from control and Alzheimer's disease brains. *Free Rad. Res. Comms.* 12-13: 571-580.

Ceballos-Picot, I., Nicole, A., Briand, P., Grimber, G., Delacourte, A., Defossez, A., Javoy-Agid, F., Lafon, M., Blouin J.L., Sinet, P.M. (1991b) Neuronal-specific expression of human copper-zinc superoxide dismutase gene in transgenic mice: animal model of gene dosage effect in Down's syndrome. *Brain Res.* 552: 198-214.

Ceballos, I., Nicole, A., Briand, P., Grimber, G., Delacourte, A., Flament, S., Thevenin, J.M., Kamoun, P., Sinet, P.M. (1991c) Expression of human CuZn superoxide dismutase gene in transgenic mice : model for gene dosage effect in Down's syndrome. *Free Rad. Res. Comms.* 12-13: 581-589.

Ceballos-Picot, I., Nicole, A., Sinet,PM. (1992a) Cellular clones and transgenic mice overexpressing copper-zinc superoxide dismutase: model for the study of free radicals metabolism and aging. *Free Radicals and Aging.*In: Emerit I., Chance B.(Eds) Birkhaüser Verlag, pp. 89-98.

Ceballos-Picot, I., Nicole, A., Sinet, P.M. (1992b) Age-related changes in antioxidant enzymes and lipid peroxidation in brains of control and transgenic mice overexpressing copper-zinc superoxide dismutase. *Mut. Res.* 275: 281-293.

Ceballos-Picot, I., Trivier, J.M., Nicole, A., Sinet , P.M. and Thevenin, M. (1992c) Age-correlated modifications of copper-zinc superoxide dismutase and glutathione-related enzyme activities in human erythrocytes. *Clin. Chem.* 38: 66-70.

Chance, B., Sies, H., Boveris, A. (1979) Hydroperoxide metabolism in mammalian organs. *Physiol. Rev.* 59: 527-533.

Chen, C.J., Warshaw, J.B., Sanadi, D.R. (1972) Regulation of mitochondrial respiration in senescence. *J. Cell Physiol.* 80: 141-148.

Chen, T.S., Richie, J.P. and Lang, C.A. (1989) The effect of aging on glutathione and cysteine levels in different rgions of the mouse brain. *Soc. Biol. Med.* 190: 399-402

Connor, J.R., Menzies, S.L., Martin, S.M., Mufson, E.J. (1990) Cellular distribution of Transferrin, Ferritin, and Iron in Normal and Aged Human Brains. *J. Neurosci. Res.* 27: 595-611.

Delacourte, A., Defossez, A., Ceballos, I., Nicole, A., Sinet , P.M.(1988) Preferential expression of copper-zinc superoxide dismutase in the vulnerable cortical neurons in Alzheimer's disease. *Neurosci. Lett.* 92: 247-253.

Di Monte, D.A., Chan, P., Sandy, M.S. (1992) Glutathione in Parkinson's disease: a link between oxidative stress and mitochondrial damage. *Ann. Neurol.* 32: S111-S115.

Dowson, J.H.(1982) Neuronal Lipofuscin Accumulation in ageing and Alzheimer dementia : a pathogenic mechanism . *Brit. J. Psychiat.* 140: 142-148.

Elroy-Stein, O., Bernstein, Y., Groner, Y.(1986) Overproduction of human CuZn superoxide dismutase in transfected cells : extenuation of paraquat-mediated cytotoxicity and enhancement of lipid peroxidation. *EMBO J.* 5: 615-622.

Elroy-Stein, O., Groner, Y.(1988) Impaired neurotransmitter uptake in PC12 cells overexpressing human Cu-Zn superoxide dismutase implications for gene dosage effect in Down's syndrome. *Cell* 52: 259-267.

Epstein, C.J., Avraham, K.B., Lovett, M., Smith, S., Elroy-Stein, O., Rotman, G., Bry, C.,Groner Y. (1987) Transgenic mice with increase Cu/Zn-superoxide dismutase activity : animal model of gene dosage effects in Down syndrome. *Proc. Natl. Acad. Sci.* 84: 8044-8048.

Evans, P.H., Klinowski, J., Yano, E., Urano, N. (1989) Alzheimer's disease: a pathogenic role for aluminosilicate-induced phagocytic free radicals. *Free Rad. Res. Comms,* 6: 317-321.

Farooqui, M.Y.H, Day, W.W., Zamorano D.M (1987) Glutathione and lipid peroxidation in the aging rat. *Comp. Biochem. Physiol.* 88B: 177-180

Freeman, B.A. and Crapo, J.D. (1982) Biology of disease: free radicals and tissue injury. *Lab. Invest.* 47: 412-427.

Fucci, L., Oliver, C.N., Coon, M.J., Stadtman, E.R. (1983) Inactivation of key metabolic enzymes by mixed function oxidation reactions: possible implication in protein turnover and ageing. *Proc. Natl. Acad. Sci.*80: 1521-1525.

Gerard-Monnier, D., Fougeat, S. and Chaudiere, J. (1992) Glutathione and cysteine depletion in rats and mice following acute intoxication with Diethylmaleate. *Biochem. Pharmacol.,* 43: 451-456.

Geremia, E., Baratta, D., Zafarana, S., Giordano, R., Pinizotto, M.R., La Rosa, M.G., Garozzo A. (1990) Antioxidant enzymatic systems in neuronal and glial cell-enriched fractions of rat brain during aging. *Neurochem. Res.* 15: 719-772.

Graham, D.G. (1978) Oxidative pathways for catecholamines in the genesis of neuromelanin and cytotoxic quinones. *Mol. Pharmacol.* 14: 633-634.

Griffith, O., and Meister, A. (1979) Glutathione: interorgan translocation, turnover and metabolism. *Proc. Natl. Acad. Sci.* 76: 5605-5610

Griffith, O. (1980) Determination of glutathione and glutathione disulfide using glutathione reductase and 2-vinylpyridine. *Anal. Biochem.* 106: 207-212.

Harman, D. (1981) The aging process. *Proc. Natl. Acad. Sci. USA* 78: 7124-7128.

Hazelton, G.A., and Lang C.A. (1980) Glutathione content of tissues in the aging mouse. *J. Biochem.* 188: 25-30

Hazelton, G.A. and Lang, C.A.(1985) Glutathione peroxidase and reductase activities in the aging mouse. *Mech. Ageing Dev.* 29: 71-78.

Hothersall, J.S., El-Hassan, A., McLean, P., Greenbaum, A.L. (1981) Age-related changes in enzymes of rat brain. 2. Redox system linked to NADPH and glutathione. *Enzyme* 26: 271-276.

Jain, A., Martensson, J., Stole, E., Auld, P.A.M., Meister, A. (1991) Glutathione deficiency leads to mitochondrial damage in brain. *Proc. Natl. Acad. Sci.*, 88: 1913-1917

Kellogg, E.W., Fridovich, I. (1976) Superoxide dismutase in the rat and mouse as a function of age and longevity. *J. Gerontol.* 31: 405-408.

Ledig, M., Fried, R., Ziessel, M., Mandel, P. (1982) Regional distribution of superoxide dismutase in rat brain during postnatal development. *Dev. Brain Res.* 4: 333-337.

Marklund, S.L., Adolfsson, R., Gottfries, C.G., Winblad, B. (1985) Superoxide dismutase isoenzymes in normal brains and in brains from patients whith dementia of Alzheimer type *J. Neurol. Sci.* 67: 319-325.

Martensson, J., Jain, A., Stole, E., Frayer, W., Auld, P.A.M., Meister, A. (1991) Inhibition of glutathione synthesis in the newborn rat: a model for endogenously produced oxidative stress. *Proc. Natl. Aced. Sci.*, 88: 9360-9364.

Marttila, R.J., Lorentz, H., Rinne, U.K (1988) Oxygen toxicity protecting enzymes in Parkinson's disease. Increase of superoxide dismutase-lie activity in the substantia nigra and basal nucleus. *J. Neurol. Sci.* 86: 321-331.

Massie, H.R., Aiello, V.R., Iodice, A.A. (1979) Changes with age in copper and superoxide dismutase levels in brains of C57BL/6J mice. *Mech. Ageing Dev.*10: 93-99.

Meister, A. and Anderson, M.E. (1983) Glutathione. *Ann. Rev. Biochem.* 52: 711-760

Mizuno, Y., Ohta, K. (1986) Regional distribution of thiobarbituric acid-reactive products, activities of enzymes regulating the metabolism of oxygen free radicals and some of the related enzymes in adult and aged rat brain. *J. Neurochem.* 46: 1344-1352.

Naryshkin, S., Miller, L., Lindeman, R. and Lang, C.A.(1981) Blood glutathione: a biochemical index of human aging. *Fed. Proc.* 40: 3179.

Norris, K.H., Hornsby, P.J. (1990) Cytotoxic effects of expression of human superoxide dismutase in bovine adrenocortical cells. *Mutation Res.* 237: 95-106.

Perry, T.L., Godin, D.V., Hansen, S. (1882) Parkinson's disease: a disorder due to nigral glutathione deficiency. *Neurosci. Lett.* 33: 305-310.

Philbert, M.A., Beiswanger, C.M., Waters, D.K., Reuhl, K.R., Lowdes, H.E.(1991) Cellular and regional distribution of reduced glutathione in the nervous system of the rat: histochemical localization by mercury orange andO-Phtaldialdehyde-induced histofluorescence. *Toxicol and Applied Pharmacol.* 107: 215-227.

Pileblad, E., and Magnusson, T. (1989) Intracerebroventricular administration of L-buthionine sulfoximine: a method for depleting brain glutathione. *J. Neurochem.* 53: 1878-1882

Przedborski, S., Jockson-Lewis, V., Kostic, V., Carlson, E., Epstein, C.J., Cadet, J.L.(1992) Superoxide dismutase, catalase and glutathione peroxidase activities in copper-zinc superoxide dismutase transgenic mice. *J. Neurochem.* 58: 1760-1767.

Raps, S.P., Lai, J.C.K., Hertz, L., Cooper, A.J.L.(1989) Glutathione is present in high concentration in cultured astrocytes but not in cultured neurons. *Brain Res.* 493: 498-501

Ravindranath, V., Shivakumar, B.R., Anandatheerthavarada, H.K. (1989) Low glutahione levels in brain regions of aged rats. *Neurosci. Lett.* 101: 187-190.

Reiss, U. and Gershon, D. (1976) Comparison of cytoplasmic superoxide dismutase in liver, heart and brain of aging rats and mice. *Biochim. Biophys. Res. Commun.* 73: 255-262.

Rosen, D.R., Siddique, T., Patterson, D. et al., (1993) Mutations in Cu/Zn superoxide dismutase gene are associated with familial amyotrophic lateral sclerosis. *Nature* 362: 59-62

Rumble, B., Tetallack, R., Hilbich, C.(1989) Amyloid A4 protein and its precursor in Down's syndrome and Alzheimer's disease. *New England J. Med.* 22: 1146-1452.

Sawada, M., Carlson, J.C. (1987) Changes in superoxide radical and lipid peroxide formation in the brain, heart and liver during the life time of the rat. *Mech. Ageing Dev.*41: 125-137.

Scarpa, M., Rigo, A., Viglino, P., Stevanato, R., Bracco, F., Battistin, L. (1987) Age-dependence of the level of the enzymes involved in the protection against active oxygen species in the rat brain. *Proc. Soc. Ex. Med.* 185: 129-133.

Schneider, D., Naryshkin, S., Lang, C.A. (1982) Blood glutathione, a biochemical index of aging women. *Fed. Proc.* 41: 7671

Scott, M.D., Meshnick, S.R., Eaton, J.W.(1987) Superoxide dismutase rich bacteria. Paradoxal increase in oxidant toxicity. *J. Biol. Chem.* 262: 3640-3645.

Scott, M.D., Meshnick, S.R., Eaton, J.W. (1989) Superoxide dismutase amplifies organismal sensitivity to ionizing radiation. *J. Biol. Chem.* 264: 2498-2501.

Semsei, I., Govinda, R.,Richardson, A. (1991) Expression of superoxide dismutase and catalase in rat brain as a function of age. *Mech. Ageing Dev.* 58: 13-19.

Sinet, P.M., Couturier, J., Dutrillaux, A., Jerome, H. (1976) Trisomie 21 et superoxide dismutase-1 (IPO-A). Tentative de localisation sur la sous-bande 21q22.1. *Exp. Cell. Res.* 97: 47-55

Sinet, P.M. (1982) Metabolism of oxygen derivatives in Down's syndrome. *Ann NY. Acad. Sci.* 386:82-94.

Sinet , P.M. and Ceballos-Picot, I. (1992) Role of free radicals in Alzheimer's disease and Down's syndrome. In: L. Packer, L. Prilipko and Y. Christen (eds.): *Free Radicals in the Brain Aging, Neurological and Mental Disorders*, Springer-Verlag, Berlin Heidelberg, pp. 91-98.

Slivska, A., Mytillineou, C., Cohen, G. (1987) Histochemical evaluation of glutathione in brain. *Brain Res.*, 209: 275-284

Slivska, A., Spina, M., Calvin, H., Cohen, G. (1988) Depletion of brain glutathione in preweaning mice by L-buthionine sulfoximine. *J. Neurochem.* 50: 1391-1393.

Smith, C.D., Carney, J.M., Starke-Reed, P.E, Oliver, C.N., Stadtman, E.R, Floyd, R.A., Markesbery, W.R.(1991) Excess brain protein oxidation and enzyme dysfunction in normal aging and in Alzheimer disease. *Proc. Natl. Acad. Sci.* 88: 10540-10543.

Sohal, R.S., Arnold, L.A., Sohal, B. (1990a) Age-related changes in antioxidant enzymes and prooxidant generation in tissues of the rat with special reference to parameters in two insect species. *Free Rad Biol. Med.*10: 495-500.

Sohal, R.S and Allen, R.G.(1990b) Oxidative stress as a causal factor in differentiation and aging: a unifying hypothesis. *Experiment. Gerontol.* 25: 499-522.

Sohal, R.S. (1991) Hydrogen peroxide production by mitochondria may be a biomarker of aging. *Mech. Ageing Dev.*60: 189-198.

Sommerville, M.J., Percy, M.E., Bergeron, C., Yoong, L.K.K., Grima, E.A. McLachlan, D.R.C. (1991) Localization and quantitation of 68 kda neurofilament and superoxide dismutase-1 mRNA in Alzheimer brain. *Molec. Brain Res.* 9: 1-8.

Svennerholm, L., Bostrom, K., Helander, C.G., Jungbjer, B. (1991) Membrane lipids in the aging human brain. *J. Neurochem.*56: 2051-2059.

Teramoto, S., Fukuchi, Y., Uejima, Y., Ito, H. and Orimo, H. (1992) Age-related changes in GSH content of eyes in mice- a comparison f senescence-accelerated mouse (SAM) and C57BL/J mice. *Comp. Biochem. Physiol.*, 102A: 693-696

Ushikima, K., Miyazaki, H., Morioka, T.(1986) Immunohistochemical localization of glutathione peroxidase in the rat brain. *Rescucitation* 13: 97-105.

Vanella, A., Villa, R.F., Gorini, A., Campisi, A., Giuffrida-Stella, A.M.(1989) Superoxide dismutase and cytochrome oxydase activities in light and heavy synaptic mitochondria from rat cerebral cortex during aging. *J. Neurosci. Res.* 22: 351-355.

Vina, J., Sastre, J., Anton, V., Bruseghini, L., Esteras, A. and Asensi, M. Effect of aging on glutathione metabolism, protection by antioxidants. (1992) In: I. Emerit and Chance B. (eds): *Free radicals and Aging*, Birhaüser Verlag, pp. 89-98

Vitoria, J., Machado, A., Satrustegui, J. (1984) Age-dependent variations in peroxide-utilizing enzymes from rat brain mitochondria and cytoplasm. *J. Neurochem.* 42: 351-356.

Wallace, D.C. (1992) Mitochondrial genetics: a paradigm for aging and degenerative diseases? *Science* 256: 628-632

Wei Soong, N., Hinton, D.R., Cortopassi, G., Arnheim, N. (1992) Mosaicism for a specificsomatic mitochondrial DNA mutation in adult human brain. *Nature Genet.* 2:318-323

Wong, G.H.W., and Goeddel, D.V (1988) Induction of manganous superoxide dismutase by tumor necrosis factor: possible protective mechanism. *Science* 242: 941-944

Yim, M. B., Chock P. B., Stadtman E.R.(1990) Copper, zinc superoxide dismutase catalyzes hydroxyl radical production from hydrogen peroxide. *Proc. Natl. Acad. Sci.*87: 5006-5010.

Oxidative Stress, Cell Activation and Viral Infection
C. Pasquier et al. (eds)
© 1994 Birkhäuser Verlag Basel/Switzerland

Protective effect of Poly(A)-Poly(U) against immune oxidative injury. Role of thiols released by activated macrophages

M. Fay[1], M. Jampy-Fay[2], K. Akarid[1] and M.A. Gougerot-Pocidalo[2]

[1] INSERM U 13 and [2] INSERM U 294 et laboratoire d'Immunologie et d'Hématologie, CHU X. Bichat, France

Summary

Various drugs with or without sulfur groups have been demonstrated to exert antioxidant properties in vivo. In this latter group immunomodulating agents such as the polyribonucleotide Poly(I)-Poly(C) have been shown to protect rats against hyperoxic pulmonary damage. However, the effects of these agents on the protection of immune cells against oxidative injury have not been described. Our work shows that the immunomodulating agent Poly(A)-Poly(U) administered in vivo to mice protected their lymphoid cells against immune oxidative injury by increasing their glutathione content. This antioxidant effect could be related to the Poly(A)-Poly(U) capacity to induce the release of thiol compounds from macrophages. The antioxidant effects of such a non toxic immunomodulating agent could be of interest in protecting lymphoid cells in pathological situations involving oxidative injury.

Introduction

Oxidative injury of immune cells has been observed in various pathological situations, such as exposure to ionizing radiation (Doria, 1982), the immunodeficiency associated with ageing (Harman, 1981), inflammatory reactions (El-Hag, 1987) and, more recently, the progression of human immunodeficiency virus infection (Halliwell, 1991). In vitro and in vivo models of immune oxidative injury have contributed to demonstrating the critical role of thiol status in the protection of lymphocyte functions against oxidative injury (Noelle, 1981; Gougerot-Pocidalo, 1985; Kraus, 1985; Gougerot-Pocidalo, 1988). Glutathione, a low-molecular-weight non-protein thiol, is one of the most important cellular antioxidant systems and plays a major role in the precise redox equilibrium of the cell. In particular, it protects lipids against peroxidation and regenerates protein sulfhydryl groups from disulfide bonds (Kosower, 1989; Meister, 1983).

Various compounds exert antioxidant properties in vivo. In particular, the synthetic polyribonucleotide poly I:poly C, has been shown to protect rats against hyperoxic lung damage (Frank, 1980; Kikkawa, 1984). Poly(A)-Poly(U) is also a double-stranded synthetic polyribonucleotide complex. Its lack of toxicity in animals and humans gives it potential for therapeutic use. The immunomodulating action of Poly(A)-Poly(U) is, at least in part, related to macrophage activation (Johnson, 1985). Since LPS-activated macrophages release cysteine, which is one of the main precursors of glutathione and regulates intracellular glutathione levels in lymphocytes (10, 29), we sought if Poly(A)-Poly(U) could afford a protective effect against an immune oxidative injury and modify the thiol content of the lymphoid cells.

Materials and Methods

Treatment with Poly(A)-Poly(U)

Poly (A)- Poly (U) was a gift from Ipsen-Biotech (Laboratoire Beaufour, France). It was dissolved in sterile 0.15 M NaCl and stored at 4°C throughout the study. Female C57/Bl6 mice 8 to 12 weeks old (Iffa-Credo, France), received 1 mg IP in 0.5 ml on seven occasions over two weeks. Control mice received 0.5 ml of sterile 0.15 M NaCl IP. The injected solutions had a pH of 7.2-7.3.

Spleen T cell preparation

Spleen cells were prepared from aseptically removed spleens and separated from red blood cells by Percoll density gradient centrifugation. After washing, spleen cells were filtered through a nylon-wool column, in order to enrich the suspension in T cells.(Campbell, 1971). After 45 min incubation at 37°C on the column, effluent cells were centrifuged and suspended in supplemented medium. Immunofluorescence staining with anti-CD3 monoclonal antibody (Caltag Lab.,USA) showed enrichment from about 30% of T cells before filtration to more than 70%. No significant difference in the percentage or number of CD3-positive cells was observed between samples from treated and control mice. These cells are referred to as spleen T cells. Trypan blue exclusion showed that viability exceeded 97%.

Ex vivo model of immune oxidative injury: O2 exposure of spleen T cells:

Spleen T cells from treated or control mice were cultured at 37°C for 24 h in a humidified normobaric 60% oxygen/5% CO_2 atmosphere ("O2 conditions") in a leak-proof chamber (20). Incubation was then continued for 24 or 48 h in 95% air/5% CO_2. Control cultures were incubated in 95% air/5% CO_2 ("air conditions") for 24, 48 or 72 h.

Proliferative response to mitogens

Spleen T cells (5×10^5 in 0.2 ml) were cultured in 96-well flat-bottomed tissue culture plates in RPMI 1640 medium (Gibco) supplemented with 5% FCS, with or without ConA (2 μg/ml). When specified, 2 mercapto-ethanol (2ME) was added at a final concentration of 5×10^{-5}M.

Cultures were incubated in O2 or air conditions as described above. Twenty-four hours before the end of the culture periods, each well was pulsed with 1 μCi methyl- [^{3}H] thymidine ([^{3}H] TdR, 2 Ci/mmole) Cells were then harvested onto glass-fiber filters with a Skatron harvester. Incorporated radioactivity was counted with a liquid scintillation counter (LKB Intruments) and data were expressed as means of triplicate cultures (cpm of stimulated wells minus cpm of unstimulated wells).

Determination of intracellular total glutathione
Total glutathione was assayed according to the method of Tietze (Tietze, 1969) after preparation of the samples as described in (Lacombe, 1985). The method used 2 mM NADPH, 3 mM DTNB and 1 U/ml glutathione reductase (Sigma). Absorbance was recorded at 412 nm using a Uvikon 810 spectrophotometer (Kontron, France). The glutathione concentration was read from a calibration curve. Results are expressed as nmoles per 10^6 viable cells.

Poly(A)-Poly(U) treatment of peritoneal macrophages in vitro
Resident peritoneal macrophages were obtained by washing the peritoneal cavity with cold PBS containing 5 U/ml heparin. Two x 10^6 macrophages, as determined by non-specific esterase staining, were plated in 24-well plates (Falcon). After adherence (90 min at 37°C), more than 95% of adherent cells were macrophages which were cultured for 48 h at 37°C in 5% CO2/95% air in the presence or absence of Poly(A)-Poly(U) at final concentrations of 0.1, 0.2 and 1 mg/ml. Extracellular total thiols were measured by a method using 5 mM DTNB (Bannai, 1980).Acid-soluble thiols were determined after sulfosalicylic acid precipitation (Bannai, 1980).

Statistical analysis
Data are presented as means ± SEM. Means were compared using Student's *t* test.

Results

Treatment of mice with Poly(A)-Poly(U) protected spleen T cells against oxidative injury ex vivo.
Treatment of mice with Poly(A)-Poly(U) gave significant protection against the decrease of proliferation caused by exposure to O2 ex vivo (Table I). After 48 h of culture, the ConA-induced proliferative response of control spleen T cells cultured in O2 conditions was about 28% that of cells cultured in air conditions, while the percentage responses (O2 cpm/air cpm) of cells from mice treated with Poly(A)-Poly(U) were about 49% (p < 0.001). Similar results were

observed after 72 h of culture. The sulfhydryl compound 2-ME (5×10^{-5}M), the antioxidant effect of which is well known, was added to cells from untreated mice in air and O_2 conditions; it increased the percentage response (O_2 cpm/air cpm) from $27.8 \pm 3.8\%$ to $95 \pm 3.5\%$, after 48 h of culture.

Table I: Treatment of mice with Poly(A)-Poly(U) protected the proliferative response of spleen T cells against ex vivo oxidative injury.[a]

Treatment	48 H Culture			72 H Culture		
	Air[b] cpmx10^{-3}	O_2[b] cpmx10^{-3}	O_2/Air[c] %	Air[b] cpmx10^{-3}	O_2[b] cpmx10^{-3}	O_2/Air[c] %
Control	117.1 ± 6.8	35.6 ± 6.1	27.8 ± 3.8	171.2 ± 10.8	80.4 ± 12.1	43.7 ± 5.0
P(A)-P(U)	117.4 ± 3.1	58.4*** ± 5.3	49.2*** ± 3.7	174.3 ± 5.1	98.6* ± 7.1	56.7* ± 4.0

[a] Spleen T cells were cultured with ConA (2μg/ml) in air (95% air and 5% CO2) or O2 conditions (60% O2 and 5% CO2 for 24 h followed by 95% air and 5% CO2 during the rest of the culture).
[b] Results are means ± s.e.m for 14 mice in 3 experiments and are expressed as cpm x 10^{-3}
[c] Percentage proliferative response O2 cpm/Air cpm was calculated by dividing ^{3}H-TdR incorporation (cpm) by spleen cells cultured in O2 conditions by the corresponding value in air conditions x 100%.
Statistical comparisons between treated and control mice : *p<0.05, ***p< 0.001.

Effect of Poly(A)-Poly(U) on the glutathione content of fresh spleen cells.
The glutathione content of spleen T cells isolated immediately from mice killed after treatment with Poly(A)-Poly(U) were significantly higher than control values (0.558 ± 0.021 and 0.462 ± 0.017 nmol/ 10^6 viable cells, respectively, p<0.05).

Effect of Poly(A)-Poly(U) treatment on glutathione content of ConA-stimulated spleen T cells in O2 and air conditions
As shown in Table II, after ConA stimulation in air and O2 conditions the glutathione content of spleen T cells from treated mice remained higher than that in controls at all times of culture. Furthermore, Poly(A)-Poly(U) treatment protected against the hyperoxia-induced decrease in glutathione content, the nadir of which was observed after 48 h culture. At this time, the percentages of the glutathione contents (O2 glutathione/air glutathione) increased from 45.4 ± 2.5 % in controls to 57.4 ± 4.0 % (p<0.05) in Poly(A)-Poly(U)-treated mice.

Table II : Kinetics of glutathione content in spleen T cells from Poly(A)-Poly(U)-treated mice after ConA stimulation in air or O2 conditions.

	24 H Culture			48 H Culture			72 H Culture		
	Air [a]	O$_2$ [a]	O$_2$/Air [b]	Air [a]	O$_2$ [a]	O$_2$/Air [b]	Air [a]	O$_2$ [a]	O$_2$/Air [b]
	n.mol/10^6 cells		%	n.mol/10^6 cells		%	n.mol/10^6 cells		%
Control	0.251 ± 0.024	0.246 ± 0.018	103.4 ± 9.4	0.551 ± 0.030	0.248 ± 0.010	45.4 ± 2.2	1.090 ± 0.040	0.766 ± 0.041	70.6 ± 3.8
Poly(A)-Poly(U)	0.379** ± 0.030	0.319*** ± 0.011	88.4 ± 6.5	0.898*** ± 0.043	0.506*** ± 0.038	57.4* ± 3.4	1.607*** ± 0.059	1.379*** ± 0.071	87.1 ± 5.8

[a] Results are means ± s.e.m. for 8 mice and are expressed as nmol/10^6 viable cells.

[b] Percentage glutathione content was calculated by dividing the glutathione value of spleen cells cultured in O2 conditions by the corresponding value in air conditions x 100%.

Statistical comparisons between treated and control mice:*p<0.05, **p<0.01, ***p<0.001.

Effect of Poly(A)-Poly(U) on the release of total thiols and acid-soluble thiols from peritoneal macrophages.

To elucidate the effect of Poly(A)-Poly(U) treatment in vivo, we measured the capacity of this compound to induce the release of thiols especially acid-soluble thiols (mainly glutathione and cysteine) from macrophages. As shown in table III, Poly(A)-Poly(U) treatment of resident peritoneal macrophages in vitro induced a significant release of total thiols and acid-soluble thiols into the extracellular medium. The effects of Poly(A)-Poly(U) (0.2 and 1 mg/ml) were similar to that obtained with LPS (E. Coli, O.55:B5, 10μg/ml).

Table III: Poly(A)-Poly(U) induces the release of total and acid-soluble thiols from macrophages

Addition to culture	Total thiols (μM)	Acid-soluble thiols (μM)
Medium	16.9 ± 0.6	9.6 ± 0.5
LPS (10μg/ml)	25.0 ± 1.5**	12.7 ± 0.5*
Poly(A)-Poly(U)		
0.1 mg/ml	25.4 ± 0.8**	12.7 ± 0.5*
0.2 mg/ml	30.4 ± 1.6**	15.7 ± 1.0**
1.0 mg/ml	30.4 ± 0.9**	14.0 ± 0.7**

*$p<0.05$, ** $p<0.001$

Discussion

Our results show that the synthetic double-stranded polyribonucleotide Poly(A)-Poly(U) protect spleen T cells against oxidative injury ex vivo when administered to C57Bl/6 mice. Administration of this non-sulfur compound increased the glutathione content in fresh spleen cells and preserved both the ConA-proliferative response and glutathione content during oxidative injury ex vivo. In addition, Poly(A)-Poly(U) induced acid-soluble-thiol release from macrophages in vitro, an observation which could account, at least in part, for the increased glutathione content of the spleen T cells.

We used an ex vivo model of oxidative damage to lymphoid cells, which permitted us to evaluate the antioxidant effects of in vivo treatment on lymphocytes. The immunological consequences of oxidative injury have been demonstrated by exposure of lymphocytes ex vivo to enzymatic systems that generate O2-derived products, as well as to various oxidizing agents and hyperoxic conditions (Chaplin, 1978; Grever, 1980; Kraus, 1985; Gougerot-Pocidalo, 1988). Models of oxidative injury in vivo have also been described involving the immune system.(Gougerot-Pocidalo, 1985, Saito, 1991) However, the ex vivo model used here allowed us to test the effects of drug administration on lymphoid cells without having to take into account the nonspecific stress of oxidative injury in vivo. The hyperoxic conditions used here have been shown previously in our laboratory to induce a significant decrease in ConA-induced proliferation and glutathione content of mouse spleen T cells after 48 h of culture.(Lacombe, 1988).

Poly(A)-Poly(U) treatment in vivo protected spleen T cells against the oxidative injury-induced reduction in both ConA-induced proliferation and glutathione content. Thiols, especially glutathione, one of the most important antioxidant systems in aerobic cells, play an important role in lymphocyte functions.(Chaplin,1978; Noelle, 1981; Messina,1989) Thiol compounds such as 2ME, cysteine and glutathione ester, which increase intracellular glutathione levels, also enhance lymphocyte functions in vitro, particularly the proliferative response to mitogens (Ohmori, 1983; Wellner, 1984; Fidelus, 1986; Gmünder, 1990). In addition, when used in vitro, these compounds protected against the consequences of oxidative injury induced in vivo and in vitro (i.e. the decrease in glutathione content and the proliferative response of spleen cells) (Noelle, 1981; Gougerot-Pocidalo, 1985; Kraus, 1985; Gougerot-Pocidalo, 1988). The non-protein thiol glutathione accounts for about 30 % of total thiols in lymphoid cells and is required for the G1/S transition of the cell cycle (Messina, 1989). We therefore measured the glutathione content of spleen cells to assess the antioxidant capacity of Poly(A)-Poly(U) administration. Since Poly(A)-Poly(U) treatment protected against the oxidative injury-induced decrease in glutathione content of spleen T cells after 48 h of culture, their protective effect on the proliferative response could involve the action of glutathione on the G1/S phase of the cell cycle. Poly(A)-Poly(U) administration increased the glutathione content in resting fresh spleen T cells, suggesting that this immunomodulating agent induces cellular events which lead to increased glutathione content in spleen T cells. LPS- and TNF-activated macrophages have been reported to release acid-soluble thiols, especially cysteine, and the addition of the latter to lymphoid cultures has been shown to increase intracellular glutathione levels and to protect against oxidative injury (Gougerot-Pocidalo, 1985; Gmunder, 1990). We therefore sought such an effect in peritoneal macrophages stimulated by Poly(A)-Poly(U) in vitro. Our results show that Poly(A)-Poly(U) had a similar effect than LPS. Gmunder et al, 1990 demonstrated that acid-soluble thiols produced by macrophages increase the intracellular glutathione content of lymphocytes, thereby regulating their proliferative response. Our results could suggest that Poly(A)-Poly(U), by stimulating the production of these thiol compounds by macrophages in vivo, protect lymphocytes against oxidative injury via an increase in the most important intracellular antioxidant, glutathione, which is in strict equilibrium with other cellular thiols. In addition, since poly(A)-Poly(U) induce the production of TNF by macrophages (Johnson, 1985) and that TNF also induces thiol release from macrophages (Gmunder, 1990), the production of TNF under Poly(A)-Poly(U) stimulation could amplify thiol release by macrophages.

In conclusion, we show that Poly(A)-Poly(U) administered to mice protects lymphoid cells against immune oxidative injury ex vivo by increasing their glutathione content. This antioxidant effect could be related, at least in part, to the ability of Poly(A)-Poly(U) to induce the release of acid-soluble thiols from macrophages. The antioxidant effects of such a non-toxic immunomodulating agent could be of interest in protecting lymphoid cells in pathological situations involving oxidative injury.

References

Bannai, S. and Ishii, T. (1980) Formation of sulfhydryl groups in the culture medium by human diploid fibroblasts. *J. Cell. Physiol.* 104: 215-223.

Campbell, P. A. and Kind, P. (1971) Bone marrow-derived cells as target cells for polynucleotide adjuvants. *J. Immunol.* 107: 1419-1423.

Chaplin, D. D. and Wedner H. J. (1978) Inhibition of lectin-induced lymphocyte activation by diamide and other sulfhydryl reagents. *Cell. Immunol.* 36: 303-311.

Doria, G., Agarossi, G. and L. Adorini (1982) Selective effects of ionizing radiations on immunoregulatory cells. *Immunol. Rev.* 65: 23-54.

El-Hag, A. and Clark R. A. (1987) Immunosuppression by activated human neutrophils. Dependence on the myeloperoxidase system. *J. Immunol.* 139: 2406-2413.

Fidelus, R. K.and Tsan M. F. (1986) Enhancement of intracellular glutathione promotes lymphocyte activation by mitogen. *Cell. Immunol.* 97: 155-163.

Frank, L. Summerville, J. and Massaro D. (1980) Protection from oxygen toxicity with endotoxin. *J. Clin. Invest.* 65: 1104-1110.

Gmunder, H., Eck, P., Benninghoff, B., Roth, S.and Droge W. (1990) Macrophages regulate intracellular glutathione levels of lymphocytes. Evidence for an immunoregulatory role of cysteine. *Cell. Immunol.* 129: 32-46.

Gougerot-Pocidalo, M. A., Fay, M., Roche, Y., Lacombe, P.and Marquetty, C. (1985) Immune oxidative injury induced in mice exposed to normobaric O2 : effects of thiol compounds on the splenic cell sulfhydryl content and Con A proliferative response. *J. Immunol.* 135: 2045-2051.

Gougerot-Pocidalo, M. A., Fay ,M., Roche, Y and Chollet-Martin,S. (1988) Mechanisms by which oxidative injury inhibits the proliferative response of human lymphocytes to PHA. Effect of the thiol compound 2-mercaptoethanol. *Immunology.* 64: 281-288.

Grever, M. R., Thompson,V. N., Balcerzack, S. P. and Sagone,A. L. (1980) The effect of oxidant stress on human lymphocyte cytotoxicity. *Blood.* 56: 284-288.

Halliwell, B., and Cross,C. E. (1991) Reactive oxygen species, antioxydants and acquired immunodeficiency syndrome. *Arch. Intern. Med.* 151: 29-31.

Harman, D. (1981) The ageing process. *Proc. Natl. Acad. Sci.* 78: 7124-7128.

Johnson, A. G. (1985) Immunomodulating effects of synthetic polyribonucleotides. *J. Biol. Resp. Modifiers.* 4: 4481-483.

Kikkawa, Y., Yano, S.and Skoza,L. (1984) Protective effect of interferon inducers against hyperoxic pulmonary damage. *Lab. Invest.* 50: 62-71.

Kosower. E.M. (1989) *Glutathione: chemical, biochemical, and medical aspects.* Vol. III, part A. In D. Dolphin, R. Poulzon, O. Avramovic (eds.) Wiley-Interscience, New York. p. 103.

Kraus, L., Gougerot-Pocidalo, M. A., Lacombe, P.and Pocidalo, J. J. (1985) Depression of Con. A proliferative response of immune cells by in vitro hyperoxic exposure. Protective effect of thiol coumpounds. *Int. J. Immunopharmac.* 7: 753-760.

Lacombe, P., Carre, I.,Fay, M. and Pocidalo, J. J. (1988) In vitro O_2-induced depression of T and B lymphocyte activation is reversed by diethyldithiocarbamate (DDC) treatment. *Immunol. Lett.* 18: 99-108.

Lacombe, P., Kraus, L,.Fay, M,. and Pocidalo, J. J. (1985) Lymphocyte glutathione status in relation to their Con A proliferative response. *FEBS.* 191: 227-230.

Meister, A. (1983) Selective modification of glutathione metabolism. *Science.* 220: 472-477.

Messina, J. P., and Lawrence, D. A. (1989) Cell cycle progression of glutathione-depleted human peripheral blood mononuclear cells is inhibited at S phase. *J. Immunol.* 143: 1974-1981.

Noelle, R. J., and Lawrence, D. A. (1981) Determination of glutathione in lymphocytes and possible association of redox state and proliferative capacity of lymphocytes. *Biochem. J.* 198: 571-579.

Ohmori, H., and Yamamoto, I. (1983) Mechanism of augmentation of the antibody response in vitro by 2-Mercaptoethanol in murine lymphocytes. II A major role of the mixed disulfide between 2-Mercaptoethanol and cysteine. *Cell. Immunol.* 79: 173-185.

Saito, K., Tanaka, Y.,Ota, T., Eto, S. and Yamashita,S. (1991) Suppressive effect of hyperbaric oxygenation on immune responses of normal and autoimmune mice. *Clin. Exp. Immunol.* 86: 322-327.

Tietze, F. (1969) Enzymatic method for quantitative determination of nanogram amounts of total and oxidized glutathione: applications to mammalian blood and other tissues. *Anal. Biochem.* 27: 502-522.

Toohey J. I. (1981) Macrophages and methylthio groups in lymphocyte proliferation. J. Supramol. Struct. *Cell. Biochem.* 17: 11-25.

Wellner, V. P., Anderson, M. E., Puri, R., N. Jensen, G. L.and Meister. A. (1984) Radioprotection by glutathione ester: transport of glutathione ester into human lymphoid cells and fibroblasts. *Proc. Natl. Acad. Sci.* USA. 81: 4732-4735.

Oxidative Stress, Cell Activation and Viral Infection
C. Pasquier et al. (eds)
© 1994 Birkhäuser Verlag Basel/Switzerland

Reactive oxygen, antioxidants, and autotoxicity in viral diseases

E. Peterhans

Institute of Veterinary Virology, University of Berne, Länggass-Str. 122, CH-3012 Berne, Switzerland

Summary

The symptoms and pathology of viral diseases are characterized by direct effects of the viruses on their host cells, leading to death of the cells in some cases but more frequently to functional impairment. In some viral diseases, the antiviral immune response significantly contributes to disease symptoms and tissue damage by hypersensitivity reactions of types I through IV. This paper deals with an additional mechanism of viral pathogenesis, referred to as "autotoxicity". This mechanisms is characterized by activation of host effector functions in the absence of a specific antiviral immune response. Influenza in mice is associated with enhanced generation of reactive oxygen species by phagocytes and by xanthine oxidase. In addition, a wide array of cytokines are detected in alveolar lavage fluid obtained from infected lungs. Collectively, the changes observed in murine influenza, and also in other viral diseases, e.g. AIDS, indicate that infection leads to oxidative stress. Antioxidants may find a place in the therapy of viral diseases in protecting against the harmful effects of oxidants. But the beneficial role of oxidants in the killing of microorganisms as well as the complex role of oxidants in metabolic regulation suggest that antioxidant therapy should be monitored carefully. The effect of antioxidant therapy could be enhanced by drugs that interfere with other pathways and mechanisms activated by viruses. Examples of possibly useful drugs include antagonists of platelet activating factor, inhibitors of lipoxygenase and nitric oxide synthase, and drugs that modulate the release and effects of certain cytokines such as tumor necrosis factor.

Introduction

Viral diseases are highly diverse with regard to the symptoms, duration and pathological changes in infected tissue. Examples include the sniffles, a harmless catarrh caused by rhinovirus, childhood diseases associated with rashes such as measles and more severe diseases like hepatitis. Viral diseases can be life-threatening and take a dramatic course, as seen in the viral hemorrhagic fevers, while others, e.g. AIDS, take a protracted course. Infections with HIV, and similarly with the lentiviruses that cause diseases in domestic animals, take years before symptoms develop, illustrating that also the incubation time of viral diseases is highly variable. Given the startling diversity, the question can be asked if there exists a common denominator in these diseases. All viruses need living cells for their replication and impair the function or kill their host cells. The irreversible neurologic consequences of polio, caused by loss of motor neurons that serve as host cells for this virus, illustrates this disease mechanism. Disease symptoms may also be caused by the immune response to the virus. The antiviral immune response helps to overcome viral infection and in many cases reliably protects against reinfection with the same virus. But it is now clear that elimination of the virus from the body can lead to symptoms and in some cases even pathology. For example, the rashes seen in measles are due to the presence of virus-immune complexes, and cytotoxic T cells have been shown to contribute to the lesions in the lungs during influenza (for a review, see Mims and White, 1984). We have proposed that "autotoxicity" may be an additional mechanism contributing to the symptoms and pathology of viral diseases. This mechanism is characterized by the activation of host effector mechanisms in the absence of antiviral antibodies or virus-specific cellular immunity (Peterhans et al., 1988a). In the following, evidence will be reviewed indicating that oxidants play a role in this mechanism of viral pathogenesis not only as potentially harmful effector molecules but also at the level of metabolic regulation.

The early biochemical changes in virus-infected cells and mitogenically activated cells are similar

As a result of replication, many viruses induce in their host cells changes that ultimately lead to cell death. In a study of the cytopathic effect induced by Semliki Forest virus, an enveloped RNA virus of the togaviridae family, cell death is observed at 10-15 hours post-infection (Peterhans et al., 1979a). Cell death is characterized by pyknosis of the nucleus, cell rounding and detachment of the cells from the substratum. Early in infection mitochondria show increased respiration and acceptor control ratios (i.e. coupling of respiration to ATP synthesis or calcium uptake) and only later is there evidence for loss of function, characterized by decreased respiration and acceptor control ratio. Impairment of mitochondrial function is suggested also by the increase in the permeability of the inner mitochondrial membrane and by the ultrastructural appearance of the mitochondria. Control experiments showed, paradoxically, that treatment with fetal calf serum induced changes identical to the early phase of viral infection, although serum promotes cell proliferation while infection with this virus results in cell death. We suggested therefore that the early phase of cytolytic virus-cell interaction was associated with cell activation (Peterhans et al., 1979a). The interpretation of these findings was that viral replication required an activated state of the host cell, a suggestion that was supported by the observation that mitogen-treated lymphocytes are more permissive for replication of diverse RNA and DNA viruses (Wheelock and Toy, 1973; Woodruff and Woodruff, 1975). It was also supported by the general experience that resting cells are inefficient in replicating viruses (Libikova et al., 1975). Several reports demonstrate that oxidants play an important role in the early phase of mitogenic lymphocyte activation (Chaudhri et al., 1986; 1988, Hunt and Fragonas, 1992). It is important to note that oxidants, depending on their concentration, can either stimulate or inhibit cell proliferation (Burdon et al., 1989). Several years before these reports, Weidemann and coworkers had shown that oxidants are generated in suspensions of rodent thymocytes stimulated with mitogenic lectins (Wrogemann et al., 1978). To demonstrate cellular oxidant generation, these authors measured luminol-dependent chemiluminescence, a technique originally introduced for the measurement of the respiratory burst in phagocytic cells (for a review, see Allen, 1986). It was found only later that the light emission observed in mitogenically stimulated thymocyte suspensions emanated from phagocytic cells that were present as a contamination accounting for less than 0.2% of the total cell number (Weidemann et al., 1987). At the time the experiments with thymocytes were published we undertook experiments using mouse spleen cells with the aim of demonstrating that viruses are capable of inducing changes similar to mitogenic lectins. Without knowing that chemiluminescence reflected respiratory burst activity of phagocytes, we observed that some viruses activated luminol-dependent chemiluminescence in cell suspensions prepared from mouse spleen. Viruses activating this cell response included Sendai virus, a paramyxovirus (Peterhans, 1979b), influenza virus (Peterhans, 1980, Mills et al., 1981) and rabbit poxvirus (for a review, see Peterhans et al., 1987). Sendai virus treated with ultraviolet light still induced chemiluminescence while treatment with heat or pronase abrogated the cell-activating capacity. Moreover, purified viral surface glycoproteins reconstituted in liposomes triggered light emission, indicating that this cell response was initiated by an interaction between the viral surface glycoproteins with the plasma membrane receptors of the virus. (Peterhans, 1980; Peterhans et al., 1983). Subsequent experiments demonstrated that some viruses, among them herpes-, lenti- and toroviruses, failed to stimulate chemiluminescence in phagocytic cells but acquired this capacity in the presence of antiviral antibodies. Experiments using $F(ab')_2$ - fragments of antibodies (i.e. antibodies that possess both antigen-binding sites but lack the part of Fc required for interaction with the Fc receptors) failed to trigger chemiluminescence, indicating that the respiratory burst was initiated via Fc receptors. Control experiments with herpes viruses showed that the failure to activate chemiluminescence in the absence of antiviral antibodies was not due to a failure to bind to the phagocytes (unpublished observation), suggesting differences in the viral surface receptors. The difference between viruses that activate the respiratory burst and viruses that lack this capacity illustrates the two levels of cell activation. Thus, herpes viruses increase the transcription of oncogenes that are also activated in mitogenically activated cells (Boldogh et al., 1991; Albrecht et al., 1992). The concentration of oxidants involved in the early stages of mitogenic cell activation is well below that produced in the respiratory burst of phagocytes. This situation is similar to that seen with nitric oxide which can serve both as a second messenger and mediate cytotoxicity (Moncada et al., 1991). The work on the virus-induced chemiluminescence opened up also interesting applications in the analysis of cell surface antigens (Figure 1).

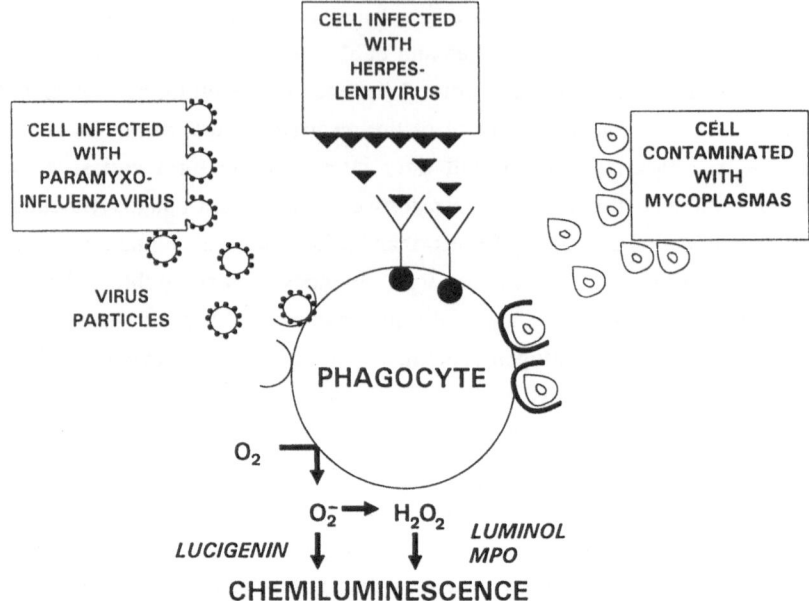

Figure 1. Induction of chemiluminescence in phagocytes. Light emission can be measured in liquid scintillation spectrometers operated in the "out-of-coincidence" mode or in specifically designed luminometers. Lucigenin reacts predominantly with O_2^-. Luminol-amplified chemiluminescence strongly depends on myeloperoxidase (MPO) or eosinophil peroxidase in the case of eosinophils. Macrophages contain little or no MPO. In these cells, luminol-amplified chemiluminescence can be amplified by addition of a peroxidase (e.g. microperoxidase).

For example, using appropriate antibodies, it is possible to do tissue typing (Descamps et al., 1980; Peterhans et al., 1981) or study Fc receptors (Peterhans et al., 1988b). In addition, since mycoplasmas stimulate the generation of oxygen species (ROS) in phagocytes, chemiluminescence measurement can be used to detect mycoplasma contamination in cultured cells (Bertoni et al., 1985).

Oxidant generation in vitro and toxicity in vivo

The generation of oxidants by phagocytic cells represents a powerful defense mechanism, acting against a wide variety of microorganisms (Belding et al., 1970; Nathan, 1982; Thomas et al., 1988; Klebanoff and Coombs, 1992) as well as against certain tumor cells (Dallegri et al., 1983; Martin and Edwards, 1993). Antibodies have a dual role in this process. By their antigenic specificity, they determine the situations in which this defense mechanism is activated. Moreover, antibodies trigger phagocytosis, thus ensuring that the particle is brought in contact with a high concentration of oxidants and other antimicrobial defense mechanisms. Since viruses that directly activate oxidant generation bypass the control and focus exerted by antiviral antibody, we hypothesized that such viruses should be toxic when injected in animals, and that this type of "autotoxicity" would contribute to the symptoms and pathology of viral diseases (Peterhans et al., 1987; 1988a). Indeed, work starting in the late 1940' had demonstrated toxic effects of paramyxo- and influenzaviruses characterized by fever and hemorrhages (Henle and Henle, 1946; Bennet et al., 1949; Mims, 1960). Similar to virus-induced chemiluminescence, toxicity was independent of viral replication but was mediated by product(s) of leukocytes. Infection with influenza virus in man, horses and swine is restricted to the airways and lungs, but is associated with very unpleasant systemic symptoms such as fever and muscle aching (Murphy and Webster, 1990). Furthermore, some of symptoms of influenza are similar to the effects (sometimes called "side-effects") seen in humans and animals treated by cytokines, in particular IL-1, TNF and interferons (Scott, 1983; Maennel et al., 1987; Remick, 1989). We concluded therefore that also cytokines contributed to autotoxicity in influenza and other viral infections (Peterhans et al., 1988a).

Oxidants, antioxidants and cytokines in murine influenza

Several human influenza viruses have been adapted to mice and cause infection of the respiratory tract, thereby serving as a model of influenza in humans (Ada and Jones, 1986). We have established in our laboratory a mouse model of infection with influenzavirus A/PR8/34 that results in massive mononuclear inflammation in the lungs and death in 5-6 days post-infection when using a virus dose of $10LD_{50}$. To assess the possible contribution of oxidants in the pathogenesis, we measured (i), generation of superoxide in inflammatory cells isolated from the lungs and superoxide production by xanthine oxidase in lung homogenates; (ii), concentrations of major antioxidants in lung homogenates. The results of these investigations can be summarized as follows:

Superoxide production by cells obtained from the lungs by alveolar lavage and stimulated with PMA reached its maximum at day 2 post-infection and thereafter declined. Taking into account the total number of cells isolated from the lungs by alveolar lavage, it can be estimated that the capacity to produce superoxide increases to over 70-fold in this compartment as a result of infection with influenza virus. As a further source of oxidants in the infected lungs, we observed an increase in xanthine oxidase activity (Christen et al., 1990). Similar observations on increased superoxide production were made in two different models of murine influenza (Akaike et al., 1990; Sharonov et al., 1991). During the course of infection, ascorbic acid, glutathione and α-

tocopherol decreased in the lungs but the ratios between the oxidized and reduced forms remained unaltered (Hennet et al., 1992a). In agreement with an earlier report (Yoshida et al., 1979), activity of indoleamine-2,3 dioxygenase increased markedly (Christen et al., 1990). Some of the metabolites of this pathway proved to be powerful antioxidants (Christen et al., 1990). As assessed by the prevention of peroxyl radical-induced loss of phycoerythrin fluorescence, the total antioxidant capacity present in lung tissue homogenate was not diminished even in terminally ill mice (Buffinton et al., 1992). Although not excluding that oxidants may cause a certain degree of tissue damage in the infected microenvironment, these results argue against the concept put forward by Maeda and Akaike (1991) that oxidants and in particular O_2^- may be the prime toxic molecule in murine influenza. The protective effect against influenza of superoxide dismutase (SOD) conjugated to pyran (Oda et al., 1989), dimerized SOD (Sharonov et al., 1991), as well as that observed in infected mice treated with the xanthine oxidase inhibitor allopurinol (Akaike et al., 1990) could be achieved in a more indirect way than prevention of toxic effects on lung tissue. For example, the release of TNF depends on oxidants and can be inhibited by antioxidants such as N-acetylcysteine and glutathione (Peristeris et al., 1992). Indeed, Sharonov and colleagues have observed that injection of SOD effectively blocks the release of this cytokine in mice infected with influenza virus (Sharonov, personal communication). Moreover, SOD could prevent the activation of proteases by protecting antiproteases against oxidative inactivation. Protease activity significantly contributes to destruction of lung tissue (Trefz et al., 1989; Jakab et al., 1990) and could also enhance the concentration of infectious virus in the lungs. The viral surface glycoprotein hemagglutinin is synthesized as a precursor and is responsible for adsorption and entry of the virus into its host cells. Only hemagglutinin cleaved into its dipeptide form is capable of mediating viral entry (Rott and Klenk, 1988). Increased protease activity on the surface of the airways and lungs could cleave the precursor into the active dipeptide, thus enhancing the concentration of infectious viruses. In a model system consisting of trypsin and protease inhibitor we have shown that the viral titer increased by four orders of magnitude after treatment of the protease inhibitor with NaOCl⁻ (Hennet et al., 1992b). That protease activity indeed plays a detrimental role is suggested also by the observation that protease inhibitors can protect mice against the lethal effect of influenza (Zhirnov et al., 1984) and that bacterial proteases can exacerbate infection (Tashiro et al., 1987).

Although influenza in humans is a local infection of the airways, it is associated with marked general symptoms such as fever and muscle aching. In mice, infection is not generalized but nevertheless leads to dramatic alterations in the physiology, including anorexia, anuria and hypothermia (Adé-Damilano, in preparation). Furthermore, decreases in glutathione and α-tocopherol were noted not only in the lungs but also in the liver, an organ not invaded by the virus (Hennet et al., 1992a). Because of earlier reports that injection of certain cytokines can provoke influenza-like symptoms in humans (Scott, 1983; Maennel et al., 1987; Remick, 1989), we assayed bronchoalveolar lavage fluid for the presence of immune mediators. Over the course of infection, we observed IL-1, IL-1, IL-6, TNF, , GM-CSF, G-CSF, M-CSF and the lipid mediators leukotriene B_4 and platelet-activating factor. (Hennet et al., 1992a). In a study based on demonstration of mRNA, additionally IL-4, IL-10 and IL-2 were observed in the lungs of mice infected with a different strain of influenza virus (Carding et al., 1993).

Antioxidants in the therapy of viral diseases

Work reviewed in this paper demonstrates that viruses are capable of stimulating ROS generation *in vitro* and that ROS play a role in the pathogenesis of viral diseases. This role is however clearly more complex than in diseases such as paraquat intoxication in which ROS are known to be key mediators of tissue damage (Bismuth et al., 1990). In murine influenza, available evidence argues against widespread oxidative damage but suggests that oxidative stress may contribute to the pathogenesis. Recently, oxidative stress has been recognized in humans infected with HIV. Dröge and coworkers noted a decrease in acid soluble thiols in plasma (Eck et al., 1989) and similar observations on this and other small molecular weight antioxidants were reported also by other groups (Buhl et al., 1989; Folkers et al., 1988). The decrease of glutathione in CD4 lymphocytes and mononuclear phagocytes seems to be a key factor promoting the replication of HIV and enhanced transcription of genes that depend on the transcription factor NFB (for a review, see Schreck et al., 1992). The initial biochemical event triggering the decrease in glutathione is unknown at present. Dröge and coworkers noted elevated glutamate in plasma of HIV-infected individuals which they proposed to decrease the uptake of cystine in macrophages (Dröge et al., 1992), leading to a decrease in the cysteine-containing tripeptide glutathione. It was suggested that drugs that replenish glutathione (e.g. N-acetylcysteine) should be useful in the therapy of infection with HIV (Fuchs et al., 1991; Dröge et al., 1992; Roederer et al., 1993).

The above observations suggest that antioxidants could find a place in the therapy of viral infections. It is necessary however to also consider the "useful" roles of oxidants, including inactivation of viruses (Belding et al., 1970; Klebanoff and Coombs, 1992) and killing of other microorganisms (Nathan, 1982; Thomas et al., 1988) . Perhaps equally important, oxidants of a certain concentration range are essential for the proliferation of lymphocytes and other cell types (Chaudhri et al., 1986; 1988, Hunt and Fragonas, 1992; Burdon et al., 1989). The inhibition of the antibody response to Newcastle disease virus in chickens receiving butylated hydroxytoluene in the feed (Brugh, 1977) shows that antioxidants can have undesirable effects if used in an uncontrolled fashion and indicates that the effect of antioxidant therapy must be monitored. To monitor for possible immunosuppressive effects, the response of lymphocytes to mitogenic stimulation and other functional parameters should be assessed and the effect of exogenous antioxidants on the concentrations and red/ox rations of major endogenous antioxidants, such as glutathione, should be monitored (De Quay et al., 1991; Baruchel and Weinberg, 1992; Ruffmann and Wendel, 1991).

It seems important to also point out that viruses stimulate a wide array of pathways not directly related to oxidant generation (Figure 2).

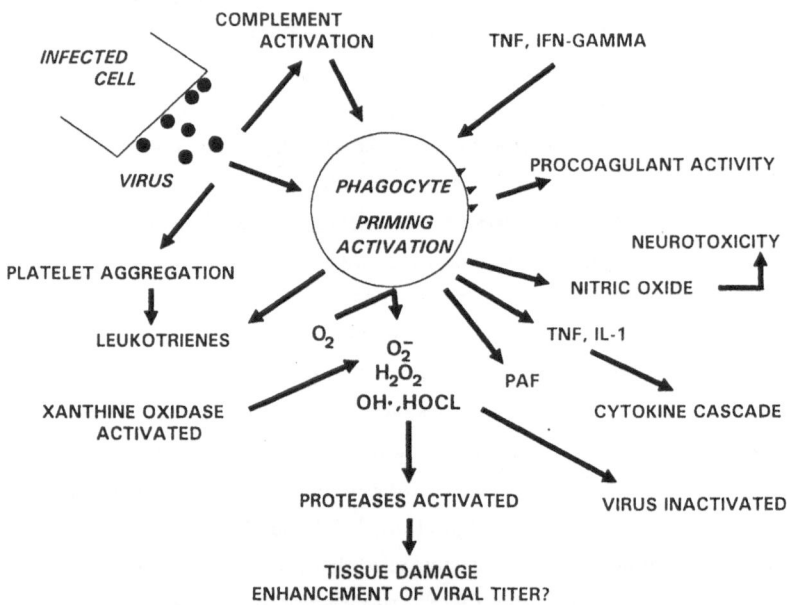

Figure 2. Autotoxic mechanisms involved in viral diseases. For details refer to text.

Certain viruses can activate the complement cascade in the absence of antiviral antibodies (Hirsch et al., 1980; Okada and Okada, 1981; Spear et al., 1990). Others induce in mononuclear phagocytes the expression of procoagulant activity (Dindzans et al., 1985; Brügger et al., 1992) or promote platelet aggregation and release of arachidonate metabolites (Scott et al., 1978, Antal et al., 1986). Priming of phagocytic cells for enhanced cytokine production (Nain et al., 1990; Gong et al., 1991) is a further potentially important mechanism that could explain why concurrent infections with endotoxin-containing bacteria could exacerbate viral infections. Very recently, stimulation of nitric oxide generation has been recognized as a pathogenic mechanism in viral encephalitis (Koprowski et al., 1993; Hayman et al., 1993). It was shown that gp120, the surface glycoprotein of HIV, mediates neurotoxicity in the presence of glutamate via a mechanism involving nitric oxide and superoxide (Dawson et al., 1993). This report is particularly interesting because glutamate is a neurotransmitter in the central nervous system (Watkins and Olverman, 1987). Excessive activation of glutaminergic neurons by HIV could not only contribute to neuronal death but also explain some of the aspects of AIDS-related neuropsychological dysfunction (Marotta and Perry, 1989). The partial inhibition of neurotoxicity by SOD also illustrates a principle that should be taken into account in the development of a therapy of viral diseases. Most of the multiple pathways that mediate the autotoxic reactions are interrelated. For example, lipoxygenase activity is involved in TNF formation (Schade et al., 1989) and TNF induces the release of leukotriene B_4 from neutrophils (Camussi et al., 1989). Targeting several pathways is therefore likely to have synergistic effects. Clearly, a therapy aiming at the alleviation of the symptoms of viral diseases does not replace but rather complements the therapy with drugs that directly interfere with the replication of viruses.

Acknowledgments

Work in the author's laboratory was supported by the Swiss National Science Fund. I thank Drs. Bertoni, B. Frech, T.W. Jungi and C. Richter for stimulating discussions.

References

Ada, G. L., and Jones, P. D. (1986) The immune responses to influenza infection. *Curr. Topics Microbiol. Immunol.* 128: 1-54.

Akaike, T., Ando, M., Odi, T., Doi, T., Ijiri, S., Araki, S., and Maeda, H. (1990) Dependence on O_2^- generation by xanthine oxidase of pathogenesis of influenza virus infection in mice. *J. Clin. Invest.* 85: 739-745.

Albrecht, T., Boldogh, I., and Fons, M. P. (1992) Receptor-initiated activation of cells and their oncogenes by herpes-family viruses. *J. Invest. Dermatol.* 98 (6 Suppl): 29S-35S.

Allen, R. C. (1986) Phagocytic leukocyte oxygenation activities and chemiluminescence: a kinetic approach to analysis. *Methods Enzymol.* 133: 449-493.

Antal, A., Gecse, A., Mojzes, L., and Foeldes, V. (1986) Die Wirkung der Influenza-Virus-Infektion auf die Arachidonsäure- Kaskade von Mäusethrombozyten. *Z. Rechtsmed.* 96: 303-308.

Baruchel, S., and Wainberg, M. A. (1992) The role of oxidative stress in disease progression in individuals infected by the human immunodeficiency Virus. *J. Leukocyte Biol.* 52: 111-114.

Belding, M. E., Klebanoff, S. J., and Ray, C. G. (1970) Peroxidase-mediated virucidal systems. *Science* 167: 195-196.

Bennet, I. L., Wagner, R. R., and Lequire, V. S. (1949) Pyrogenicity of influenza virus in rabbits. *Proc. Soc. Exp. Biol. Med.* 71: 132-133.

Bertoni, G., Keist, R., Groscurth, P., Wyler, R., Nicolet, J., and Peterhans, E. (1985) A chemiluminescent assay for mycoplasmas in cell cultures. *J. Immunol. Methods.* 78: 123-133.

Bismuth, C., Garnier, R., Baud, F.J., Muszynski, J., Keyes, C. (1990) Paraquat poisoning. An overview of the current status. *Drug saf.* 5: 243-251.

Boldogh, I., Abubakar, S., Millinoff, D., Deng, C. Z., and Albrecht, T. (1991) Cellular oncogene activation by human cytomegalovirus. Lack of correlation with virus infectivity and immediate early gene expression. *Arch. Virol.* 118: 163-177.

Brugh Jr, M. (1977) Butylated hydroxytoluene protects chickens exposed to newcastle disease virus. *Science* 197: 1291-1292.

Brügger, M., Jungi, T. W., Zurbriggen, A., and Vandevelde, M. (1992) Canine distemper virus increases procoagulant activity of macrophages. *Virology* 190: 616-623.

Buffinton, G. D., Christen, S., Peterhans, E., and Stocker, R. (1992) Oxidative stress in lungs of mice infected with influenza-A virus. *Free Radical Res. Commun.* 16: 99-110.

Buhl, R., Holroyd, K. J., Mastrangeli, A., Cantin, A. M., Jaffe, H. A., Wells, F., Saltini, C., and Crystal, R. G. (1989) Systematic glutathione deficiency in symptom-free seropositive individuals. *Lancet*, Dezember: 1294-1297.

Burdon, R. H., Gill, V., and Rice Evans, C. (1989) Cell proliferation and oxidative stress. *Free Radical Res. Commun.* 7: 149-159.

Camussi, G., Tetta, C., Bussolino, F., and Baglioni, C. (1989). Tumor necrosis factor stimulates human neutrophils to release leukotriene-B_4 and platelet-activating factor. Induction of phospholipase-A_2 and acetyl-CoA:1-alkyl-*sn*-glycero-3-phosphocholine O_2-acetyltransferase activity and inhibition by antiproteinase. *Eur. J. Biochem.* 182, 661-666.

Carding, S.R., Allan, W., Mcmickle, A., Doherty, P.C. (1993) Activation of cytokine genes during primary and secondary influenza pneumonia. *J. Exp. Med.* 177: 475-482.

Chaudhri, G., Clark, I. A., Hunt, N. H., Cowden, W. B., and Ceredig, R. (1986) Effect of antioxidants on primary alloantigen-induced T cell activation and proliferation. *J. Immunol.* 137: 2646-2652.

Chaudhri, G., Hunt, N. H., Clark, I. A., and Ceredig, R. (1988) Antioxidants inhibit proliferation and cell surface expression of receptors for interleukin-2 and transferrin in T lymphocytes stimulated with phorbol myristate acetate and ionomycin. *Cell. Immunol.* 115: 204-213.

Christen, S., Peterhans, E., and Stocker, R. (1990) Antioxidant activities of some tryptophan metabolites: possible implication for inflammatory diseases. *Proc. Natl. Acad. Sci. USA* 87: 2506-2510.

Dallegri, F., Frumento, G., and Patrone, F. (1983) Mechanisms of tumour cell destruction by PMA-activated human neutrophils. *Immunology* 48: 273-279.

Dawson, V. L., Dawson, T. M., Uhl, G. R., and Snyder, S. H. (1993) Human immunodeficiency virus type-1 coat protein neurotoxicity mediated by nitric oxide in primary cortical cultures. *Proc. Natl. Acad. Sci. USA* 90, 3256-3259.

de Quay, B., Malinverni, R., and Lauterburg, B. H. (1992) Glutathione depletion in HIV-infected patients: role of cysteine deficiency and effect of oral N-acetylcysteine. *Aids* 6: 815-819.

Descamps, B., Nguyen, A. T., and Feuillet-Fieux, M. N. (1980) Flash detection of anti-H-2 antibodies using chemiluminescence without complement. *Annal. Immunol. (Inst. Pasteur)* 131D: 337-346.

Dindzans, V. J., MacPhee, P. J., Sum Fung, L., Leibowitz, J. L. and Levy, G.A. (1985) The immune response to mouse hepatitis virus: expression of monocyte procoagulant activity and plasminogen activator during infection *in vivo. J. Immunol.* 135: 4189-4197.

Dröge, W., Eck, H. P., and Mihm, S. (1992) HIV-induced cysteine deficiency and T-cell dysfunction - A rationale for treatment with N-acetylcysteine. *Immunol. Today* 13: 211- 214.

Eck, H. P., Gmünder, H., Hartmann, M., Petzold, D., Daniel, V., and Dröge, W. (1989) Low Concentrations of acid-Soluble Thiol (Cysteine) in the blood plasma of HIV-1-Infected Patients. *Biol. Chem. Hoppe Seyler* 370: 101-108.

Folkers, K., Langsjoen, P., Nara, Y., Muratsu, K., Komorowski, J., Richardson, P. C., and Smith, T. H. (1988) Biochemical deficiencies of coenzyme Q10 in HIV-infection and exploratory treatment. *Biochem. Biophys. Res. Commun.* 153: 888-896.

Fuchs, J., Ochsendorf, F., Schofer, H., Milbradt, R., and Rübsamen-Waigmann, H. (1991) Oxidative imbalance in HIV infected patients. *Med. Hypoth.* 36: 60-64.

Gong, J. H., Sprenger, H., Hinder, F., Bender, A., Schmidt, A., Horch, S., Nain, M., and Gemsa, D. (1991) Influenza-A virus infection of macrophages - enhanced tumor necrosis factor-(TNF-) gene expression and lipopolysaccharide-triggered TNF- release. *J. Immunol.* 147: 3507-3513.

Hayman, M., Arbuthnott, G., Harkiss, G., Brace, H., Filippi, P., Philippon, V., Thomson, D., Vigne, R., and Wright, A. (1993) Neurotoxicity of peptide analogues of the transactivating protein tat from Maedi-Visna virus and human immunodeficiency virus. *Neuroscience 53:* 1-6.

Henle, W., and Henle, G. (1946) Studies on the toxicity of influenza viruses. II. The effect of intra-abdominal and intravenous injection of influenza viruses. *J. Exp. Med.* 84: 639-660.

Hennet, T., Peterhans, E., and Stocker, R. (1992a) Alterations in antioxidant defences in lung and liver of mice infected with influenza A virus.*J. Gen. Virol.* 73: 39-46.

Hennet, T., Ziltener, H. J., Frei, K., and Peterhans, E. (1992b) A kinetic study of immune mediators in the lungs of mice infected with influenza-A virus. *J. Immunol.* 149: 932-939.

Hirsch, R. L., Winkelstein, J. A., and Griffin, D. E. (1980) The role of complement in viral infections. III. Activation of the classical and alternative complement pathways by Sindbis virus. *J. Immunol.* 124: 2507-2510.

Hunt, N. H., and Fragonas, J. C. (1992) Effects of anti-oxidants on ornithine decarboxylase in mitogenically-activated T lymphocytes. *Biochim. Biophys. Acta* 1133: 261-267.

Jakab, G. J. (1990) Sequential virus infections, bacterial superinfections, and fibrogenesis. *Am. Rev. Resp. Dis.* 142: 374-379.

Klebanoff, S. J., and Coombs, R. W. (1992) Virucidal effect of polymorphonuclear leukocytes on human immunodeficiency virus-1. Role of the myeloperoxidase system. *J. Clin. Invest.* 89: 2014-2017.

Koprowski, H., Zheng, Y. M., Heber Katz, E., Fraser, N., Rorke, L., Fu, Z. F., Hanlon, C., and Dietzschold, B. (1993) *In vivo* expression of inducible nitric oxide synthase in experimentally induced neurologic diseases. *Proc. Natl. Acad. Sci. U S A* 90: 3024-3027.

Libikova, H. (1975) Viral infection and interferon in cell cultures aged *in vitro. Adv. Exp. Med. Biol* 53: 469-479.

Maeda, H., and Akaike, T. (1991) Oxygen free radicals as pathogenic molecules in viral diseases. *Proc. Soc. Exp. Biol. Med.* 198: 721-727.

Maennel, D. N., Northoff, H., Bauss, F., and Falk, W. (1987) Tumor necrosis factor: A cytokine involved in toxic effects of endotoxin. *Rev. Inf. Dis.* 9,supplement 5: 602- 606.

Marotta, R. and Perry, S. (1989) Early neuropsychological dysfunction caused by human immunodeficiency virus. *J. Neuropsychiatry Clin. Neurosci.* 1: 255-235-

Martin, J.H.J., and Edwards, S.W. (1993) Changes in mechanisms of monocyte/macrophage-mediated cytotoxicity during culture - reactive oxygen intermediates are involved in monocyte-mediated cytotoxicity, whereas reactive nitrogen intermediates are employed by macrophages in tumor cell killing. *J. Immunol.* 150: 3478-3486.

Mills, E. L., Debets-Ossenkopp, Y., Verbrugh, H. A., and Verhoef, J. (1981) Initiation of the respiratory burst of human neutrophils by influenza virus. *Infect. Immunol.* 32: 1200-1205.

Mims, A. A., and White, D. O. (1984) "Viral Pathogenesis and Immunology". Blackwell Scientific Publications, Oxford,

Mims, C. A. (1960) An analysis of the toxicity for mice of influenza virus. I. Intracerebral toxicity. *Br. J. Exp. Pathol.* 41: 586-592.

Moncada, S., Palmer, R.M.J., and Higgs, E.A. (1991) Nitric oxide - physiology, pathophysiology, and pharmacology. *Pharmacol. Rev.* 43, 109-142.

Murphy, B. R., and Webster, R. G. (1990) Chapter 40. Orthomyxoviruses. in: "Fields Virology" Second Edition. B.N. Fields, D.M. Knipe, R.M. Chanock, M.S. Hirsch, J.L. Melnick, T.P. Monath, B. Roizman, Vol. 1, pp.1091-1153. Raven Press, New York.

Nain, M., Hinder, F., Gong, J. H., Schmidt, A., Bender, A., Sprenger, H., and Gemsa, D. (1990) Tumor necrosis factor- production of influenza-A virus-infected macrophages and potentiating effect of lipopolysaccharides. *J. Immunol.* 145: 1921-1928.

Nathan, C. F. (1982) Secretion of oxygen intermediates: role in effector functions of activated macrophages. *Fed. Proc.* 41: 2206-2211.

Oda, T., Akaike, T., Hamamoto, T., Suzuki, F., Hirano, T., and Maeda, H. (1989) Oxygen radicals in influenza-induced pathogenesis and treatment with pyran polymer-conjugated SOD. *Science* 244: 974-976.

Okada, H., and Okada, N. (1981) Sendai virus infected cells are readily cytolysed by guinea-pig complement without antibody. *Immunology* 43: 337-343.

Peristeris, P., Clark, B. D., Gatti, S., Faggioni, R., Mantovani, A., Mengozzi, M., Orencole, S. F., Sironi, M., and Ghezzi, P. (1992) N-acetylcysteine and glutathione as inhibitors of tumor necrosis factor production. *Cell. Immunol.* 140: 390-399.

Peterhans, E., Hänggeli, E., Wild, P., and Wyler, R. (1979a) Mitochondrial calcium uptake during infection of chicken embryo fibroblasts with Semliki Forest virus. *J. Virol.* 29: 143-152.

Peterhans, E. (1979b) Sendai virus stimulates chemiluminescence in mouse spleen cells. *Biochem. Biophys. Res. Commun.* 91: 383-392.

Peterhans, E. (1980) Chemiluminescence: an early event in the interaction of Sendai and influenza viruses with mouse spleen cells. I. The role of the envelope glycoproteins in the stimulation of chemiluminescence. *Virology* 105: 445-455.

Peterhans, E., Albrecht, H., and Wyler, R. (1981) Detection of H-2 and Sendai virus antigens by chemiluminescence. *J. Immunol. Methods.* 47: 295-302.

Peterhans, E., Bächi, T., and Yewdell, J. (1983) Evidence for different receptor sites in mouse spleen cells for the Sendai virus hemagglutinin-neuraminidase (HN) and fusion (F) glycoproteins. *Virology* 128: 366-376.

Peterhans, E., Grob, M., Bürge, T., and Zanoni, R. (1987) Virus- induced formation of reactive oxygen intermediates in phagocytic cells. *Free Radical. Res. Commun.* 3: 39-46.

Peterhans, E., Jungi, T. W., and Stocker, R. (1988a) Autotoxicity and reactive oxygen in viral disease. in: "Oxy-Radicals in Molecular Biology and Pathology", UCLA Symposia on Molecular and Cellular Biology, New Series. (P.A. Cerutti, I. Fridovich, J.M. McCord, eds.) Vol. 82, pp. 543-562.

Peterhans, E., Jungi, T.W., Bürge, Th., Grob, M., Jörg, A. (1988b), Phagocyte Chemiluminescence pp 309-334 in: Free Radicals. Methodology and Concepts. eds C. Rice-Evans and B. Halliwell, Richelieu Press, London, .

Remick, D. G., and Kunkel, S.L. (1989) Toxic effects of cytokines in vivo. *Lab. Invest.* 60: 317-319.

Roederer, M., Staal, F. J. T., Ela, S. W., Herzenberg, L. A., and Herzenberg, L. A. (1993) N-acetylcysteine - potential for AIDS therapy. *Pharmacology* 46: 121-129.

Rott, R., and Klenk, H. D. (1988) The molecular biology of influenza virus pathogenicity. *Adv. Virus Res.* 34: 247-281.

Ruffmann, R., and Wendel, A. (1991) GSH Rescue by N-acetylcysteine. *Klinische Wochenschrift* 69: 857-862.

Schade, U. F., Burmeister, I., Engel, R., Reinke, M., and Wolter, D. T. (1989) lipoxygenase inhibitors suppress formation of tumor necrosis factor *In vitro* and *In vivo. Lymphokine Res.* 8: 245-250.

Schreck, R., Albermann, K., and Baeuerle, P. A. (1992) Nuclear factor kappaB - An oxidative stress-responsive transcription factor of eukaryotic cells. *Free Radical Res. Commun.* 17: 221-237.

Scott, G. M. (1983) The toxic effects of interferon in man. *Interferon* 5: 85-114.

Scott, S., Reimers, H. J., Chernesky, M. A., Greenberg, J. P., Kinolugh-Rathbone, R. L., Packham, M. A., and Mustard, J. F. (1978) Effect of viruses on platelet aggregation and platelet survival in rabbits. *Blood* 52: 47-55.

Sharonov, B. P., Dolganova, A. V., and Kiselev, O. I. (1991) Effective application of superoxide dismutase on later stages influenza. *Voprosi Virusologii (Russia)* 36: 477-481.

Spear, G. T., Landay, A. L., Sullivan, B. L., Dittel, B., and Lint, T. F. (1990) Activation of complement on the surface of cells infected by human immunodeficiency virus. *J. Immunol.* 144: 1490-1496.

Tashiro, M, Ciborowski, P., Reinacher, M., Pulverer, G., Klenk, H.D., and Rott, R. (1987) Synergistic role of staphylococcal proteases in the inductions of influenza pathogenicity. *Virology* 157: 412-430.

Thomas, E. L., Lehrer, R. I., and Rest, R. F. (1988) Human neutrophil antimicrobial activity. *Rev. Infect. Dis.* 10 Suppl 2: S450-S456.

Trefz, G., Heck, B., Schulz, V., and Ebert, W. (1989) Functional activity of the 1-proteinase inhibitor in serum and bronchoalveolar lavage fluid in congenital lung emphysema. *Pneumologie* 43: 446-451.

Watkins, J.C., and Olvermann, H.J. (1987) Agonists and antagonists for excitatory amino acids. *Trends in Neurosciences* 10: 265-272

Weidemann, M. J., Yin Fo, D., Kolbuch-Braddon, M. E., Mitchell, S. V., Domzig, W., and Peterhans, E. (1987) Characterization of chemiluminescence-generating subpopulations within the rodent thymus. in: "Bioluminescence and Chemiluminescence. New Perspectives", (J. Schölmerich, R. Andreesen, A. Kapp, M. Ernst, and W.G. Woods, eds), pp. 109-112. John Wiley and Sons, Chichester.

Wheelock, E. F., and Toy, S. T. (1973) Participation of lymphocytes in viral infections. *Adv. Immunol.* 16: 123-184.

Wrogemann, K., Weidemann, M.J., Peskar, B.A., Staudinger, H., Rietschel, E.T., Fischer, H. (1978) Chemiluminescence and immune cell activation. I. Early activation of rat thymocytes can be monitored by chemiluminescence measurements. *Eur. J. Immunol.* 8: 749-752.

Woodruff, J. F., and Woodruff, J. J. (1975) The effect of viral infections on the function of the immune system. In: "Viral Immunology and Immunopathology" (A. L. Notkins, ed.), pp. 393-418. Academic Press, New York.

Yoshida, R., Urade, Y., Tokuda, M., and Hayaishi, O. (1979) Induction of indoleamine 2,3-dioxygenase in mouse lung during virus infection. *Proc. Natl. Acad. Sci. USA* 76: 4084-4086.

Zhirnov, O. P., Ovcharenko, A. V., and Bukrinskaya, A. G. (1984) Suppression of influenza virus replication in infected mice by protease inhibitors. *J. Gen. Virol.* 65: 191-196.

Oxidative Stress, Cell Activation and Viral Infection
C. Pasquier et al. (eds)
© 1994 Birkhäuser Verlag Basel/Switzerland

Redox Control of Gene Expression by Eukaryotic Transcription Factors NF-κB, AP-1 and SRF/TCF

M. Meyer*, R. Schreck+, J. M. Müller and P.A. Baeuerle

Laboratory for Molecular Biology, Gene Center, Am Klopferspitz 18a, W-8033 Martinsried, Germany
*Current address: European Molecular Biology Laboratory, Meyerhofstr. 1, W-6900 Heidelberg 1, Germany
+Current address: Howard Hughes Medical Institute, Fred Hutchinson Center, 1124 Columbia Street, Seattle, Washington 98104, USA

Summary

In higher eukaryotic cells, three unrelated transcription factor systems have been identified as regulators of gene transcription in response to alterations of the intracellular level of reactive oxygen intermediates. One is NF-κB, an inducible protein complex that activates gene transcription specifically in response to peroxides. A second one is AP-1, a factor newly synthesized upon stimulation of cells by both H_2O_2 as well as antioxidants. A third one is SRF/TCF, a nuclear transcription factor responsible for transcriptional activation of the gene encoding the AP-1 subunit c-fos in response to both H_2O_2 and antioxidants. In this article, we will review the current knowledge about redox regulation of gene expression by NF-κB, AP-1 and SRF/TCF and discuss the different molecular mechanisms by which these factors induce gene expression.

Introduction

Cells in eukaryotic organisms can be subject to dramatic changes in their exposure to dioxygen and reactive oxygen intermediates (ROIs) (reviewed by Halliwell and Gutteridge, 1989). Because high levels of oxygen and its partially reduced species are cytotoxic, cells have evolved mechanisms to sense increased levels of ROIs and to efficiently protect themselves from oxidative damage to DNA, proteins and lipids. Because hypoxia or anoxia are also harmful to aerobic organisms, cells have likewise evolved mechanisms to sense below normal concentrations of oxygen and reactive intermediates and to adapt to conditions caused by hypoxia. Adaptations to changes in the concentration of ROIs are achieved by a sophisticated apparatus composed of (1) various enzymes interconverting and eliminating ROI species, (2) enzymes repairing damaged DNA and proteins and (3) a variety of antioxidative compounds, such as glutathione (GSH), amino acids and vitamines, which can directly neutralize radicals.

An important regulatory mechanism allowing cells to respond to extracellular stimuli is the transcriptional activation of genes via inducible transcriptional activator proteins. In the past decade, many such proteins have been identified and characterized in detail, including factors responding to steroid hormones, heat shock, heavy metals, hormones, cytokines, growth factors and virus infection. The transcription factors can be grouped into pre-existing or "primary" factors and those requiring *de novo* synthesis, here called "secondary" transcription factors. The primary factors are usually present in a latent form in the nucleus or cytoplasm and require a posttranslational modification or interaction with a ligand in order to bind de novo to regulatory DNA sequences, or, should they already be bound to DNA, to acquire transcription activating potential. The secondary factors depend on primary factors for transcriptional activation of their genes. Because in prokaryotes and eukaryotes the adaptation to an altererd redox state involves transcriptional control of gene expression, redox-controlled primary and secondary transcription factors must exist.

In bacteria, the molecular mechanisms controlling ROI homeostasis at the level of transcription are fairly well understood. oxyR, a primary transcription factor, is specifically activated by H_2O_2 and controls the expression of many genes encoding antioxidative enzymes in response to oxidative stress (reviewed in Storz et al., 1990). oxyR is already bound to promoters and is activated by a covalent modification, presumably a direct amino acid oxidation. A second system, comprising the two transcription factors soxR and soxS, is specifically activated by superoxide-producing agents (reviewed in Demple, 1991). soxR is a primary factor which activates transcription of soxS, a secondary factor. sox2, like oxyR, induces the transcription of enzymes protecting bacterial cells from oxidative damage.

It is important to realize that cells continuously produce ROIs from dioxygen during electron transfer reactions occuring at the plasma membrane, the cytosol, mitochondria, the endoplasmic reticulum and in peroxisomes. For this reason, a certain intracellular level of ROIs must be recognized as normal by a cell. We refer to this level as "normoxic". The normoxic level, which might vary with the cell type, explains why cells can exhibit a response to both antioxidants as well as to oxidants. In cell cultures, a transcriptional response can be triggered by addition of antioxidative chemicals, such as quinones, flavones and agents containing reduced sulfur. In the organism, an antioxidant response might also be induced by antioxidants but more relevant are conditions which locally decrease the oxygen supply, such as injury, embolism and ischemia. Reperfusion of hypoxic tissue is known to lead to severe oxidative damage (reviewed in Korthuis and Granger, 1986). This indicates the existence of an adaptive process which downregulates mechanisms protecting cells from ROIs under the normoxic condition. It seems advantagous if during hypoxia cells would induce genes whose products quench the oxidative stress occuring upon a subsequent reperfusion. In fact, antioxidants induce expression of the antioxidative enzymes glutathione S-transferase (Y_a subunit) (Rushmore et al., 1991) and NAD(P)H-quinone reductase in cell cultures (Li and Jaiswal, 1992). Cells also induce genes encoding antioxidative enzymes upon oxidative stress. Examples are the genes of the mitochondrial Mn-dependent superoxide dismutase (MnSOD) (Wong and Goeddel, 1988) and thioredoxin/ADF (reviewed in Yodoi and Uchiyama, 1992) which are newly expressed if cells are exposed to conditions known to elicit moderate oxidative stress, such as the inflammatory cytokines tumor necrosis factor-a (TNF-α) and interleukin-1 (IL-1).

Compared to prokaryotic systems, the knowledge about eukaryotic transcription factors involved in redox-dependent gene expression is still scarce. However, within the past few years, eukaryotic factors have been identified which control gene expression in response to H_2O_2 and antioxidants. These factors, NF-κB, AP-1 and SRF/TCF, did not need to be newly discovered. On the contrary, they were among the best characterized inducible transcription factors of higher eukaryotes. All three factors have in common that they are activated by a large variety of stimuli. In view of their broad, almost non-specific panel of inducers, the responsiveness to redox changes received little attention so far. In this review, we will describe the properties of each transcription factor system and compare the mechanisms by which the three factors respond to changes of the cellular redox status.

Redox-Dependent Gene Control by NF-κB

Several recent reviews have covered the biochemistry and physiology of the primary transcription factor NF-κB (Grimm and Baeuerle, 1993; Baeuerle and Baltimore, 1988; Baeuerle, 1991; Blank et al., 1992). We will therefore only briefly address some important biochemical aspects of NF-κB. A remarkable feature of NF-κB is an extreme variety of inducing conditions (Fig. 1).

This prompts the question whether these diverse stimuli use many, a few or only one intracellular mechanism to ultimately activate NF-κB. Some activators of NF-κB use protein kinase C (PKC), while other activators including TNF-α, IL-1 and LPS seem to induce the factor independently of PKC. Classical second messenger systems such as cAMP, cGMP and calcium so far failed to detectably activate NF-κB on their own. Thus, rather non-classical messenger systems are involved in the activation of NF-κB.Another hallmark of NF-κB is its cytoplasmic occurence in nonstimulated cells. The cytoplasmic form of NF-κB is unable to bind to DNA and enter the nucleus. Upon stimulation of cells, NF-κB aquires DNA-binding activity and migrates into the nucleus. There, it binds to regulatory DNA elements containing consensus binding sites, an event initiating gene transcription in synergy with other transcription factors. The molecular basis for the subcellular redistribution of NF-κB and gain of DNA-binding in response to extracellular stimuli is the release of an inhibitory subunit, called IκB, from the cytoplasmic form of NF-κB. It seems that all inducers of NF-κB can trigger this posttranslational event. Very recent results suggest that IκB is rapidly degraded following

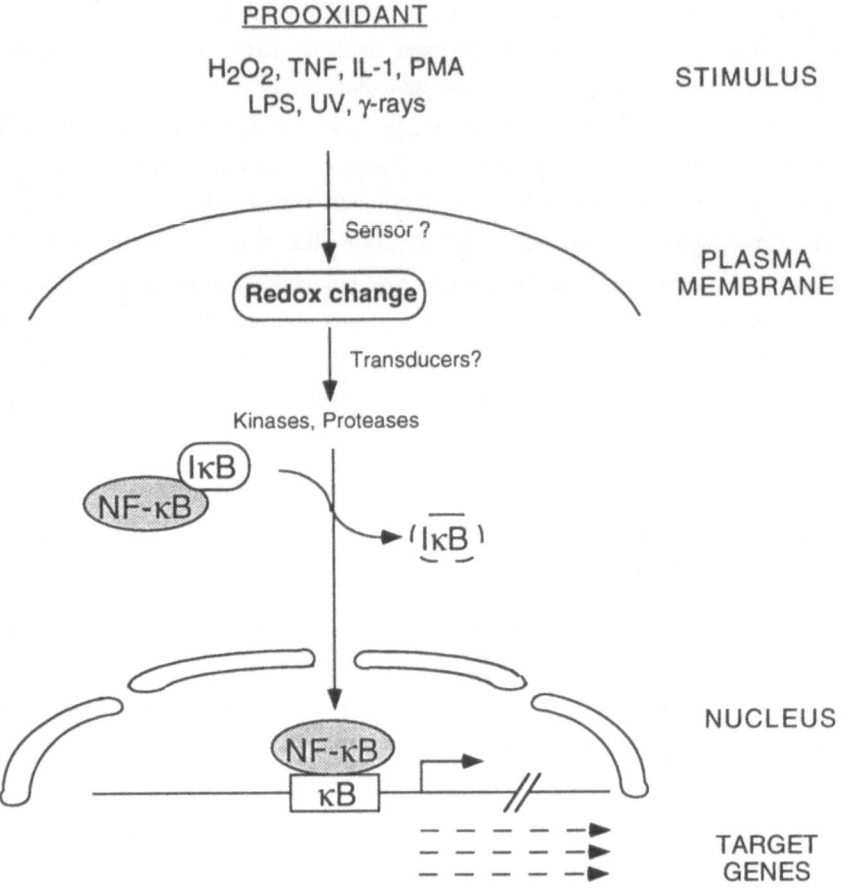

Fig. 1. A model illustrating the activation of NF-kB by prooxidant stimuli.

Tab. I. Conditions activating transcription factor NF-κB. It should be pointed out that most activating conditions are cell type-specific. Only if a cell has a specific receptor for an agent can it activate NF-κB in reponse References are listed in Schreck et al. (1992b) with the exception of leukemia inhibitory factor (Gruss et al., 1992) and Epstein-Barr-Virus transactivating proteins EBNA2 and LMP (Scalal et al., 1993; Hammarskjöld and Sinmurda, 1992).

Class	Agent
Bacterial Products	Lipopolysaccharide
	Exotoxins B
	Toxic shock syndrome toxin 1
	Muramyl peptides
Viruses	Human immunodeficiency virus type 1
	Human T cell leukemia virus type 1 (HTLV-l)
	Hepatitis B virus (HBV)
	Herpes simplex virus type 1
	Human herpes virus-6
	Newcastle disease virus
	Sendai virus
	Epstein-Barr virus (EBV)
Viral products	Double-stranded RNA
	Tax (from HTLV-l)
	HBx (from HBV)
	MHBst (from HBV)
	EBNA2 (EBV)
	Latent membrane protein (EBV)
Parasite	Theileria parva
Inflammatory cytokines	Tumor necrosis factor-α
	Lymphotoxin
	Interleukin- 1
	Interleukin-2
	Leukotriene B4
	Leukemia inhibitory factor
T Cell mitogens	Antigen
	Lectins (+ phorbol ester)
	Calcium ionophores (+ phorbol ester)
	Anti-CD3
	Anti-CD2
	Anti-CD28
B cell mitogen	Anti-surface IgM
Protein synthesis inhibitors	Cycloheximide
	Anisomycin
Physical stress	UV-A, -B and -C light
	γ rays
Drugs	Okadaic acid
	Phorbolesters
	4-Nitroquinolineoxide
Oxidants	Hydrogen peroxide
	Butyl peroxide

stimulation of cells (Brown et al., 1993; Sun et al., 1993; T. Henkel and P. Baeuerle, submitted). It is, however, still unclear whether proteolysis actually releases IκB from NF-κB, or if IκB is dissociated from NF-κB by another modification and subsequently degraded. The DNA-binding nuclear form of NF-κB is a heterodimer composed of two structurally related subunits, called p50 and Rel-A (formerly p65). Rel-A is the subunit interacting directly with IκB and is responsible for transactivating genes. p50 might be solely required in the dimeric complex with Rel-A for high-affinity DNA-binding. In vertebrates, three more proteins with homology to p50 and Rel-A have been found called c-rel, p52 and Rel-B. In vitro, extensive heterotypic dimerization is observed among the five members of the NF-κB transcription factor family but it is unclear to what extent it can occur under physiological conditions. The cDNAs for all NF-κB subunits have been cloned (reviewed in Blank et al., 1992).

Most of the inducers of NF-κB are either exogenous pathogens or inflammatory cytokines produced by the organism in response to pathogens (Tab. I). The activated NF-κB primarily initiates transcription of genes whose products serve to counteract the effects of pathogen exposure and to coordinately activate the immune system. Major groups of target genes are encoding cytokines, cell surface receptors, cell adhesion molecules, acute phase proteins and hematopoetic growth factors (listed in Baeuerle, 1991). Also, a few viruses such as HIV-1 and CMV have regulatory NF-κB binding sites in their control regions.

NF-κB functions in cells as a sensor for pathogenic conditions, signal transducer between cytoplasm and nucleus and as an immediate-early activator of gene transcription. With these properties NF-κB might play an important role in T cell, macrophage and B cell activation upon inflammation and infection. In endothelial and immune cell activation may occur during graft versus host reactions, in liver cells during the acute phase response and in keratinocytes upon UV-A irridation of skin.

The first indirect indication that NF-κB is involved in redox control of gene expression came from the observation that mM-amounts of N-acetyl-L-cysteine (NAC), a precursor for glutathione, suppresses activation of the factor (Staal et al., 1990). Another hint came from the observation that many conditions activating NF-κB are known to induce oxidative stress including TNF-α, IL-1, PMA, UV-A and γ rays (reviewed in Schreck and Baeuerle, 1991; 1992a). These observations prompted us to test directly whether NF-κB is activated by H_2O_2, a relatively stable ROI that can be added to cultured cells. Micromolar amounts of H_2O_2 have been found to activate NF-κB in the human T lymphoma line Jurkat and the human carcinoma cell line HeLa (Schreck et al., 1991; Meyer et al., 1993). The activation by H_2O_2 was relatively fast and independent of protein synthesis, suggesting that H_2O_2 caused the release of IκB from NF-κB in the cytoplasm.

We have tested a variety of other conditions known to produce or to deplete ROI species in intact cells (Tab. II) (reviewed in Schreck et al., 1992a). NF-κB could not be activated in Jurkat cells by agents producing superoxide, singlett oxygen, nitrous oxide or chemical stress. Equally ineffective were a peroxisome proliferator, heat shock and a lipid peroxidation product, suggesting that NF-κB is very specifically activated by peroxides. This is further supported by induction of NF-κB using 300 μM butylperoxide. The peroxide-specific activation of NF-κB is remeniscent of the prokaryotic factor oxyR (Storz et al., 1990). The inhibitory effect of metal chelators indicates a requirement for conversion of H_2O_2 into hydroxyl radicals as catalyzed by free Fe and Cu in the Fenton reaction (Schreck et al., 1992b).

As might have been expected, the activation of NF-κB by H_2O_2 was suppressed by preincubation of cultured cells with antioxidants such as NAC, dithiocarbamates and iron chelators (Schreck et al, 1991; 1992b). We then used antioxidants as pharmacological tools to investigate whether the activation of NF-κB by other stimuli also relies on the production of ROIs. The NF-κB-activating effect of TNF, IL-1, PMA (alone and in combination with lectins), three viral transactivator proteins, muramyl peptides, LPS, γ rays, UV-A, leukotriene-B4, double-stranded RNA and cycloheximde could be blocked by various antioxidants. This strongly suggests that H_2O_2 and OH· serve as common messenger in the activation of NF-κB by a great variety of pathogens. Furthermore, increased production of these ROIs seems to be a very general response of cells to pathogen stimulation.

In the meantime, many antioxidative substances (listed in Tab. III) have been successfully tested for their inhibition of NF-κB activation in various cell types using various stimuli. The agents are so diverse in their chemical properties such that the only common denominator of their inhibiting activity towards NF-κB might be their antioxidative potential. Table IV lists some manipulations and drugs that could not prevent activation of NF-κB.

Tab. II. Stress factors that fail to activate NF-κB. References are listed in Schreck et al. (1992b). The failure of photosensitization reactions to activate NF-κB has been reported by Legrand-Poels et al. (1993). It should be noted that most of these agents were tested with a Jurkat T cell line highly responsive to H2O2. We cannot exclude that the conditions listed here activate NF-κB in another cell type.

Agent	Class Tested	Concentrations	Incubation Period
Paraquat	O2⁻ generating	0.1 μM - 5 mM	3.5 h, 18 h
Doxorubicin	O2⁻ generating	10-50 μM	2 h
Mitomycin C	O2⁻ generating	1-50 μg/ml	2 h
Menadione	O2⁻ generating	1-100 μM	2 h
Sodium nitroprusside	NO generating	1-250 μM	1 h
Endoperoxide of dipropionate 3,3 '-(1,4-naphtylidene)	Singlett oxygen generating	0.1 - 8 mM	1 h
Photosensitization reaction	Singlett oxygen generating		
4-Hydroxynonenal product	Lipid peroxidation	0.1-100 μM	1 h
Clofibrate	Peroxisome proliferator	1-1000 μM	18 h
4-Nitroquinolineoxide	DNA damaging	1-50 μM	2 h
1-β-D-arabinofuranosyl-cytosine (Ara-C)	DNA damaging	0.1-100 μM	2 h
CuS04	Chemical stress	100-750 μM	4 h
CdS04	Chemical stress	50 μM	4 h
Sodium arsenite	Chemical stress	50 μM	4 h
42°C	Heatshock		1 h

Tab. III. Antioxidant inhibitors of NF-κB activation. A most recent list of studies reporting on the inhibition of NF-κB by antioxidants.

Agent	Class	Effective Concentration (ID90)	Reference
N-Acetyl-L-cysteine	Scavenger (-SH)	20 - 30 mM	Staal et al., 1990 Schreck et al., 1991
L-Cysteine	Scavenger (-SH)	300 μM	Mihm et al., 1991
2-Mercaptoethanol	Scavenger (-SH)	14 mM	Schreck et al., 1991
α-Lipoic acid	Scavenger (-SH)	4 mM	Suzuki et al., 1993
Aurothioglucose	Scavenger (-SH)	100 μM	Williams et al., 1992
Pyrrolidine dithio-carbamate	Scavenger (>NCS2)	60-200 μM	Schreck et al., 1991; 1992b
Diethyldithio-carbamate	Scavenger (>NCS2)	100 μM	Schreck et al., 1992b
Disulfiram	Scavenger (>NCS2)	100 μM	Schreck et al., 1992b
Butylated hydroxy-anisol	Scavenger (non S)	300 μM	Israel et al., 1992; Schulze-Osthoff et al., 1993
Nordihydroguairetic acid	Scavenger (non S)	10 μM	Schulze-Osthoff et al., 1993
Rotenone	Electron transport inhibitor	10 μM	Schulze-Osthoff et al., 1993
Orthophenanthroline	Metal chelator (Cu)	100 μM	Schreck et al., 1992a
Desferroxamine	Metal chelator (Fe)	100 μM	Schreck et al., 1992a
α-Tocopherol	Vitamine	>300 μM	Israel et al., 1992

It is presently unknown how the various NF-κB inducers cause the production of H_2O_2. During inflammation, the H_2O_2 produced from activated neutrophils and macrophages can itself be an exogenous factor which can diffuse through the plasma membrane of a target cell. Other sources of H_2O_2 can be the mitochondria. In the case of TNF, it has been shown by organelle depletion experiments and the use of electron transport inhibitors that mitochondria are a main source for the ROIs required to activate NF-κB and to elicit TNF cytotoxicity (Schulze-Osthoff et al., 1993). Another possible source for H_2O_2 are NADPH-oxidases. These plasma membrane-associated enzymes can inducibly produce ROIs and might not be restricted to specialized blood cells (reviewed in Schreck et al., 1991). Less active apparently cytokine-inducible forms of NADPH oxidases have been described in a variety of non-immune cells. Other enzymes which can produce H_2O_2 or $O_2 \cdot^-$ are xanthine oxidase, glucose oxidase, enzymes involved in arachidonic acid metabolism and proteins transporting electrons in mitochondria, peroxisomes and the endoplasmic reticulum. Future pharmacological experiments are required to investigate the sources for the ROIs activating NF-κB.

Tab. IV.   Treatments that fail to prevent the activation of NF-κB by TNF.

Agent	Class	Concentrations Tested	Incubation Period
Overexpression of Mn-superoxidedismutase	O_2^- eliminating	-	constitutive
Enucleation	Removal of DNA	-	-
Diphenylene iodonium	NADPH oxidase inhibitor	1 - 20 μM	1.5 h
Nω-Nitro-L-arginine methylester	NO synthesis inhibitor	5 -10 mM	16 h
Diclofenac	Cyclooxygenase inhibitor	1- 50 μM	1.5 h
Quinacrine	PLA$_2$ inhibitor	5 μM	2 h

We could not observe that the NF-κB--IκB complex is activated in purified or crude fractions upon direct treatment with oxidants (Schreck et al., 1991). On the contrary, there are reports demonstrating direct oxidation of p50 and the related c-rel protein on cysteine residues under cell-free conditions, which results in decreased DNA-binding of NF-κB complexes (Toledano and Leonard, 1991; Matthews et al., 1992). We doubt that these *in vitro* reactions are of physiological significance based on the fact that in intact cells NF-κB is activated but not inhibited under prooxidant conditions. Furthermore, it is questionable whether mM-amounts of glutathione will allow a selective oxidation of proteins in the cytosol or nucleus. These considerations suggest that increased concentrations of peroxides must be sensed in a cellular compartment facing more dramatic changes in ROI levels than the redox-buffered cytosol. In this compartment a messenger must be produced or activated by $H_2O_2/OH\cdot$ which can indirectly or directly communicate with NF-κB--IκB in the cytosol. A possible scenario is the activation of a membrane-associated tyrosine protein kinase or inactivation of a tyrosine protein phosphatase by ROI produced from a membrane-associated enzyme. There are, in fact, several reports in the literature showing increased tyrosine phosphorylation in reponse to H_2O_2 treatment of cells (Heffetz et al., 1990). The effects were potentiated by the tyrosine protein phosphatase inhibitor vanadate. Furthermore, an endoplasmic reticulum-associated tyrosine kinase was described which was activated in intact cells by oxidation through diamide (Bauskin et al., 1991). It is tempting to speculate that tyrosine kinases and tyrosine phosphatases or serine/threonine-specific kinases and phosphatases are sensors as well as messengers in the activation of NF-κB by oxidative stress. A similar scenario has recently been proposed for the activation of AP-1 in response to UV light (Devary et al., 1992). The induction required both tyrosine and serine/ threonine-specific kinases and was blocked by NAC.

Under all conditions tested so far, NF-κB behaved as an oxidative stress-responsive transcription factor. Because hypoxia and antioxidants are also known to induce novel gene expression, we became interested in transcription factors induced in eukaryotes under antioxidant conditions (that would prevent the activation of NF-κB).

Redox Regulation of Gene Expression by AP-1

Like NF-κB, the transcription activator AP-1 binds to DNA as heterodimer and its two DNA-binding subunits belong to a multigene transcription factor family (reviewed in Karin, 1991). However, unlike NF-κB, AP-1 relies predominantly on new synthesis for its activation. Two well-studied subunits of AP-1 are c-jun and c-fos. Both are encoded by proto-oncogenes showing an important role of AP-1 in the control of cell proliferation. The two subunits dimerize via hydrophobic interaction of a-helices forming the "leucine zipper" motif and bind to symmetric DNA sequences with the consensus 5´-TGANTCA-3´. The activation of AP-1 was predominantly investigated by studying the induction of c-fos and c-jun mRNAs. Numerous stimuli have been reported to induce the genes. This can in part be explained by various primary factors participating in transcriptional regulation of the genes. Shibanuma et al. (1988) were the first to report induction of c-fos and c-jun genes upon H_2O_2-treatment of cells. This finding was confirmed by several other investigators (Stein et al., 1989; Devary et al., 1991; Nose et al., 1991; Amstad et al., 1992). Many other prooxidant stimuli have been found to induce c-fos and c-jun mRNAs including UV (Devary et al., 1991; Stein et al., 1989), ionizing radiation (Datta et al., 1992) and IL-1 (Munoz et al., 1992). The DNA binding and transactivating potential of AP-1 is also augmented upon H_2O_2 treatment of cells (Devary et al., 1991). However, the relatively weak activation of AP-1 activity by H_2O_2 is in apparent contrast to the strong increase of c-fos and c-jun mRNAs by the oxidant. This finding prompted us to compare and contrast the activation of AP-1 and NF-κB by H_2O_2 and the effects of antioxidants within the same cell type under identical conditions.

As reported earlier for Jurkat T cells (Schreck et al., 1991), NF-κB behaved as oxidative stress-responsive factor in HeLa cells under all conditions (Meyer et al., 1993). Micromolar amounts of H_2O_2 rapidly increased the DNA binding and transactivating potential of NF-κB 20-30-fold for several hours. Under identical conditions, AP-1 was only very transiently activated 2.5-4-fold in its DNA binding and transactivating potential by H_2O_2, while the mRNA levels for c-jun and c-fos showed a much more dramatic increase. It thus seems that there is an additional step controlling the activity of AP-1 following its *de novo* expression upon a prooxidant stimulus. Such an additional step could involve the oxidization of c-fos and/or c-jun proteins. From cell-free experiments, there is very good evidence that conserved cysteine residues in the DNA-binding domains of the subunits can undergo reversible oxidation (Abate et al., 1990). Oxidation will prevent DNA-binding of AP-1. An enzymatic activity, called Ref-1, was purified which can reactivate in vitro oxidized AP-1 (Xanthoudakis et al., 1992). cDNA cloning and antibody studies showed that Ref-1 encoded a nuclear protein with homology to DNA repair enzymes.

At present, there are no data supporting the idea that AP-1 can undergo reversible oxidiation within intact cells. The high levels of antioxidants and antioxidative enzymes in the cytoplasm and nucleus make it rather unlikely that AP-1 is (selectively) damaged in the course of a prooxidant traetment of cells. We could still detect AP-1 DNA binding activity in HeLa cells treated with 500 μM H_2O_2 while, under this condition, the DNA binding of NF-κB, which was maximally activated at 250 μM H_2O_2, was already impaired (Meyer et al., 1993). NF-κB has also been reported to have oxidiation-sensitive cysteine residues in the DNA binding domains of its subunits which, in an oxidized form, suppress the DNA binding activity (Toledano and Leonard, 1991; Matthews et al., 1992). However, within living cells, NF-κB is clearly not inactivated but, on the contrary, it is strongly activated under mild prooxidant conditions.

The cysteine residues in the DNA-binding domains of p50, Rel-A, c-fos and c-jun are all conserved and flanked by positively charged residues. The latter property would render the cysteines very vulnerable to oxidation. Future studies have to address the question whether an oxidation of these cysteine residues can occur within intact cells and whether this is part of a regulatory event. From a physiological point of view, it would make sense to inactivate a mitogenic factor such as AP-1 under prooxidant conditions. Oxidative stress increases the number of DNA lesions which should be excluded from replication. Only when the lesions are repaired and normoxic (or hypoxic) conditions re-established, should AP-1 become active and trigger DNA synthesis.

Consistent with the notion that PMA can act as a prooxidant stimulus (reviewed in Cerutti, 1985), the induction of NF-κB by PMA in HeLa cells was potently suppressed by the antioxidants PDTC and NAC (Meyer et al., 1993). In sharp contrast, the activation of AP-1 by PMA was not blocked but rather enhanced by the antioxidants. This shows that PMA-activated protein kinase C uses distinct intracellular signalling pathways for the activation of NF-κB and AP-1. While PKC-induced production of peroxides might be important for NF-κB, it is the dephosphorylation of pre-existing c-jun homodimers which triggers the activation of AP-1 in HeLa cells by the phorbolester PMA (Boyle et al., 1991; Papavassiliou et al., 1992).

In the course of our investigations, we noted in control experiments that PDTC and, to a lesser extent, NAC could on their own activate the DNA binding and transactivating potential of AP-1. Data from the laboratories of Pickett and Jaiswal on so-called antioxidant response elements (AREs) strongly support our finding that AP-1 is activated by antioxidants (Rushmore et al., 1991; Li and Jaiswal, 1992). In the case of the human NAD(P) quinone reductase gene, the ARE was identical with an AP-1 binding site. However, in a consensus ARE only one of the two AP-1 half sites is conserved. It is possible that DNA binding subunits of AP-1 use these sites together with an yet unidentified heterodimerization partner of distinct DNA binding specificity. The ARE studies used as antioxidants butylated hydroxyanisol (BHA) and β−naphtoflavone (β-NF). It would be interesting to test whether PDTC and NAC can also activate these ARE sites. In our laboratory, we used the classical canonical AP-1 binding site but we do not know yet whether BHA or β-NF stimulate binding of AP-1 to this site.

Another antioxidant condition which strongly activates AP-1 is transient overexpression of the enzyme thioredoxin (Meyer et al., 1993). At the maximum of the biphasic dose response curve, transactivation by AP-1 was stimulated 25-fold upon expression of human thioredoxin in HeLa cells. Expression of the antioxidative repair enzyme might be considered a physiological antioxidative response since it occurs when cells are exposed to prooxidant stimuli such as TNF, IL-1 and PMA. Antioxidant stimulation with PDTC or by thioredoxin expression gave an induction of AP-1 in HeLa cells which was stronger than seen with PMA and much stronger

230

than with H_2O_2. We thus consider antioxidants as novel important and physiological inducers of the AP-1 transcription factor.

The mechanism by which antioxidants and H_2O_2 activate AP-1 involves transcription of c-fos and c-jun genes (Shibanuma et al., 1988). PDTC could not activate AP-1 in the presence of a protein synthesis inhibitor, and the antioxidant as well as H_2O_2 caused a rapid increase in c-fos and c-jun mRNA levels. Thus, AP-1 must be considered a secondary redox-controlled transcription factor relying on redox-controlled primary transcription factors. If in addition AP-1 activity is regulated by direct oxidation/reduction the factor would be subject to a dual redox control. As will be discussed in the following chapter, the serum response element in the upstream promoter of the c-fos gene is responsible for both antioxidant as well as prooxidant induction of AP-1/c-fos.

Why should the mitogenic factor AP-1 be activated under hypoxic conditions? Hypoxia can be a consequence of injury and other conditions perturbing the normal oxygen supply of tissue. A consequence of reperfusing hypoxic tissue is severe oxidative stress and damage. AP-1 activated by hypoxia would have the potential to induce genes counteracting such reperfusion effects. Interestingely, two genes with ARE/AP-1 sites encode antioxidative enzymes (see above). Injury requires wound healing, a process involving cell proliferation. It is conceivable that hypoxia, in addition to peptide growth factors, is an important stimulus to activate the mitogenic transcription factor AP-1 upon tissue damage.

Redox Regulation of Gene Expression by SRF/TCF

Earlier studies showed that the cis-acting element responsible for inducing the c-fos gene in response to H_2O_2 (and UV light) is identical to the serum response element (SRE; Shibanuma et al., 1988; Stein et al., 1989). We recently found that the antioxidant (PDTC) response element in the c-fos gene is also identical to the SRE (Meyer et al., 1993), suggesting that the same element is capable of conferring transcriptional activity in response to prooxidant as well as antioxidant stimuli.

Devary et al. (1991) reported that the c-jun gene requires the AP-1 binding site in its promoter in order to respond to H_2O_2. Because the c-jun mRNA increases slightly later in response to H_2O_2 and PDTC than the c-fos mRNA, it is possible that the AP-1 produced from newly synthesized c-fos and small amounts of pre-existing c-jun is responsible for turning on c-jun transcription. The initital event of AP-1 activation would then be c-fos gene induction via the SRE. The proposed sequence of these events is depicted in a model (Fig. 2).

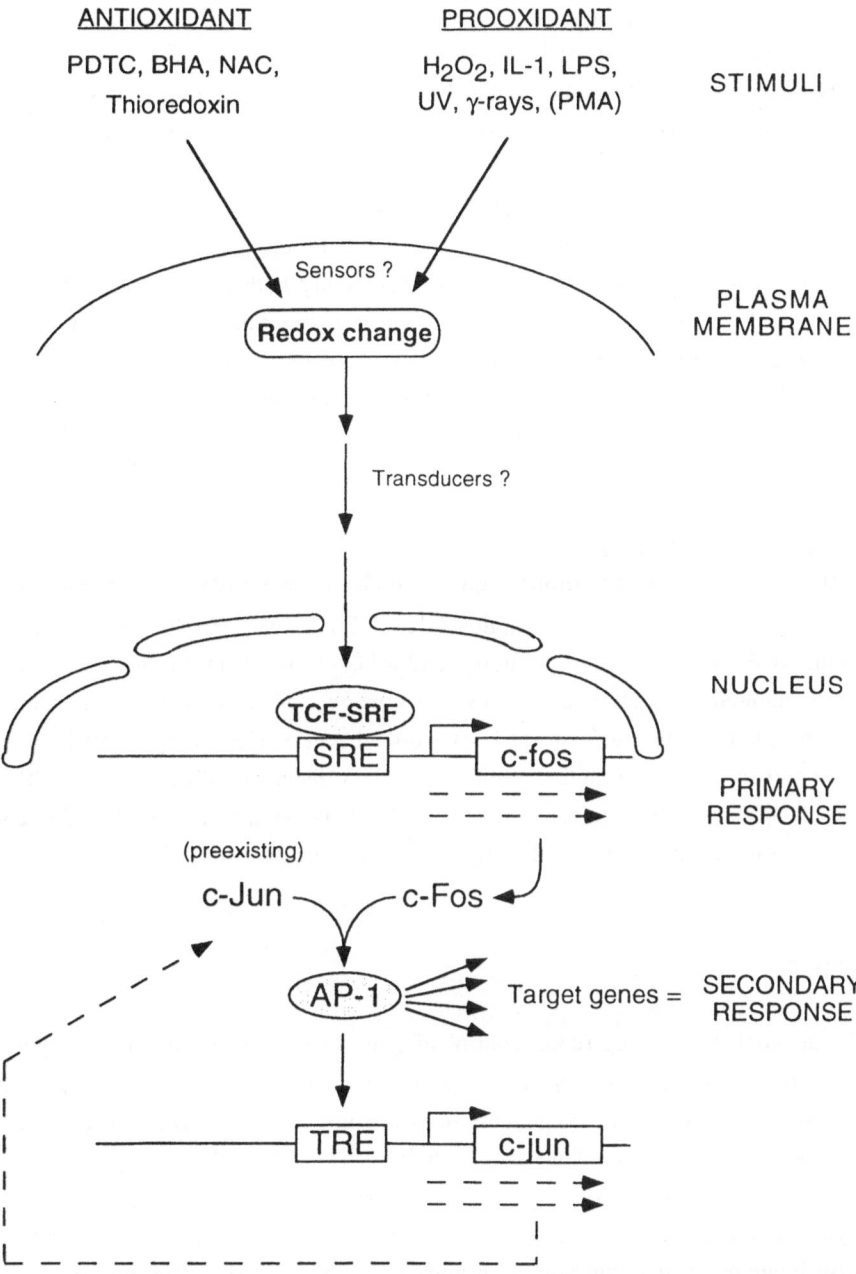

Fig. 2. A model illustrating the activation of AP-1 and SRF/TCF by prooxidant as well as antioxidant stimuli.

The SRE and its associated proteins are fairly well characterized (reviewed in Treisman, 1993). In contrast to NF-κB and AP-1, SRE-binding proteins already contact DNA before stimulation of cells (Herrera et al., 1989) suggesting they have to undergo a posttranslational modification in order to acquire transcription activating potential rather than *de novo* DNA binding. The first SRE-binding protein identified and cloned is called serum response factor (SRF). This protein contacts a palindromic sequence in the SRE as a homodimer. Interestingly, SRF is not directly involved in activating the SRE in response to serum stimulation. Nordheim and colleagues (Shaw et al., 1989) identified a second factor contacting both an upstream DNA sequence flanking the SRF-binding site as well as SRF in a ternary complex. It is this protein, called ternary complex factor (TCF), which is primarily required for serum responsiveness of the c-fos promoter. TCF again belongs to a multigene transcription factor family including the proto-oncogene *ets*. cDNA clones encoding TCF proteins have recently been isolated (Dalton and Treisman, 1992; Hipskind et al., 1991). There is evidence that the TCF protein elk-1 undergoes phosphorylation in response to serum stimulation and that this modification is required to activate elk-1 (Gille et al., 1992).

So far, there is very limited information available as to whether H_2O_2 and antioxidant treatments act on SRF or TCF transcription factors. Preliminary results suggest that the SRE requires the TCF binding site in order to respond to redox signals (J. Müller, A. Nordheim and P. Baeuerle, unpublished observation). This raises the possibility that redox signals, like serum, alter the phosphorylation status of TCF. It would be very interesting to find out whether prooxidant and antioxidant stimuli result in the same or distinct modifications of TCF proteins. If protein phosphorylation is critically involved it should be possible to identify redox-controlled protein kinases or phosphatases and their upstream regulators.

Perspectives

An ultimate goal in studying redox control of gene expression in higher eukaryotes is the identification of biomolecules capable of sensing intracellular changes of the redox status and the other signalling pathway components involved in translating redox changes into novel patterns of gene expression in the nucleus. With NF-κB, AP-1 and SRF/TCF three transcription factors have been identified that serve as final nuclear targets for redox-controlled signalling pathways. There are certainly more to be identified. The pivotal importance of these three factors for inducible gene expression in response to mitogenic and pathogenic signals suggests that redox signals are central and general messengers in fundmental regulatory mechanisms involving immediate early responses to proliferation and pathogenic signals. It is tempting to speculate that redox signals have been evolutionary conserved and further evolved to control such processes when ~10^9 years ago all organisms on our planet were exposed to a life-threatening change from a reducing to an oxidating atmosphere.

Acknowledgements

We are grateful to Dr. Heike L. Pahl for critical reading of the manuscript. This work was supported by grants from the Deutsche Forschungsgemeinschaft (SFB 217) and Bundesministerium für Forschung und Technologie (BMFT) awarded to P.A.B. and the Irene-Vogler-Award to M.M.

References

Abate, C., Patel., L., Rauscher, F.J. III, and Curran, T. (1990). Redox regulation of Fos and Jun DNA-binding activity in vitro. *Science* 249: 1157-1161.

Amstad, P.A., Krupitza, G., and Cerutti, P.A. (1992). Mechanism of c-fos induction by active oxygen. *Cancer Res.* 52: 3952-3960.

Baeuerle, P.A. (1991). The inducible transcription activator NF-kappa B: regulation by distinct protein subunits. *Biochem. Biophys. Acta* 1072: 63-80.

Baeuerle, P.A., and Baltimore, D. (1988). I kappa B: a specific inhibitor of the NF-kappa B transcription factor. *Science* 242: 540-546.

Bauskin, A.R., Alkalay, I. and Ben-Neriah, Y. (1991). Redox regulation of a protein tyrosine kinase in the endoplasmatic reticulum. *Cell* 66: 685-696.

Blank, V., Kourilsky, P., and Israel, A. (1992). NF-kappa B and related proteins: Rel/dorsal homologies meet ankyrin-like repeats. *Trends Biochem. Sci.* 17: 135-140.

Boyle, W.J., Smeal, T., Defize, L.H.K., Angel, P., Woodgett, J.R., Karin, M., and Hunter, T. (1991). Activation of protein kinase C decreases phosphorylation of c-jun at sites that negatively regulate its DNA-binding activity. *Cell* 64: 573-584.

Brown, K., Park, S., Kanno, T., Franzoso, G., and Siebenlist, U. (1993). Mutual regulation of the transcriptional activator NF-kB and its inhibitor IκB-a. *Proc. natl. Acad. Sci. USA* 90: 2532-2536.

Cerutti, P.A. (1985). Prooxidant states and tumor promotion. *Science* 277: 375-381.

Dalton, S., and Treisman, R. (1992). Characterization of SAP-1, a protein recruited by serum response factor to the c-fos serum response element. *Cell* 68: 597-612.

Datta, R., Hallahan, D.E., Kharbanda, S.M., Rubin, E., Sherman, M.L., Huberman, E., Weichselbaum, R.R., and Kufe, D.W. (1992). Involvement of reactive oxygen intermediates in the induction of c-jun gene transcription by ionizing radiation. *Biochemistry* 31: 8300-8306.

Demple, B. (1991). Regulation of bacterial oxidative stress genes. *Annu. Rev. Genet.* 25: 315-337.

Devary, Y., Gottlieb, R.A., Lau, L.F., and Karin, M. (1991). Rapid and preferential activation of the c-jun gene during the mammalian UV response. *Mol. Cell. Biol.* 11: 2804-2811.

Devary, Y., Gottlieb, R.A., Smeal, T., and Karin, M. (1992). The mammalian ultraviolet response is triggered by activation of Src tyrosine kinases. *Cell* 71: 1081-1092.

Gille, H., Sharrocks, A.D., and Shaw, P.E. (1992). Phosphorylation of transcription factor p62^{TCF} by MAP kinase stimulates ternary complex formation at c-fos promoter. *Nature* 358: 414-417.

234

Grimm, S. and Baeuerle, P.A. (1993). The inducible transcription factor NF-κB: structure - function relationship of its protein subunits. *Biochem. J.* 290: 297-308.

Gruss, H.J., Brach, M.A., and Herrmann, F. (1992). Involvement of nuclear factor-kappa B in induction of the interleukin-6 gene by leukemia inhibitory factor. *Blood* 80: 2563-2570.

Halliwell, B., and Gutteridge, J.M.C. (1989) *Free Radicals in Biology and Medicine*. Second 1Edition. Clarendon Press, Oxford.

Hammarskjöld, M.-L., Simurda, M.C. (1992). Epstein Barr virus latent membrane protein transactivates the the human immmunodeficiency virus type 1 long terminal repeat through induction of NF-κB activity. *J.Virol.* 66: 6496-6501.

Heffetz, D., Bushkin, J., Dror, R. and Zick, Y. (1990). H_2O_2 potentiates phosphorylation of nvel putative substrates for the insulin receptor kinase in intact Fao cells. *J. Biol. Chem.* 265: 2896-2902.

Herrera, R., Shaw, P. and Nordheim, A. (1989). Occupation of the c-fos serum response element in vivo by a multi-protein complex is unaltered by growth factor induction. *Nature* 340: 68-70.

Hipskind, R.A., Rao, V.N., Mueller, C.G.F., Reddy, E.S.P., and Nordheim, A. (1991). Ets-related protein Elk-1 is homologous to the c-fos regulatory factor $p62^{TCF}$. *Nature* 354: 531-534.

Israel, N., Gougerot-Pocidalo, M.A., Aillet, F., and Virelizier, J.L. (1992). Redox status of cells influences constitutive and induced NF-κB translocation and HIV long terminal repeat activity in human T and monocytic cell lines. *J. Immunol.* 149: 3386-3393.

Karin, M. (1991). Signal transduction and gene control. *Curr. Opin. Cell Biol.* 3: 467-473.

Korthuis, R.J., and Granger, D.N. (1986) in *Physiology of Oxygen Radicals*, Taylor, A.E., Matalon, S. and Wrad, P.A. (eds.), Baltimore: Williams and Wilkins, pp. 217-249.

Legrand-Poels, S., Hoebeke, M., Vaira, D., Rentier, B., and Piette, J. (1993). HIV-1 promoter activation following an oxidative stree mediated by singlet oxygen. *J. Photochem. Photobiol.* 17: 229-237.

Li, Y. and Jaiswal, A.K. (1992). Regulation of human NAD(P)H:quinone oxidoreductase gene. Role of AP1 binding site contained within human antioxidant response element. *J. Biol. Chem.* 267: 15097-15104.

Matthews, J.R., Wakasugi, N., Virelizier, J.-L., Yodoi, Y., and Hay, R.T. (1992). Thioredoxin regulates the DNA binding activity of NF-kappa B by reduction of a disulphide bond involving cysteine 62. *Nucl. Acids Res.* 20: 3821-3830.

Meyer, M., Schreck, R., and Baeuerle, P.A. (1993). H_2O_2 and antioxidants have opposite effects on activation of NF-κB and AP-1 in intact cells: AP-1 as secondary antioxidant- responsive factor. *EMBO J.* 12: 2005-2015.

Mihm, S., Ennen, J., Pessara, U., Kurth, R., and Dröge, W. (1991). Inhibition of HIV-1 replication and NF-kappa B activity by cysteine and cysteine derivatives. *AIDS* 5: 497-503.

Munoz, E., Zubiaga, A.M., Huang, C.-K., and Huber, B.T. (1992). Interleukin-1 induces c-fos and c-jun gene expression in T helper type II cells through different signal transmission pathways. *Eur. J. Immunol.* 22: 1391-1396.

Nose, K., Shibanuma, M., Kikuchi, K., Kageyama, H., Sakiyama, S., and Kuroki, T. (1991). Transcriptional activation of early-response genes by hydrogen peroxide in a mouse osteoblastic cell line. *Eur. J. Biochem.* 201: 99-106.

Papavassiliou, A.G., Bohmann, K., and Bohmann, D. (1992). Determining the effect of inducible protein phosphorylation on the DNA-binding activity of transcription factors. *Anal. Biochem.* 203: 302-309.

Rushmore, T.H., Morton, M.R., and Pickett, C.B. (1991). The antioxidant responsive element. Activation by oxidative stress and identification of the DNA consensus sequence required for functional activity. *J. Biol. Chem.* 266: 11632-11639.

Scala, G., Quinto, I., Ruocco, M.R., Mallardo, M., Ambrosino, C., Squitieri, B., Tassone, P., and Venuta, S. (1993). Epstein-Barr virus nuclear antigen 2 transactivates the long terminal repeat of human immunodeficiency virus type 1. *J. Virol.* 67: 2853-2861.

Schreck, R. and Baeuerle, P.A. (1991). A role for oxygen radicals as second messengers. *Trends Cell Biol.* 1: 39-42.

Schreck, R., Rieber, P., and Baeuerle, P.A. (1991). Reactive oxygen intermediates as apparently widely used messengers in the activation of the NF-kappa B transcription factor and HIV-1. *EMBO J.* 10: 2247-2258.

Schreck, R., Albermann, K., and Baeuerle, P.A. (1992a). Nuclear factor kappa B: an oxidative stress-responsive transcription factor of eukaryotic cells (a review). *Free Rad. Res. Comms.* 17: 221-237.

Schreck, R., Meier, B., Männel, D., Dröge, W., and Baeuerle, P.A. (1992b). Dithiocarbamates as potent inhibitors of nuclear factor kappa B activation in intact cells. *J. Exp. Med.* 175: 1181-1194.

Schulze-Osthoff, K., Beyaert, R., Vandevoorde, V., Haegeman, G., and Fiers, W. (1993) *EMBO J.*, in press.

Shaw, P.E., Schröter, H., and Nordheim, A. (1989). The ability of a ternary complex to form over the serum response element correlates with serum inducibility of the human c-fos promoter. *Cell* 56: 563-572.

Shibanuma, M., Kuroki, T., and Nose, K. (1988). Superoxide as a signal for increase in intracellular pH. *Oncogene* 3: 17-21.

Staal, F.J.T., Roederer, M., Herzenberg, L.A., and Herzenberg, L.A. (1990). Intracellular thiols regulate activation of nuclear factor kappa B and transcription of human immunodeficiency virus. *Proc. Natl. Acad. Sci. USA* 87: 9943-9947.

Stein, B., Rahmsdorf, H.J., Steffen, A., Litfin, M., and Herrlich, P. (1989). UV-induced DNA damage is an intermediate step in UV-induced expression of human immunodeficiency virus type 1, collagenase, c-fos, and metallothionein. *Mol. Cell. Biol.* 9: 5169-5181.

Storz, G., Tartaglia, L.A., Farr, S.B., and Ames, B.N. (1990). Bacterial defenses against oxidative stress. *Trends Genet.* 6: 363-368.

Sun, S.-C., Ganchi, P.A., Ballard, D.W., and Greene, W.C. (1993) NF-κB controls expression of inhibitor IκB-a: Evidence for an inducible autoregulatory pathway. *Science* 259: 1912-1915.

Suzuki, Y.J., Aggarwal, B.B., and Packer, L. (1993). Alpha-lipoic acid is a potent inhibitor of NF-kappa B activation in human T cells. *Biochem. Biophys. Res. Comms.* 189: 1709-1715.

Toledano, M.B. and Leonard, W.J. (1991). Modulation of transcription factor NF-κB binding activity by oxidation-reduction in vitro. *Proc. Natl. Acad. Sci. USA* 88: 4328-4332.

Treisman, R. (1993). The serum response element. *Trends Biochem. Sci.* 17: 423-426.

Williams, D.H., Jeffrey, L.J., and Murray, E.J. (1992). Aurothioglucose inhibits induced NF-kB and AP-1 activity by acting as an IL-1 functional antagonist. *Biochim. Biophys. Acta* 1180: 9-14.

Wong, G.H.W.and Goeddel, D.V. (1988). Induction of manganeous superoxide dismutase by tumor necrosis factor: possible protective mechanism. *Science* 241: 941-943.

Xanthoudakis, S., Miao, G., Wang, F., Pan, Y.C., and Curran, T. (1992). Redox activation of Fos-Jun DNA binding activity is mediated by a DNA repair enzyme. *EMBO J.* 11: 3323-3335.

Yodoi, J. and Uchiyama, T. (1992). Diseases associated with HTLV-I virus, Il-2 receptor dysregulation and redox regulation. *Immunol. Today* 13: 405-411.

Oxidative Stress, Cell Activation and Viral Infection
C. Pasquier et al. (eds)
© 1994 Birkhäuser Verlag Basel/Switzerland

Redox Mechanisms in T Cell Activation

N.H. Hunt, D.M. van Reyk, J.C. Fragonas, T.M. Jeitner and S.D. Goldstone

Department of Pathology, University of Sydney, N.S.W. 2006, Australia

Summary

A variety of anti-oxidants, including free radical scavengers and iron chelators, inhibit the proliferation of mitogen-stimulated T lymphocytes in vitro. These agents do not operate by inhibiting the induction of ornithine decarboxylase activity. Phorbol myristate acetate induces formation of reactive oxygen species in T cell-enriched populations. Flow cytometry has identified this reactive oxygen within the T lymphocytes themselves, though other cells may contribute indirectly. The binding to DNA of the transcription factors AP-1 and NF-kB is increased during the commitment period, 2-4 hours after mitogen addition to T lymphocytes, and this binding is inhibited by the anti-oxidant aminothiol compound, cysteamine. Intracellular oxidant formation in mitogen-stimulated T cells might activate transcription factors, allowing their translocation into the nucleus and subsequent reduction and binding to DNA, thereby inducing early gene expression. Since the replication of the human immunodeficiency virus-1 is linked to NF-kB and can be inhibited by some anti-oxidants, it is possible that this virus is stimulated to replicate by the intracellular oxidative events that we have shown to be an obligatory step in T lymphocyte recruitment into the cell cycle.

Introduction

The potentially damaging reactions of reactive oxygen species (ROS) in biological systems, particularly mammalian cells, have been studied extensively. However, there are a number of circumstances in which ROS, or free radicals not centred on oxygen, modulate certain normal biological processes (discussed by Halliwell and Gutteridge, 1989). In this article we present evidence that activation of T lymphocytes is one such system in which ROS may play a positive, modulatory role.

It was shown some years ago that rat thymocytes respond to addition of concanavalin A (con A; a plant lectin that stimulates proliferation of T cells) or Sendai virus with a burst of luminol-dependent chemiluminescence (Wrogemann *et al.*, 1978; Kolbuch-Braddon *et al.*, 1984). This chemiluminescence was largely extracellular and non-mitochondrial in origin and had the characteristics of being induced by superoxide, or ROS derived from it. It was suggested that T cells themselves might be the source of the ROS (Wrogemann *et al.*, 1978). Rat thymocyte populations largely consist of T lymphocytes, but there are also appreciable numbers of other cells that are known to be capable of generating ROS, such as macrophages and

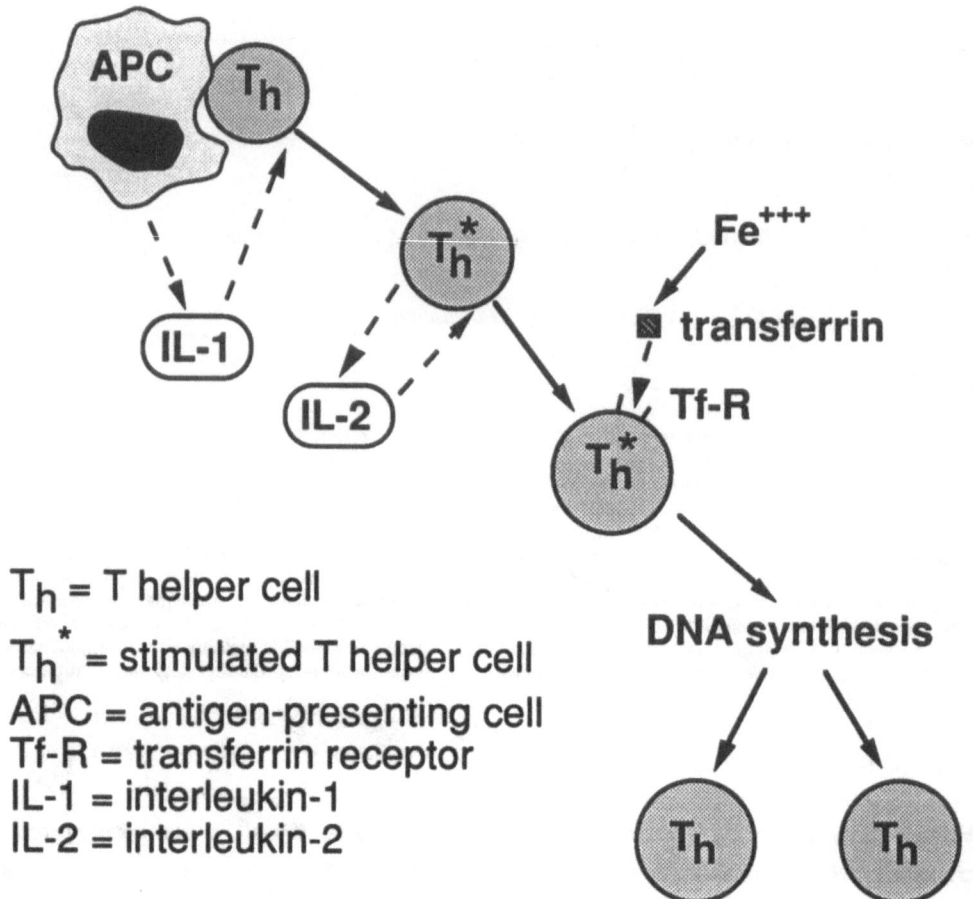

Figure 1. Simplified diagram of early events in the activation of T lymphocytes, of the "helper" sub-class, by antigen. Foreign antigen is processed by APC and presented at the cell surface, in the context of the major histocompatibility complex, to T cells. This delivers a signal that is translated through a complex set of biochemical pathways leading to gene expression and commitment to the cell cycle.

polymorphonuclear leukocytes, and subsequent studies by these workers showed that phagocytes were the most likely source of the oxidants in this system (Weidemann *et al.*, 1987). However, these studies did suggest that it might be worthwhile investigating a link between oxidative phenomena and T cell activation.

It is well established that lymphocytes, like other cells, are susceptible to oxidative damage (Sagone *et al.*, 1984; Smit and Anderson, 1990). Thus, anti-oxidants can protect them from damage under circumstances where they are exposed to large fluxes of ROS, for example when in contact with phagocytes artificially exposed to sustained stimulation by activators of the oxidative burst (Sagone *et al.*, 1984; Zoschke and Staite, 1987). However, whether such circumstances are likely to pertain in lymphoid organs *in vivo* is debatable. Novogrodsky and colleagues (1982) showed that hydroxyl radical scavengers, for example dimethyl sulphoxide, thiourea, dimethylurea and mannitol, inhibited the mitogenic response of human peripheral blood lymphocytes (HPBL) to phorbol myristate acetate (PMA), con A and phytohaemagglutinin (PHA, another mitogenic plant lectin), suggesting that hydroxyl radicals might be involved in mediating the signal(s) that triggers T cell activation. These findings prompted us to study further the possible roles of ROS in T lymphocyte activation and proliferation (Chaudhri *et al.*, 1986, 1988; Hunt *et al.*, 1989, 1991, 1992; Hunt and Fragonas, 1992; van Reyk *et al.*, 1992).

The proliferation of T lymphocytes is a pivotal event in the cell-mediated immune response. Particulate foreign antigens are partially degraded by antigen-processing cells, for example dendritic cells, macrophages or B lymphocytes, and presented on the surface of these cells, in the context of the Major Histocompatibility Complex, to T cells bearing the cell suface marker CD4 (Unanue, 1992). This initiates a complex series of events, involving the *de novo* expression of over 100 genes (Ullman *et al.*, 1990), that results in DNA synthesis and mitosis. Key events in this process include the production by accessory cells of interleukin (IL)-1, and of IL-2 by CD4[+] T cells themselves (Figure 1). The T cells also express cell surface receptors for IL-2 and the iron transport molecule transferrin. The CD4[+] T "helper" cells, of two classes, produce cytokines that modulate the activities of other cells, for example B lymphocytes and macrophages.

Inhibition of T cell activation by anti-oxidants

Agents that interfere with oxidative processes may do so by preventing formation of radicals (or their precursors), or by preferentially reacting with them to form stable, non-oxidizing products. For convenience we will refer to such agents as "anti-oxidants". Radical scavengers and iron chelators both fit this definition, the former by reacting preferentially with radicals and the latter by binding iron and preventing its involvement in reactions that lead to the formation of radical species. A more detailed discussion of these concepts can be found in Halliwell and Gutteridge (1989).

A range of anti-oxidants was found to inhibit, in a dose-dependent manner, the antigen-driven proliferation associated with the mixed lymphocyte reaction of murine lymph node cells (Chaudhri *et al.*, 1986). These agents were: the resonance-stabilized free radical scavenger butylated hydroxyanisole (BHA); iron chelators, including desferrioxamine (DES) and desferrithiocin (DFT); and the non-permeant electron acceptor ferricyanide. Limit dilution analysis showed that proliferation of every potential responder cell in the mixed lymphocyte cultures was inhibited. These compounds did not inhibit the production of IL-1 by accessory cells or that of IL-2 by the T cells themselves, but did inhibit the cell-surface expression of receptors for IL-2 (IL-2R). Since the only obvious common feature of the anti-oxidant compounds was their ability to interfere with oxidative processes, we proposed that "free radicals are involved in the activation of T lymphocytes as positive mediators" (Chaudhri *et al.*, 1986). Others later came to the same conclusion (Sekkat *et al.*, 1988; Fidelus, 1988).

We subsequently showed that BHA, DES and ferricyanide inhibited the proliferation of various types of murine T cell induced to proliferate by a combination of PMA and the calcium ionophore ionomycin but did not inhibit IL-2 production (Chaudhri *et al.*, 1988). We also confirmed that these agents inhibited cell surface expression of IL-2R and transferrin receptors, whereas the ribonucleotide reductase inhibitor hydroxyurea did not. The inhibition of IL-2R expression by DES was seen at somewhat higher concentrations than were required to inhibit entry into DNA synthesis, suggesting that the latter phenomenon was not a consequence of the former. We have never found inhibition by DES or BHA of IL-2 production in mitogen-stimulated T cells, either murine or human, in agreement with some reports (Carotenuto *et al.*, 1986) but not others (Dornand and Gerber, 1989). However, at least one anti-oxidant, the aminothiol cysteamine, does inhibit IL-2 production in human peripheral blood lymphocytes (HPBL) stimulated with mitogens (S.D. Goldstone and N.H. Hunt, unpublished observations).

Our original observations on murine T cells later were extended to HPBLs, with broadly similar results. DES, DFT, BHA and ferricyanide all inhibited DNA synthesis induced by the lectin phytohaemagglutinin (PHA) or by PMA/ionomycin (Hunt *et al.*, 1991). The selenocompound ebselen, which has glutathione peroxidase-like activity (Parnham and Graf, 1987), the anti-oxidant nordihydroguaiaretic acid (NDGA), and diphenylene iodonium, which inhibits NADPH oxidase (Maly *et al.*, 1989), were also anti-proliferative in the HPBL system. These observations are consistent with the idea that intracellular ROS formation is part of a signalling pathway in mouse and human T cells, but studies with inhibitors clearly must be complimented by more direct approaches, as discussed below. A common error of interpretation is to ascribe the actions of NDGA to its ability to inhibit arachidonate lipoxygenase, when it is just as likely that its antioxidant properties (Darley-Usmar *et al.*, 1989) are relevant. Another factor that needs always to be considered is that some of the anti-oxidants are cytotoxic in the longer term experiments of 2-3 days required for studies of DNA synthesis, and this necessitates careful controls. Some of these agents, for example NDGA, BHA or pyrrolidine dithiocarbamate, are of most use in short-term studies of early G_1 phase events, though this can raise the problem of relating such events to ones occurring later in the cell cycle.

The intracellular milieu of cells is, on the whole, extremely reduced in chemical terms, as indicated by the cytosolic redox ratio of glutathione where GSH/GSSG is 100:1 or greater, though there are cellular compartments where this ratio does not prevail (Hwang *et al.*, 1992). If an oxidative event were critical in T lymphocyte activation, it would not be surprising if the intracellular GSH status could influence proliferation. This is indeed so, with depletion of intracellular GSH having been shown to inhibit proliferation of T cells, and augmentation to enhance it (Fidelus and Tsan, 1986; Hamilos *et al.*, 1989; Suthanthiran *et al.*, 1990). These observations have attracted great interest in light of the marked decreases in GSH levels in blood mononuclear cells observed in HIV-seropositive individuals (Dröge *et al.*, 1988), who at late stages of their disease are immunocompromised. GSH and N-acetylcysteine inhibit the induction of HIV replication in U1 monocytic cells (Kalebic *et al.*, 1991), whereas H_2O_2 stimulates HIV-1 replication in Jurkat cells (Schreck *et al.*, 1991). These observations have led to proposals to treat HIV-infected patients with various anti-oxidants, including thiols (Roederer *et al.*, 1990; Chinnadurai, 1991; Harakeh and Jariwalla, 1991).

Ornithine decarboxylase

Polyamines are essential for cellular proliferation, for example by modulating the transcription of growth-associated genes in some cell types (Celano *et al.*, 1989). Ornithine decarboxylase (ODC) is the rate-limiting enzyme in polyamine biosynthesis. The activity of the enzyme is very low in resting T lymphocytes but is induced within a few hours after mitogen addition (White *et al.*, 1987; Hunt and Fragonas, 1992). Anti-oxidants (DES, NDGA, ferricyanide) were found to strongly, but not completely, inhibit the increase in ODC activity seen after addition of con A or PMA/ionomycin to murine T cells (Hunt and Fragonas, 1992). However, these agents did not influence intracellular levels of ODC mRNA, suggesting that their effect was exerted post-transcriptionally, which is consistent with what is known about the regulation of ODC production in T cells (White *et al.*, 1987). Addition of the polyamine putrescine to activated T cells was not sufficient to overcome the inhibition of DNA synthesis induced by DES, though it could reverse the anti-proliferative effect of the ODC inhibitor difluoromethylornithine (Table I). This probably indicates that the anti-oxidants do not exert their inhibitory actions through inhibition of ODC activity, implying that the residual enzyme activity is sufficient for cell cycle progression. This interpretation is consistent with our recent observations that DES exerts its effects late in G_1 (T.M. Jeitner, D.M. van Reyk and N.H. Hunt, unpublished observations).

Table I. Effect of exogenous putrescine on inhibition of DNA synthesis induced by desferrioxamine (DES) and difluoromethylornithine (DFMO) in mitogen-activated T lymphocytes.

Treatment	Without putrescine	With putrescine
Unstimulated	2883 ± 137	ND
PMA/IoM	27,725 ± 2349	ND
PMA/IoM + DES	1849 ± 270	1376 ± 51
PMA/IoM + DFMO	2566 ± 371	27,306 ± 167

Murine lymph node T cells were activated by addition of PMA/ionomycin (IoM) in the presence or absence of DES (20 μM), DFMO (2.5 mM) or putrescine (1 mM), and DNA synthesis was measured 48 hours later by [^{3}H]-thymidine incorporation into DNA. ND = not determined (putrescine alone does not affect DNA synthesis, either basal or stimulated, in this system: unpublished observations). Values are mean ± SEM from triplicate determinations.

Iron chelators

Several structurally-distinct iron chelators have been shown to inhibit the proliferation of murine or human T cells *in vitro* (Table II). These effects are prevented by pre-complexing the chelators with iron. DES cannot remove iron from the iron transport protein transferrin (Pollack *et al.*, 1976) and thus presumably does not operate by chelating all available extracellular iron. Although DES is generally considered not to enter most cells readily, it has been shown to enter HPBL (Polson *et al.*, 1990). Furthermore, certain high molecular weight polymeric iron chelators do not inhibit T cell proliferation (Hunt *et al.*, 1989). Thus it seems likely that iron chelators exert their anti-proliferative effects through interacting with intracellular iron pools. One way that this might influence cellular activities is by preventing the known (Halliwell and Gutteridge, 1989) involvement of iron in the catalysis of hydroperoxides, though this is speculative. It has long been believed that iron chelators block cell proliferation through inhibition of ribonucleotide reductase, an enzyme required for DNA synthesis, and this possibility has not been ruled out in the case of T lymphocytes though there are some arguments against this idea (Chaudhri *et al.*, 1986). It has been suggested that small traces of iron are

Table II. Inhibition of T lymphocyte proliferation by iron chelators.

Chelator	Species	Reference
Desferrioxamine	Mouse	Chaudhri et al. 1986, 1988; Dornand & Gerber, 1989; van Reyk et al. 1992; Hunt & Fragonas, 1992
	Human	Hoffbrand et al. 1976; Ganeshaguru et al. 1980; Lederman et al. 1984; Carotenuto et al. 1986; Hunt et al. 1989, 1991; Polson et al. 1990
Desferrithiocin	Mouse	Chaudhri et al. 1986, 1988
	Human	Hunt et al. 1989, 1991; Bierer & Nathan, 1990
Octanohydroxamic acid	Mouse	Chaudhri et al. 1986
Pyridoxal isonicotinoyl hydrazone	Mouse	Chaudhri et al. 1986
SAG-15	Mouse	van Reyk et al. 1992

necessary for the production of the cell-cycle regulatory protein kinase p34^{cdc2} (Terada et al., 1993). Our early work suggested that DES might act early in G_1 but our most recent studies have shown that it exerts its crucial anti-proliferative action close to the onset of DNA synthesis (T.M. Jeitner, D.M. van Reyk and N.H. Hunt, unpublished observations).

The anti-proliferative ability of iron chelators suggested that they might indicate some new directions for the development of immunosuppressive or anti-leukaemic agents (Hunt et al., 1989). DES has been used successfully in the treatment of leukaemia (Estrov et al., 1987), but the short half-life of the drug and lack of activity by oral administration makes it far from ideal clinically. For this reason we examined the effects of a new iron chelator, 1-[N-ethoxycarbonylmethyl-pyridoxylidenium]-2-[2'-pyrimidyl] hydrazine bromide (SAG-15), in comparison with two established agents, d-razoxane (ADR-529 or ICRF-187) and 1-hydroxypyridine-2-thione (omadine) on lymphoproliferation. Although all three agents inhibited T cell proliferation, only SAG-15 acted in an iron-dependent manner (van Reyk et al., 1992). SAG-15 arrested cells in G_1 and ADR-529 blocked in G_2/M. Omadine, an agent suggested as a potential anti-cancer agent (Blatt et al., 1989), was extremely toxic to lymphocytes at concentrations as low as 100 nM.

Cysteamine

Aminothiol compounds have long been known for their radioprotective abilities (Yuhas, 1980). Some of these compounds, for example cysteamine, have anti-oxidant activity through mechanisms such as direct radical scavenging, reduction of oxidised sulphydryls, or by increasing intracellular GSH. Cysteamine inhibits the growth of several experimental tumours, including hyperprolactinaemic rat pituitary adenomas (Jeitner and Oliver, 1990), the proliferation of HPBL (Jeitner and Hunt, 1990; Hunt *et al.*, 1992) and that of the human leukaemic cell line CCRF-CEM (Jeitner and Hunt, 1990). The anti-oxidant activity of cysteamine could be the basis of its anti-proliferative effect, in T cells at least. In HPBL, cysteamine acts within the first 2 hours after mitogen addition, that is during the commitment period, but in the continuously-cycling CCRF-CEM line it arrests cell progression during S phase. The commitment period is the time during which the cell requires continual exposure to a mitogen or growth factor to maintain progression through the cell cycle (Pardee, 1989). In light of the current interest in using thiols for treatment of HIV-infected patients *(vide supra)*, it should be pointed out that cysteamine is used clinically, for example in nephropathic cystinosis (Yudhoff *et al.*, 1981), and is much more active than *N*-acetylcysteine at inhibiting lymphoproliferation *in vitro* (Jeitner and Hunt, unpublished observations).

Intracellular ROS production

As previously discussed, there were early indications in the literature that lymphocytes could produce ROS. It later was established that B lymphocytes, the proliferation of which is also inhibited by anti-oxidants (Hunt *et al.*, 1991), possess a functional NADPH oxidase (Maly *et al.*, 1989) but T lymphocytes do not (Pick and Gadba, 1988). Notwithstanding this latter observation, PHA and PMA were shown to stimulate the intracellular oxidation of dichlorofluorescin (DCFH) to dichlorofluorescein in T-enriched HPBL (Sekkat *et al.*, 1988). The T lymphocyte component of the mixed cell population was not directly identified but, since some clones of a T cell line (Jurkat cells) responded to the same agonists in a similar way, it was argued that the intracellular ROS formation demonstrated by DCFH oxidation was T cell-associated. These workers have reported recently that other leukocytes provide "help" necessary for T lymphocytes to produce ROS (Rabesandratana *et al.*, 1992).

Positive identification of the responding cells as T lymphocytes is necessary for the reasons briefly alluded to earlier; for example, the T cell mitogen con A induces the oxidative burst in human monocytes and neutrophils (Lacal *et al.*, 1990). We approached this problem by using concurrent identification, using antibodies against cell-specific markers, of DCFH-oxidising cells in studies employing flow cytometry (Figure 2). PMA induced an increase in DCFH oxidation within mouse cells identified as T lymphocytes by possession of the specific cell-surface marker Thy-1. Consistent results have been obtained in HPBL (C.L. Kneale and N.H. Hunt, unpublished data).

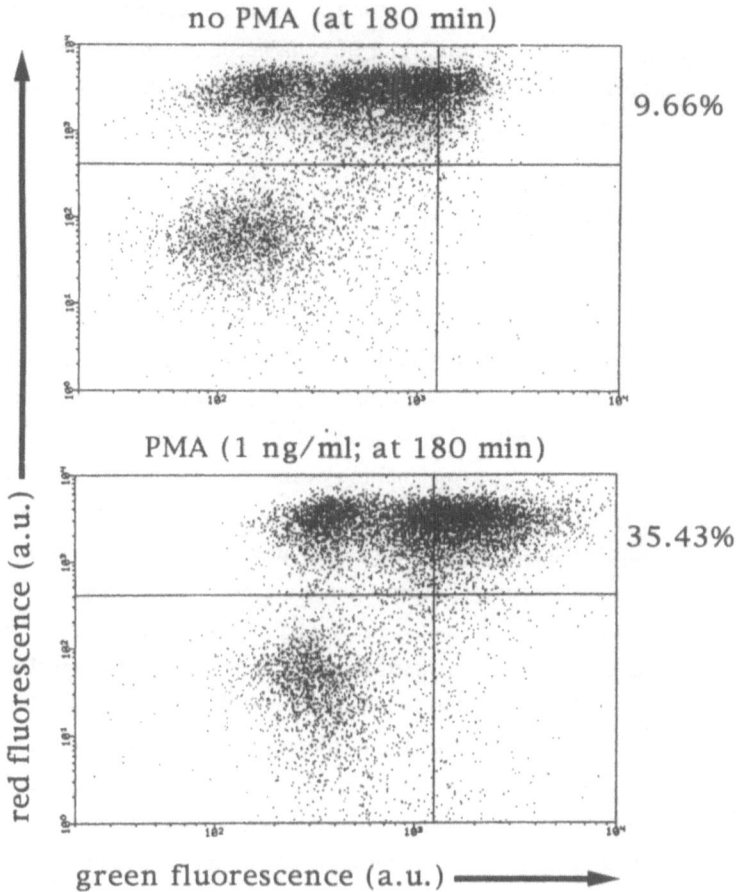

no PMA (at 180 min)

9.66%

PMA (1 ng/ml; at 180 min)

35.43%

red fluorescence (a.u.)

green fluorescence (a.u.)

Figure 2. Dichlorofluorescin (DCFH) oxidation in mouse lymph node T cells. The cells were sequentially stained with biotin-conjugated anti-mouse Thy-1.2 and phycoerythrin-streptavidin (red fluorescence) prior to loading with DCFH (green fluorescence) in the form of the diacetate (Bass *et al.*, 1983) for 15 min. Flow cytometric analysis was carried out in unstimulated cells or those exposed to PMA for 60 minutes.

Given the lack of a functional NADPH oxidase in T lymphocytes (Pick and Gadba, 1988), the source of the intracellular ROS demonstrated by us (Figure 2) and others (Sekkat *et al.*, 1988; Rabesandratana *et al.*, 1992) is not clear. A number of candidate mechanisms have been canvassed, including arachidonate metabolism (Chaudhri *et al.*, 1986; Dornand and Gerber, 1989), but the evidence is not definitive. Whether or not this intracellular ROS formation is related to mitogenic induction is not established but, as shown below, the anti-oxidant cysteamine concurrently inhibits DCF oxidation, transcription factor binding to DNA, IL-2 production and DNA synthesis in T cells, suggesting that such a relationship does exist. Others have suggested that the basal redox equilibrium of cells needs to tend towards oxidation to facilitate, but not induce, activation of the transcription factor NF-κB and the HIV long terminal repeat, in studies carried out in cell lines (Israel et al., 1992).

Transcription factors

The time after mitogenic stimulation of T cells that anti-oxidants act, and by inference the time at which the putative oxidative event occurs, has been established for some agents. It is clear that not all operate through the same mechanism (Sekkat et al., 1988). Cysteamine acts within 2 hours of mitogenic activation in HPBLs, but its mode of action has not been clearly identified. One possibility is that it inhibits an early signalling event, for example the activity of one or more protein kinases, and this possibility is under investigation in our laboratory. Given the important role of transcription factors in the early events after T cell activation that lead to gene expression and cell cycle progression (Lenardo and Baltimore, 1989; Granelli-Piperno and Nolan, 1991; Carter et al., 1991), and since some contain cysteine groups essential for their activity, they are obvious candidates for oxidative modification as part of a signalling cascade. Exogenous ROS have indeed been shown to activate the transcription factor NF-κB and thereby induce replication of latent HIV in certain cell lines (Schreck et al., 1991). Some anti-oxidants, notably N-acetylcysteine and pyrrolidine dithiocarbamate, inhibit these processes (Staal et al., 1990; Schreck et al., 1991, 1992). Cysteamine could interfere with such processes by reacting directly with ROS, or by reducing oxidised groups on transcripition factors or other components of signalling pathways.

We have found that cysteamine blocks the increased DNA binding of AP-1 and NF-κB that follows mitogen addition to murine and human T cells (S.D. Goldstone and N.H. Hunt, unpublished observations). However, the aminothiol does not affect the binding of the constitutive transcription factor SP-1. Under the same conditions, cysteamine inhibited intracellular DCFH oxidation, IL-2 mRNA and IL-2 protein production, and DNA synthesis. A speculative interpretation of these results is that mitogens induce in T cells the intracellular formation of ROS and that this leads to oxidation of AP-1 and/or NF-κB, which might be necessary for their transport into the nucleus. Binding of transcription factors to DNA only occurs under reducing conditions (Bannister et al., 1991), and there is a redox factor, Ref-1, in the nucleus capable of reducing oxidised transcription factors (Xanthoudakis et al., 1992). Among the genes regulated by AP-1 is that for IL-2 (Granelli-Piperno and Nolan, 1991), production of which is essential for T cell passage through G_1 into DNA synthesis. As alluded to earlier, it also is quite possible that cysteamine inhibits an upstream event in T cell activation and thereby prevents activation of transcription factors. These possibilities are under investigation in our laboratory.

Concluding remarks

Our work and that of others, notably Dornand and Gerber, has shown that an intracellular oxidative event is essential for T cell commitment to the cell cycle. This may well be linked to gene expression through redox modulation of transcription factor activity. A critical factor in our discovery of this process has been our decision to work with resting T lymphocytes. Work on cell cycle control in continuously-cycling cell lines is difficult without effective methods for synchronisation. This need is obviated in resting T cells which are positioned in G_0, awaiting recruitment into the cycle. Another important aspect of our experimental approach has been to study only effects related to intracellular oxidative events induced in T lymphocytes by mitogenic stimuli. Addition of exogenous ROS is likely to provoke a number of responses, including the expression of genes for anti-oxidant enzymes and other inter-related responses to environmental stress, which may be difficult to dissect away from those responses related to mitogenic activation.

A number of studies, including presentations at this conference, have established that the replication of some viruses, such as HIV and SV40, may be induced by oxidants. One possible source of such oxidants *in vivo* could be formation by phagocytes as a consequence of the autotoxic reactions provoked by certain viruses (Peterhans *et al.*, 1989). Another possibility relates to our own work, where we have shown that intracellular oxidant formation is a feature of T lymphocyte activation. T cells are infected by HIV, and it may be that some such viruses have developed replication cues linked to the signalling mechanisms of T cells, and that ROS formation may be one of the most fundamental of these cues. The ability of certain ROS to activate NF-κB and HIV replication (Schreck *et al.*, 1991) is consistent with this possibility. For many years our thinking in free radical research has been dominated by consideration of the biologically damaging effects of radicals. It is ironic that just as we are coming to appreciate that they also may serve a regulatory role in such a fundamental process as T cell activation it is also becoming clear that this role is linked to the survival of a devastating infectious agent like HIV.

248

Acknowledgements

This work was supported by the National Health & Medical Research Council of Australia and the University of Sydney Cancer Research Fund.

References

Bannister, A.J., Cook, A., and Kouzarides, T. (1991) *In vitro* DNA binding activity of Fos/Jun and BZLF1 but not C/EBP is affected by redox changes. *Oncogene* 6: 1243-1250.

Bass, D.A., Parce, J.W., Dechatelet, L.R., Szedja, P., Seeds, M.C., and Thomas, M. (1983) Flow cytometric studies of oxidative product formation by neutrophils: a graded response to membrane stimulation. *J. Immunol.* 130: 1910-1917.

Bierer, B.E., and Nathan, D.G. (1990) The effect of desferrithiocin, an oral iron chelator, on T-cell function. *Blood* 76: 2052-2059.

Blatt, J., Taylor, S.R., and Kontoghiorghes, G.J. (1989) Comparison of activity of deferoxamine with that of oral iron chelators against human neuroblastoma cell lines. *Cancer Res.* 49: 2925-2927.

Carotenuto, P., Pontesilli, O., Cambier, J.C., and Hayward, A.R. (1986) Desferoxamine blocks IL 2 receptor expression on human T lymphocytes. *J. Immunol.* 136: 2342-2347.

Carter, R., Cosenza, S.C., Pena, A., Lipson, K., Soprano, D.R., and Soprano, K.J. (1991) A potential role for c-jun in cell cycle progression through late G_1 and S. *Oncogene* 6: 229-235.

Celano, P., Baylin, S.B., and Casero, R.A.,Jr. (1989) Polyamines differentially modulate the trnscription of growth-associated genes in human colon carcinoma cells. *J. Biol. Chem.* 264: 8922-8927.

Chaudhri, G., Clark, I.A., Hunt, N.H., Cowden, W.B., and Ceredig, R. (1986) Effect of antioxidants on primary alloantigen-induced T cell activation and proliferation. *J. Immunol.* 137: 2646-2652.

Chaudhri, G., Hunt, N.H., Clark, I.A., and Ceredig, R.H. (1988) Antioxidants inhibit proliferation and cell surface expression of receptors for interleukin 2 and transferrin in T lymphocytes stimulated with PMA and ionomycin. *Cell. Immunol.* 115: 204-213.

Chinnadurai, G. (1991) Modulation of HIV-enhancer activity by heterologous agents: A minireview. *Gene* 101: 165-170.

Darley-Usmar, V.M., Hersey, A., and Garland, L.G. (1989) A method for the comparative assessment of antioxidants as peroxyl radical scavengers. *Biochem. Pharmacol.* 38: 1465-1469.

Dornand, J. and Gerber, M. (1989) Inhibition of murine T-cell responses by anti-oxidants: The targets of lipo-oxygenase pathway inhibitors. *Immunology* 68: 384-391.

Dröge, W., Eck, H.-P., Näher, H., Pekar, U., and Daniel, V. (1988) Abnormal amino-acid concentrations in the blood of patients with acquired immunodeficiency syndrome (AIDS) may contribute to the immunological defect. *Biol. Chem. Hoppe-Seyler* 369: 143-148.

Estrov, Z., Tawa, A., Wang, X-H., Dubé, I.D., Sulh, H., Cohen, A., Gelfand, E.W., and Freedman, M.H. (1987) In vitro and in vivo effects of deferoxamine in neonatal acute leukemia. *Blood* 69: 757-761.

Fidelus, R.K. and Tsan, M.-F. (1986) Enhancement of intracellular glutathione promotes lymphocyte activation by mitogen. *Cell. Immunol.* 97: 155-163.

Fidelus, R.K. (1988) The generation of oxygen radicals: a positive signal for lymphocyte activation. *Cell. Immunol.* 113: 175-182.

Ganeshaguru, K., Hoffbrand, A.V., Grady, R.W., and Cerami, A. (1980) Effect of various iron chelating agents on DNA synthesis in human cells. *Biochem. Pharmacol.* 29: 1275-1279.

Granelli-Piperno, A. and Nolan, P. (1991) Nuclear transcription factors that bind to elements of the IL-2 promoter: Induction requirements in primary human T cells. *J. Immunol.* 147: 2734-2739.

Halliwell, B. and Gutteridge, J.M.C. (1989) *Free radicals in biology and medicine*, Oxford, Clarendon Press.

Hamilos, D.L., Zelarney, P., and Mascali, J.J. (1989) Lymphocyte proliferation in glutathione-depleted lymphocytes: Direct relationship between glutathione availability and the proliferative response. *Immunopharmacology* 18: 223-235.

Harakeh, S. and Jariwalla, R.J. (1991) Comparative study of the anti-HIV activities of ascorbate and thiol-containing reducing agents in chronically HIV-infected cells. *Am. J. Clin. Nutr.* 54 Suppl.: 1231S-1235S.

Hoffbrand, A.V., Ganeshaguru, K., Hooton, J.W.L., and Tattersall, M.N.H. (1976) Effect of iron deficiency and desferrioxamine on DNA synthesis in human cells. *Br. J. Haematol.* 33: 515-552.

Hunt, N.H., Chaudhri, G., Buffinton, G.D., Cook, E., Cowden, W.B., Clark, I.A., and Dorsch, S.E. (1989) Iron and infectious disease. In: *Free Radicals, Metal Ions and Biopolymers*, 233-247. Edited by Beaumont, P., Deeble, D., Parsons, B., and Rice-Evans, C., London, Richelieu Press.

Hunt, N.H., Cook, E.P., and Fragonas, J.-C. (1991) Interference with oxidative processes inhibits proliferation of human peripheral blood lymphocytes and murine B-lymphocytes. *Int. J. Immunopharmacol.* 13: 1019-1026.

Hunt, N.H. and Fragonas, J.-C. (1992) Effects of anti-oxidants on ornithine decarboxylase in mitogenically-activated T lymphocytes. *Biochim. Biophys. Acta* 1133: 261-267.

Hunt, N.H., Fragonas, J.-C., Kneale, C.L., Jeitner, T.M., and Van Reyk, D.M. (1992) Anti-oxidants and T cell activation. In: *Active Oxygens, Lipid Peroxides and Antioxidants*, 719-722. Edited by Yagi, K., Kondo, M., Niki, E., and Yoshikawa, T., Amsterdam, Elsevier Excerpta Medica.

Hwang, C., Sinskey, A.J., and Lodish, H.F. (1992) Oxidized redox state of glutathione in the endoplasmic reticulum. *Science* 257: 1496-1502.

Israel, N., Gougerot-Pocidalo, M.-A., Aillet, F., and Virelizier, J.L. (1992) Redox status of cells influences constitutive or induced NF-kB translocation and HIV long terminal repeat activity in human T and monocytic cell lines. *J. Immunol.* 149: 3386-3393.

Jeitner, T.M. and Hunt, N.H. (1990) Cysteamine and its phenolic analogues inhibit the proliferation of normal, leukemic and drug resistant leukemic T cells. *Free Radic. Biol. Med.* 9: 172.

Jeitner, T.M. and Oliver, J.R. (1990) Possible oncostatic action of cysteamine on the pituitary glands of oestrogen-primed hyperprolactinaemic rats. *J. Endocrinol.* 127: 119-127.

Kalebic, T., Kinter, A., Poli, G., Anderson, M.E., Meister, A., and Fauci, A.S. (1991) Suppression of human immunodeficiency virus expression in chronically infected monocytic cells by glutathione, glutathione ester, and *N*-acetylcysteine. *Proc. Natl. Acad. Sci. USA* 88: 986-990.

Kolbuch-Braddon, M.E., Peterhans, E., Stocker, R., and Weidemann, M.J. (1984) Oxygen uptake associated with Sendai-virus-stimulated chemiluminescence in rat thymocytes contains a significant non-mitochondrial component. *Biochem. J.* 222: 541-551.

Lacal, P.M., Balsinde, J., Cabañas, C., Bernabeu, C., Sánchez-Madrid, F., and Mollinedo, F. (1990) The CD11c antigen couples concanavalin A binding to generation of superoxide anion in human phagocytes. *Biochem. J.* 268: 707-712.

Lederman, H.M., Cohen, A., Lee, J.W.W., Freedman, M.H., and Gelfand, E.W. (1984) *Blood* 64: 748-753.

Lenardo, M.J. and Baltimore, D. (1989) NF-kB: a pleiotropic mediator of inducible and tissue-specific gene control. *Cell* 58: 227-229.

Maly, F.-E., Nakamura, M., Gauchat, J.-F., Urwyler, A., Walker, C., Dahinden, C.A., Cross, A.R., Jones, O.T.G., and De Weck, A.L. (1989) Superoxide-dependent nitroblue tetrazolium reduction and expression of cytochrome b-245 components by human tonsillar B lymphocytes and B cell lines. *J. Immunol.* 142: 1260-1267.

Novogrodsky, A., Ravid, A., Rubin, A.L., and Stenzel, K.H. (1982) Hydroxyl radical scavengers inhibit lymphocyte mitogenesis. *Proc. Natl. Acad. Sci. USA* 79: 1171-1174.

Parnham, M.J. and Graf, E. (1987) Seleno-organic compounds and the therapy of hydroperoxide-linked pathological conditions. *Biochem. Pharmacol.* 36: 3095-3102.

Pardee, A.B. (1989) G_1 events and regulation of cell proliferation. Science 246: 603-608.

Peterhans, E., Jungi, T.W., and Stocker, R.S. (1989) Autotoxicity and reactive oxygen in viral disease. In: *Oxy-Radicals in Molecular Biology and Pathology*, 543-562. Edited by Cerutti, P.A., Fridovich, I., and McCord, J.M., New York, Alan R. Liss.

Pick, E. and Gadba, R. (1988) Certain lymphoid cells contain the membrane-associated component of the phagocyte-specific NADPH oxidase. *J. Immunol.* 140: 1611-1617.

Pollack, S., Aisen, P., Lasky, F.D., and Vanderhoff, G. (1976) Chelate mediated transfer of iron from transferrin to desferrioxamine. *Brit. J. Haematol.* 34: 231-235.

Polson, R.J., Jenkins, R., Lombard, M., Williams, A.C., Roberts, S., Nouri-Aria, K., Williams, R., and Bomford, A. (1990) Mechanisms of inhibition of mononuclear cell activation by the iron-chelating agent desferrioxamine. *Immunology* 71: 176-181.

Rabesandratana, H., Fournier, A.-M., Chateau, M.-T., Serre, A., and Dornand, J. (1992) Increased oxidative metabolism in PMA-activated lymphocytes: a flow cytometric study. *Int. J. Immunopharmacol.* 14: 895-902.

Roederer, M., Staal, F.J.T., Raju, P.A., Ela, W.S., and Herzenberg, L.A. (1990) Cytokine-stimulated human immunodeficiency virus replication is inhibited by N-acetyl-L-cysteine. *Proc. Natl. Acad. Sci. USA* 87: 4884-4888.

Sagone, A.L.,Jr., Husney, R., Guter, H., and Clark, L. (1984) Effect of catalase on the proliferation of human lymphocytes to phorbol myristate acetate. *J. Immunol.* 133: 1488-1494.

Schreck, R., Meier, B., Mannel, D.N., Droge, W., and Baeuerle, P.A. (1992) Dithiocarbamates as potent inhibitors of nuclear factor kB activation in intact cells. *J. Exp. Med.* 175: 1181-1194.

Schreck, R., Rieber, P., and Baeuerle, P.A. (1991) Reactive oxygen intermediates as apparently widely used messengers in the activation of the NF-kB transcription factor and HIV-1. *EMBO J.* 10: 2247-2258.

Sekkat, C., Dornand, J., and Gerber, M. (1988) Oxidative phenomena are implicated in human T-cell stimulation. *Immunology* 63: 431-437.

Smit, M.J. and Anderson, R. (1990) Inhibition of mitogen-activated proliferation of human lymphocytes by hypochlorous acid *in vitro*: Protection and reversal by ascorbate and cysteine. *Agents and Actions* 30: 338-343.

Staal, F.J.T., Roederer, M., Herzenberg, L.A., and Herzenberg, L.X. (1990) Intracellular thiols regulate activation of nuclear factor kB and transcription of human immunodeficiency virus. *Proc. Natl. Acad. Sci. (USA)* 87: 9943-9947.

Suthanthiran, M., Anderson, M.E., Sharma, V.K., and Meister, A. (1990) Glutathione regulates activation-dependent DNA synthesis in highly purified normal human T lymphocytes stimulated via the CD2 and CD3 antigens. *Proc. Natl. Acad. Sci. USA* 87: 3343-3347.

Terada, N., Or, R., Szepesi, A., Lucas, J.J., and Gelfand, E.W. (1993) Definition of the roles for iron and essential fatty acids in cell cycle progression of normal human T lymphocytes. *Exp. Cell Res.* 204: 260-267.

Ullman, K.S., Northrop, J.P., Verweij, C.L., and Crabtree, G.R. (1990) Transmission of signals from the T lymphocyte antigen receptor to the genes responsible for cell proliferation and immune function: The missing link. *Annu. Rev. Immunol.* 8 421-452.

Unanue, E.R. (1992) Cellular studies on antigen presentation by class II MHC molecules. *Curr. Opinion Immunol.* 4: 63-69.

van Reyk, D.M., Sarel, S., and Hunt, N.H. (1992) Inhibition of lymphocyte proliferation by three putative iron chelators: Identification of differences in their mechanisms of action. *Int. J. Immunopharmacol.* 14: 925-932.

Weidemann, M.J., Yin Foo, D., Kolbuch-Braddon, M.E., Mitchell, S.V., Domzig, W., and Peterhans, E. (1987) Characterisation of chemiluminescence-generating subpopulations within the rodent thymus. In: *Proceedings of the Fourth International Bioluminescence Symposium*, 109-112. Edited by Schoelmerich, J., New York, Wiley.

White, M.W., Kameji, T., Pegg, A.E., and Morris, D.R. (1987) Increased efficiency of translation of ornithine decarboxylase mRNA in mitogen-activated lymphocytes. *Eur. J. Biochem*. 170: 87-92.

Wrogemann, K., Weidemann, M.J., Peskar, B.A., Staudinger, H., Rietschel, E.T., and Fischer, H. (1978) Chemiluminescence and immune activation. 1. Early activation of rat thymocytes can be monitored by chemiluminescence measurements. *Eur. J. Immunol*. 8: 749-752.

Xanthoudakis, S., Miao, G., Wang, F., Pan, Y.-C.E., and Curran, T. (1992) Redox activation of Fos-Jun DNA binding activity is mediated by a DNA repair enzyme. *EMBO J*. 11: 3323-3335.

Yudkoff, M., Foreman, J.W. and Segal, S. (1981) Effects of cysteamine therapy in nephropathic cystinosis. *N. Engl. J. Med*. 304: 141-145.

Yuhas, J.M. (1980) On the potential application of radioprotective drugs in solid tumor radiotherapy. In: *Radiation-Drug Interactions in the TReatment of Cancer*, 113-134. Edited by Sokol, G.H. and Maickel, R.P., New York, Wiley.

Zoschke, D.C. and Staite, N.D. (1987) Suppression of human lymphocyte proliferation by activated neutrophils or H_2O_2: surviving cells have an altered T helper/T suppressor ratio and an increased resistance to secondary oxidant exposure. *Clin. Immunol. Immunopathol*. 42: 160-170.

Oxidative Stress, Cell Activation and Viral Infection
C. Pasquier et al. (eds)
© 1994 Birkhäuser Verlag Basel/Switzerland

Implication of Oxydative Phenomena in T Cell Activation

M. Gerber[1] and J. Dornand[2]

[1]INSERM, INSERM-C.R.L.C., 34094 Montpellier Cedex 5, France.
[2]INSERM U 65, U.S.T.L., 34095 Montpellier Cedex 5, France.

Summary

We have shown that free radical scavengers inhibit IL2 synthesis and that reactive oxygen species are induced upon T cell activation. Because of their physiological relevance, we concentrated our study on lipoxygenase (LO) inhibitors. We investigated the mechanisms by which LO inhibitors decrease IL2 production in Jurkat cells. We demonstrated that the inhibition, linked to blockade of the [Ca2+]i rise involving TCR triggering, resulted from the action of these compounds on the signal transduction, upstream from inositol-triphosphate synthesis. When the breakdown of inositol phospholipids induced by the ligand-receptor interaction was bypassed, e.g. after PMA+A23187 stimulation, IL2 secretion was also suppressed by LO inhibitors, but at much higher concentrations. This indicates that the event depending upon ligand receptor binding are the main target of LO inhibitors. None of three PKC-dependent events investigated were affected in Jurkat cells, stimulated in the presence of LO inhibitors. Furthermore, these compounds did not inhibit IL2 production in PMA -treated Jurkat cells cultured with vanadate, which maintains protein phosphorylation on tyrosine residues and induces IL2 secretion. This suggests that LO inhibitors might affect the tyrosine kinase pathway in TCR- activated Jurkat cells. These results are consistent with a role for LO metabolite(s) in signal transduction pathways. However, some other redox mechanism might be involved at other steps of IL2 synthesis, as indicated by the partial effect of LO inhibitors when the ligand-receptor binding is by-passed, and by the effect of other types of anti-oxidants.

Introduction

Several authors reported that irradiation enhanced IL2 production by T cells (rev in Gerber et al., 1985). The explanation which proposed that suppressor CD8 cells were more radiosensitive than IL2-CD4 producing cells was non-tenable radiobiologically (Gerber, 1984 and Gerber et al, 1989). The other explanation based on the IL2-non-consumption of non-proliferating cells was inadequate since the kinetics of IL2 production did not coincide with the kinetics of cell proliferation. Therefore, another explanation had to be found: Novogrodsky (1982) showed that some free radical scavengers (FRS) inhibited PMA-induced mitogenesis and suggested that hydroxyl radical was a signal for cell proliferation. Other authors also suggested that reactive oxygen species were involved in T cell activation, (Wrongemann et al., 1978). Thus, we decided to investigate this hypothesis in our model. We showed first that dimethyl-sulfoxide (DMSO) an hydroxyl radical scavenger and nor-dihydro-guaraïetic acid (NDGA), a peroxidation chain blocking agent with an effect more specifically directed at the lipoxygenase (LO) pathway impaired IL2 production (Gerber et al., 1985). We decided then to look for for foot-prints of reactive oxygen species (ROS) and to focus on lipoxygenase inhibitors because of their physiological relevance and also because several authors suggested their involvment in cytokines production (Farrar and Humes, 1985; Kato et al,1986; Rola-Pleszczynski, 1988).

Materials and Methods

Chemicals and reagents

Phorbol-12 myristate-13 (PMA), indomethacin, orthovanadate, Quin 2-AM, H7 and calcium ionophore (A23187) were purchased from Sigma, Phytohemagglutinin (PHA) from Difco, RPMI 1640 and fetal calf serum (FCS) from GIBCO, tritiated thymidine from CEA (France) and myo-^{3}[H]inositol from Amersham. Staurosporine was obtained from Ciba-Geigy, Basel Switzerland; 2',7'-dichlorofluoresceine diacetate (DCFH-DA) from eastman Kodak; BCECF (2-7bis [carboxyethyl] 5-6 carboxyfluorescein) is a product of HSC Research Development Corporation, Toronto, Canada. Leukotrienes were kindly supplied by Dr. Rokach (Merck Frost Laboratories, Dorval,Canada).

Free radical scavengers and LO inhibitors

The hydroxyl anion scavenger dimethyl sulfoxide (DMSO) and the LO inhibitors NDGA, BHA, esculetin were purchased from Sigma. Other compounds were obtained from various compagnies: Desferal from Ciba-Geigy, Basel, Switzerland; CBS1108 from Chauvin-Blache, Montpellier, France (Bertez et al.,1984); ETYA from Hoffmann-Laroche, Nutley, New Jersey USA; AA861 from Takeda, Osaka, Japon; BW A4C from Wellcome laboratories, Beckenham, England.

When drugs were present in the cultures throughout the activation, the absence of toxicity was verified by trypan blue exclusion.

Measurement of oxidative products

Cell suspensions were processed by the method of Bass et al (1983) using DCFH-DA. EDTA was added after the incubation with the fluorchrome to avoid cell aggregation which precludes the use of a cytofluorimeter. PHA and PMA were added to the cell suspension immediately before cytometry. In some experiments irradiation (^{60}Co gammatron at a dose rate of 1.10 Gy/mn in air at time of the experimentation)was performed 35mn prior to stimulation and 140mM DMSO was added prior to irradiation and stimulation. At least 10 000 cells were examined in each sample and the number of fluorescent cells was recorded.

Cells

The Jurkat cell line and the IL2 -dependent cytotoxic T cell line (CTLL) were previously used in our laboratory (Gerber et al, 1985). They were all maintained in RPMI supplemented with 5% FCS; CTLL cells were cultured in the presence of 50 U/ml recombinant IL2 (r-IL2), a gift from P. Casellas , SANOFI, Montpellier .

IL2 production and determination

Jurkat cells (10^6/ml in FCS-RPMI) were stimulated for 24 h at 37°C with 5 µg/ml PHA, or 60 ng/ml A23187, or 500 µM $VO_4^=$, in the presence of 10 ng / ml PMA. The cells were then centrifuged, the supernatants filtered through O.2 µm Millipore filters and extensively dialysed . Their IL2 content was determined in a bioassay with the CTLL cells (Gerber et al , 1985). A reference standard for IL2 unitage, established by the Biological Response Modifier Program (BRMP) of the National Cancer Institute (USA) (Rossio et al., 1985), was used to standardize the IL2 assays. When IL2 production was performed in the presence of different drugs, control experiments of CTLL proliferation showed that these compounds did not affect proliferation at the concentrations used in the assayed supernatant dilutions (data not shown).

Phosphoinositide breakdown

Jurkat cells prepared in an inositol free -RPMI medium, were labelled for 18-20 h with 5 µCi myo-2^3[H]inositol. They were washed twice and resuspended in the same medium, then incubated for 20 min at 37°C in the absence or presence of indicated concentrations of different LO inhibitors with 10mM lithium . Aliquots (15×10^6 cells/ml) were treated with 5 µg/ml ConcanavalinA (ConA). The reaction was stopped 20 min later with perchloric acid, and the inositol phosphate level of each sample (IPs= IP1+IP2+IP3) was measured after chromatography on an anion exchange Dowex 1X-8, as described (El Moatassim et al., 1989).

Modulation of the CD3/TCR complex

Jurkat cells (10^6/ml) were cultured for 2 h at 37°C in culture medium supplemented or not with 10 ng/ ml PMA in the presence or absence of different LO inhibitors, FRS or PKC inhibitors. They were then centrifuged, and incubated on ice for 45min in 100µl RPMI with 5 % FCS, 0.1 % NaN3 supplemented or not with 10 µg FITC-conjugated anti-CD3 antibody (Ortho), washed twice with PBS-0.1 % NaN3, and analyzed on an Epics flow cytometer (Coulter Electronics).

Evaluation of IL2 receptor 55Kd chain (IL2R) expression by flow cytometry

Jurkat cells (2×10^6) were cultured in FCS-RPMI for 24 h with or without the different activators, in the presence or absence of various compounds at the indicated concentrations . They were washed twice and incubated on ice for 45 min in 100µl RPMI with 5 % FCS, 0.1 % NaN3 supplemented or not with 2 µg anti-IL2R antibody (anti-Tac provided by Dr. T.A.Waldmann, Bethesda). After two washings, the cells were incubated with FITC- conjugated anti-mouse IgG for 45min at 4°C, washed twice with PBS, 0.1 % NaN3 and analyzed .

Measurement of cytoplasmic pH

Cytoplasmic pH in Jurkat cells was measured fluorimetrically with the BCECF fluorescent probe, exactly as described in Grinstein et al (1985). Briefly, cell suspensions in PBS were incubated (50×10^6/ml) with 2.5µM BCECF for 30 min at 37°C, then extensively washed and resuspended at 5×10^6 /ml in buffer pH 7.2 (140mM NaCl, 1mM KCl, 1mM CaCl2,1 mM MgCl2, 10 mM glucose, 20 mM Tris /2 [N-morpholino] ethane sulfonic acid). Fluorescence of the suspension was measured (excitation wavelength 492 nm, emission wavelength 528 nm) and intra-cellular pH was determined as in Grinstein et al, (1985) before and after addition of PMA.

LTB4 detection

The Amersham kit was used for the RIA. The technics of radio-chemical analysis and thin layer chromatography have been described in Gerber et al (1985) and Gerber et al (1988).

Results

Measurement of oxidative products in Jurkat cells

DCFH-DA rapidly diffuses into the cells and is hydrolyzed in 2', 7' dichlorofluorescein (DCF-H), which remains stable for few hours. Non-fluorescent DCF-H becomes fluorescent upon oxidation by ROS (Figure 1). When Jurkat cells were pre-treated with DCF-DA, the number of fluorescent cells increased about twofold after PHA+PMA stimulation (2897 ± 389 versus 5746 ± 171, p=0.05). Irradiation increased the number (8632 ± 250, p<0.01) which remains comparable after irradiation plus stimulation. DMSO decreased the number of fluorescent cells to 916 ± 755 for the baseline to 180 ± 223 for the stimulated cells, 1507 ± 640 for the irradiated cells and 2403 ± 1071 for the irradiated and stimulated cells.

Effect of various LO inhibitors on IL2 production by Jurkat cells submitted to different treatments

We had shown before (Gerber et al 1985; Dornand et al, 1987) that NDGA inhibited IL2 production by PHA-PMA treated Jurkat cells. Since NDGA possesses unspecific FR scavenging properties beside blocking LO, it was compared to different LO inhibitors. At non-toxic concentrations, all of the different drugs assayed dose-dependently inhibited IL2 production in PHA+PMA -treated Jurkat cells. Esculetin appeared the less potent of all compounds tested and AA861 the most specific, the IC50s of both products being 20μM and 1μM respectively (Fig. 2 and 3); the other drugs had IC50s ranging from 5 to15μM.

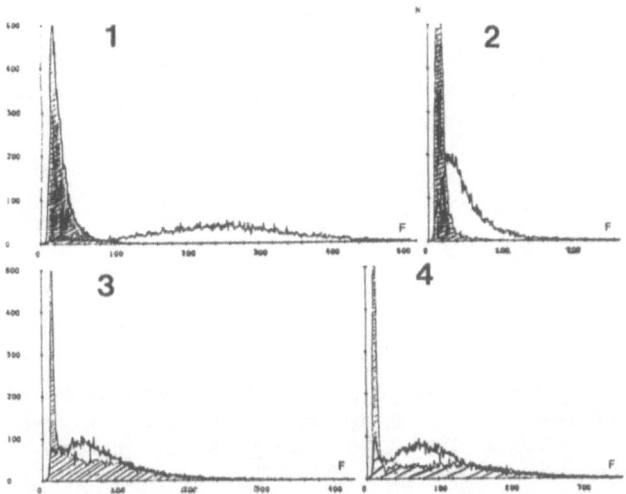

Figure 1. 1-Fluorescence profile of 10 000 DCFH-DA loaded Jurkat cells 35 mn after 20Gy irradiation (hatched profile, non irradiated cells; open profile, irradiated cells). 2- Fluorescence profile of 10 000 DCFH-DA loaded Jurkat cells 5 mn after PHA (20μg) and PMA (10ng); (hatched profile, non stimulated cells; open profile, stimulated cells). 3-Fluorescence profile of 10 000 DCFH-DA loaded Jurkat cells 35 mn after 20Gy irradiation (hatched profile DCFH-DA loaded Jurkat cells treated with 140mM DMSO prior to irradiation; open profile, no treatment). 4- Fluorescence profile of 10 000 DCFH-DA loaded Jurkat cells 20 mn after PHA (20μg) and PMA (10ng) and 35 mn after irradiation; (hatched profile DCFH-DA loaded Jurkat cells treated with 140mM DMSO prior to stimulation and irradiation; open profile, no treatment).

When Jurkat cells were activated by PMA+A23187, the LO inhibitors still inhibited IL2 production, but, higher concentrations of these compouds were required. This is shown for AA861 in Fig. 3. Similar results were obtained with the other compounds. The FRSs DMSO and Desferal also suppressed IL2 production whatever the activation pathway (not shown).

Kinetics of the effect of LO inhibitors on IL2 production

In this experiment, ConA was used instead of PHA, because it can be removed from the cell surface with α-methyl mannoside (α-MM); the inhibitory effects of NDGA, (or of other LO inhibitors, not shown) on IL2 production, were strongly significant only when these drugs were present at the onset of the activation, and during the first hours required for optimal transduction of the membrane signals (Weiss et al., 1987), (Table I). When added after this period of time,

258

they did not affect IL2 production. Furthermore, this experiment confirmed that the drug effect did not result from toxicity since Jurkat cells were still able to produce IL2 when the drugs were added to the cell cultures 5 h after stimulation, and to the cells cultured for further 19 h.

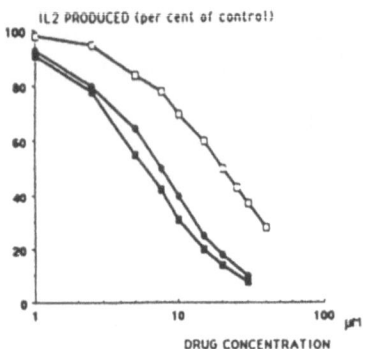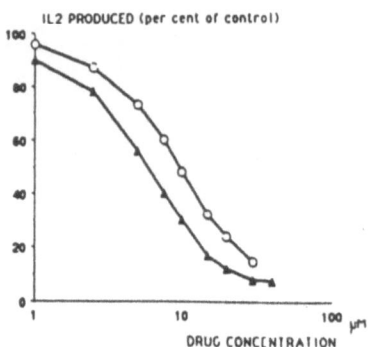

Figure 2: Effect of increasing concentrations of LO inhibitors on IL2 production in Jurkat cells.
PMA treated Jurkat cells were stimulated with 10μg/ ml PHA for 24h in the presence of NDGA (■), ETYA (▲),esculetin (□), CBS 1108 (○), and BW A4C (●) (x-axis). The IL2 level of the supernatants is expressed as percentage of the control (y-axis) (IL2 control value: 36 ± 5 U/ml).

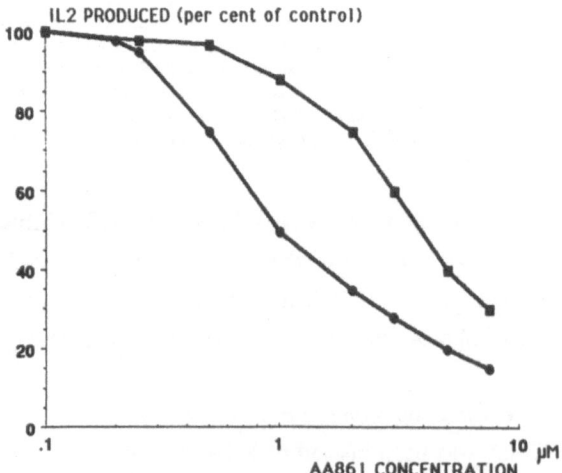

Figure 3: Effect of increasing concentrations of AA861 (x-axis) on IL2 production in PMA-treated Jurkat cells activated with10μg/ml PHA (●) or 200ng/ml A23187 (■). The IL2 level of the supernatants is expressed as percentage of the controls (y-axis); IL2 control values were 36 ± 5 U/ml and 48 ± 7 U/ml respectively for PHA- and A23187- activated cells.

Similar results were obtained when A23187 was substituted for ConA, and removed from the culture by washing.

These results established that LO inhibitors affected the early steps of the cascade of events leading to IL2 production.

Table I. Kinetics of the effect of LO inhibitors on IL2 (units/ml) production.

	Incubation time of stimuli and inhibitors			
Stimulus 1	0-24 h	0-5 h[*]	0-24 h	0-5 h[*]
Inhibitors 2	0-24 h	0-5 h[*]	0-5 h	5-24 h
1 ConA-PMA 2 None	24 ±2	15 ±3	17 ±4	15 ±3
1 ConA-PMA 2 NDGA 20μM	5 ±1	4 ±1	13±1	13 ±1
1 ConA-PMA 2 NDGA 30μM	2.5 ±0.7	<1	13 ±2	11.5±1.5
1 A23187-PMA 2 None	32 ±3	25 ±4	N.D.	25±4
1 A23187-PMA 2 NDGA 30μM	17 ±1	10 ±2	N.D.	22 ±2
1 A23187-PMA 2 NDGA 40μM	5±1	3 ±1	N.D.	20 ±1

Jurkat cells were activated by the indicated stimuli at the initiation of the culture in presence or absence of NDGA and DMSO. In some experiments*, the cells were washed with α-MM 5 h after initiation of the culture and cultured in fresh medium containing α–MM with or without the indicated stimuli 1 and inhibitors 2. IL2 produced in the supernatant was measured 24 h after the first stimulus and expressed as units/ml.

Effect of LO inhibitors on $[Ca^{++}]i$ rise

Since LO inhibitors affect an early event (s) of the activation pathway, we tested their effect on the $[Ca^{++}]i$ rise resulting from the interaction of a specific ligand with the T cell receptor (TCR). Figure 4 shows that NDGA and BHA impaired $[Ca^{++}]i$ rise in a dose-dependent way after PHA stimulation. NDGA at 10 μM completely blocked $[Ca^{++}]i$ rise after triggering by anti-CD3 antibody; this inhibition is independent from extracellular calcium since it was still observed in calcium-free medium.

260

Table II shows that there was no effect of NDGA and BHA when [Ca++]i rise was induced by calcium ionophore instead of PHA, by-passing the ligand-receptor binding. DMSO and desferal had no effect on [Ca^{2+}]i whatever the type of activation (Fig. 5).

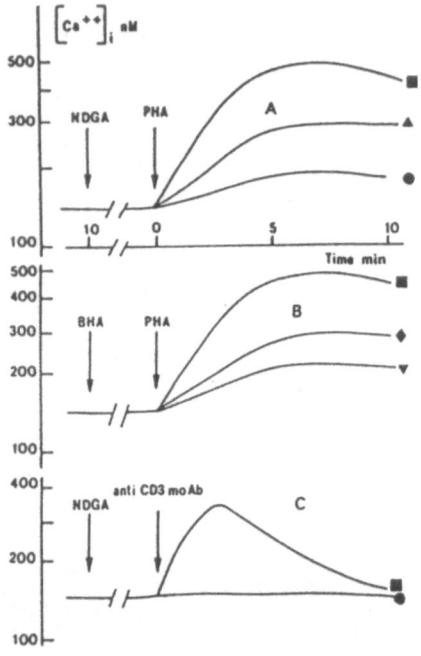

Figure 4. Inhibitory effect of NDGA and BHA on [Ca++]i rise induced in Jurkat cells. Cells were loaded with Quin2-AM and their fluorescence monitored. Where indicated (arrow) 10µg/ml PHA or 10ng/ml anti-CD3 MoAb was added to the cell suspension (time0). [Ca^{2+}]i rise was calculated and plotted as a function of time (■). In some experiments 7µM (▲) or 20µM(●)NDGA curve A and 50mM (◆) or 100µM BHA (▼) (curveB) was added to the suspension 10mn before the stimulus. In curve C, the anti-CD3 MoAb triggered [Ca^{2+}]i rise was determined in calcium-free medium (■) and NDGA 10µM (●) was added to the suspension 10mn before the stimulus.

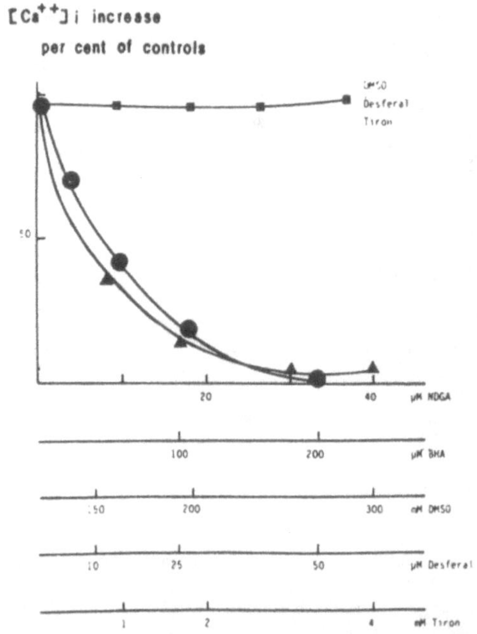

Figure 5. Inhibitory effect of NDGA(▲),BHA (●), DMSO, Desferral and Tiron on [Ca++]i rise induced in Jurkat cells. Same method as in Fig. 4

Table II. Effect of LO inhibitors on [Ca^{++}]i rise after A23187 triggering.

Preincubation	[Ca^{++}]i (nM)	
	Pre-ionophore	Post-ionophore
Mesium	172	612
NDGA(10µM)	157	587
NDGA(40µM)	162	550
BHA(75µM)	192	540
BHA(150µM)	170	431

[Ca^{++}]i was determined as in Fig3. Cells were incubated in medium with or without LO inhibitors for 10 mn and baseline (pre-ionophore) was measured. A23187 (0.5 µM) was then added and fluorescence determination was recorded 10mn later, when a plateau was reached. Results represent one experiment representative of many others.

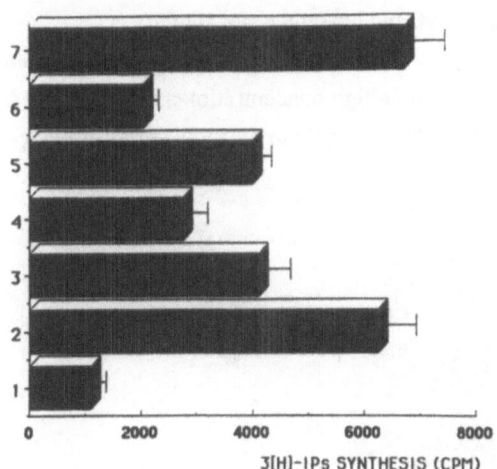

3[H]-IPs SYNTHESIS (CPM)

Figure 6: ConA stimulated generation of inositol phosphates in ^{3}[H]inositol loaded Jurkat cells pretreated with LO inhibitors or DMSO. Jurkat cells were preincubated for 20 min with medium alone (1, 2), or supplemented with different LO inhibitors (3: 10µM NDGA; 4: 20µM NDGA; 5: 10 µM ETYA; 6: 20µM ETYA) or with 150mM DMSO (7). They were then activated for a further 20 min with ConA (2-7), or not (1); the IPs associated radioactivity of each sample was measured (x-axis), after chromatography as previously described (EL Moatassim et al.,1989). Results are expressed as means ± SD of triplicate determinations in one representative experiment out of three.

262

Effect of various LO inhibitors and DMSO on phospholipase C activation

The $[Ca^{2+}]i$ rise triggered by ligand-receptor binding results from inositol triphosphate (IP3) synthesis after inositol-diphosphate hydrolysis by the activated phospholipase C (Imboden and Stobo, 1985).

We thus measured the effect of NDGA, ETYA or DMSO on IPs formation in Jurkat cells stimulated with Con A. As we had previously observed in thymocytes (El Moatassim et al.,1989), ConA was shown to induce a significant increase (400-500 %) of the IPs level in Jurkat cells. NDGA (or ETYA), which did not affect the basal level itself (not shown), dose-dependently inhibited the ConA-induced IPs increase, whereas DMSO did not modify it (Fig. 4).

Effect of LO inhibitors and DMSO on PKC dependent-events

Effect of LO inhibitors on CD3/TCR modulation by PMA

Although not conclusively established, there is some evidence suggesting that CD3/TCR modulation by PMA is related to its phosphorylation by PKC (Cantrell et al., 1985). The PKC inhibitors H7, and staurosporine, prevented this modulation when added at 10 µM or 1µM respectively (Thuiller et al., 1990 and ourselves, not shown).

Fig. 7 demonstrates that in 2 h, PMA treatment negatively modulated the expression of the CD3/TCR complex and that even at high concentrations, neither NDGA, AA861, the other LO inhibitors (not shown), nor DMSO, were able to antagonize this effect.

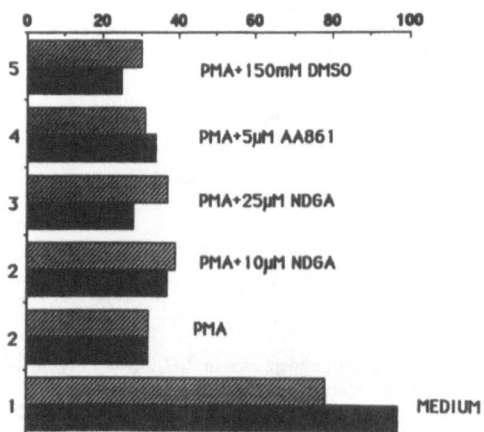

Figure 7. LO inhibitors and DMSO do not modify PMA-induced CD3 modulation in Jurkat cells. Jurkat cells were incubated at 37°C for 2 h in the presence of medium alone, 10 ng/ml PMA, or PMA + different additives, as indicated; the density of the CD3 antigen on the cell surface was then evaluated in terms of the percentage of fluorescent cells (filled bars) and the measurement of the median fluorescence intensity (arbitrary unit) of the positive population (hatched bars).

Effect of LO inhibitors on PMA-induced pH change

Activation of the Na^+H^+ antiport results in an intra-cellular pH change and requires PKC activation (Grinstein et al., 1985; 1986; 1987). Thus, we studied the effects of LO inhibitors on PMA-induced intra-cellular pH change. Jurkat cells responded to the addition of 100 ng/ml PMA with a significant cytoplasmic alkalinization, resulting from activation of the Na^+H^+ antiport, as demonstrated by the inhibitory effect of 100µM dimethyl- amiloride. The maximal intra-cellular pH change was reached 10-15 min after the addition of PMA and averaged 0.2 pH unit (Fig. 8). When cells were preincubated for 10 min with 15µM NDGA (or with other LO ihibitors, not shown) before PMA addition , the intra- cellular pH was not affected, and upon subsequent PMA addition, the expected pH change was observed (Fig6) , indicating that there was normal activation of the Na^+H^+ antiport.

Effect of NDGA and DMSO on IL2R expression

IL2R expression is an early PKC-dependent event (Nel et al., 1987). Therefore, we evaluated how it was affected by the different drugs. In the absence of any activating agent, less than 1% of Jurkat cells expressed IL2R. When they were activated for 24 h, by PHA+PMA or PMA+A23187 or PMA alone, 56.6 %, 56.4 %and 34.6 % of them respectively expressed IL2R. Table III shows that NDGA had little effect on the induction of IL2R in activated cells, even at concentrations that inhibited IL2 production. DMSO appeared to have a stronger effect.

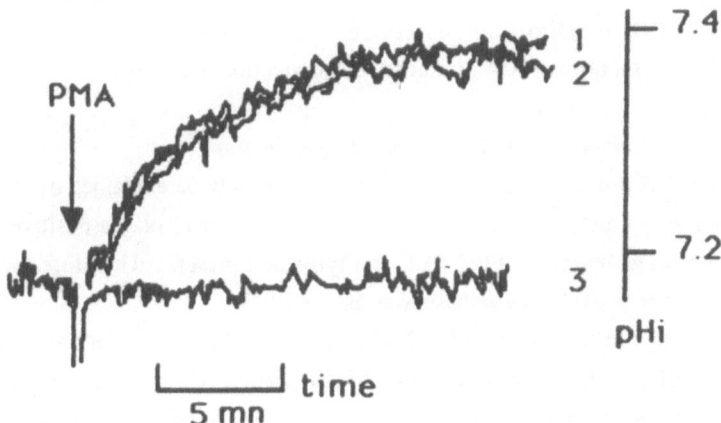

Figure 8: Effect of NDGA on intracellular pH change in Jurkat cells triggered by PMA. BCECF loaded Jurkat cells (1) were suspended in buffer pH 7.2 at 37°C . In some experiments 15µM NDGA (2) or 100µM dimethyl-amiloride (3) were present; when indicated 100 ng/ml PMA was added to the cell suspension, 10 min later. Fluorescence of the suspension was monitored at 528 nm and the Jurkat intracellular pH change determined (Grinstein et al., 1985).

Table III. Effect of NDGA or DMSO on IL2 receptor expression

Antioxidant treatment				
Stimuli	None	NDGA 25µM	NDGA 40µM	DMSO 150mM
PHA-PMA	56.6	40.6	30	N.D.
A23187-PMA	56.4	36.4	29	17.1
PMA	34.6	31.6	24	15.2

The % of fluorescent positive cells was evaluated by flow cytometry after treatment for 24 h with various stimuli in the presence or absence of different antioxidants at the indicated concentrations. The number of fluorescent cells in the unstimulated control was < 1%.

Effect of various LO inhibitors and DMSO on VO4= induced IL2 production

VO4= is an inhibitor of the phosphatase which specifically dephosphorylates tyrosine residues. It is able to induce IL2 production in PMA-treated Jurkat cells, since the phosphorylation of proteins on tyrosine kinase is involved in the synthesis of IL2 by these cells (Peyron et al.,1989a,1989b). Fig. 9 shows that PMA-activated Jurkat cells, cultured with optimal concentrations of VO4=, were able to produce up to one third of the IL2 produced in the presence of PHA. In these experimental condition, DMSO still abolished IL2 production, whereas NDGA was not inhibitory. Conversely, NDGA strongly potentiated IL2 production when used at higher concentrations. Its effect was comparable to that of indomethacin, a cyclooxygenase (CO) inhibitor. Similar results were obtained with ETYA and esculetin, whereas AA861 did not affect VO4=-induced IL2 production in any way .

Absence of effect of leukotrienes and HETE on IL2 production

The inhibition of IL2 production by LO inhibitors can possibly be explained by inhibition of the synthesis of some LO metabolite(s) of arachidonic acid that may play a positive role in signal transduction pathways involving PIP2-PLC and tyrosine kinase(s). Therefore the effect of the addition of leukotrienes onto IL2 synthesis was assayed.

There was no change in the amount of IL2 when A4, B4, D4 leukotrienes and 15 HETE were added into the Jurkat cell cultures at the time of stimulation- (Fig. 10). Moreover, none of the leukotriene tested can substitute for the signal given by PHA and PMA (data not shown).

Absence of effect of leukotrienes and 15-HETE on NDGA-induced inhibition of IL2 production

LTA4, LTB4, LTC4, and LTD4 or 15-HETE were added back at various concentrations (10^{-10} to10^{-7} M) to Jurkat cells activated by PHA (or A23187) + PMA in the presence of different inhibitors. In all cases we did not observe any reversal of the inhibitory effect of the drugs on IL2 production by these LO metabolites. One representative experiment involving NDGA and some other AA metabolite(s) is shown in Fig.11.

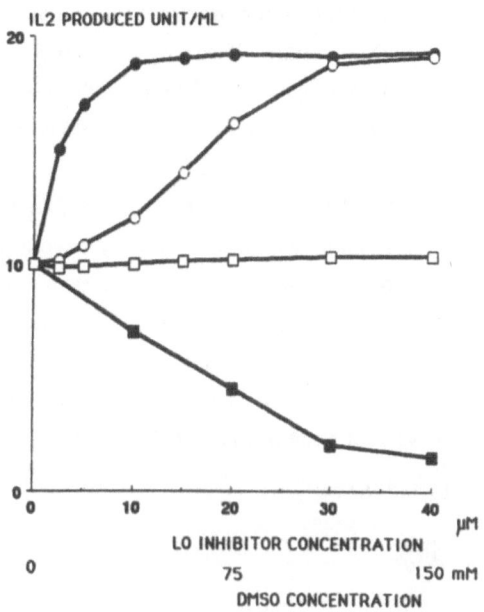

Figure 9: Effect of increasing concentrations of NDGA (O), AA861 (□) and indomethacin(●)DMSO(■) (x-axis) on IL2 production (y-axis) in PMA + VO4= treated Jurkat cells. The cells were cultured for 24 h at 37°C with 10ng/ml PMA and 500μM VO4=, in the presence or absence of the different drugs; the IL2 content of their supernatants was then measured and expressed in units/ml.

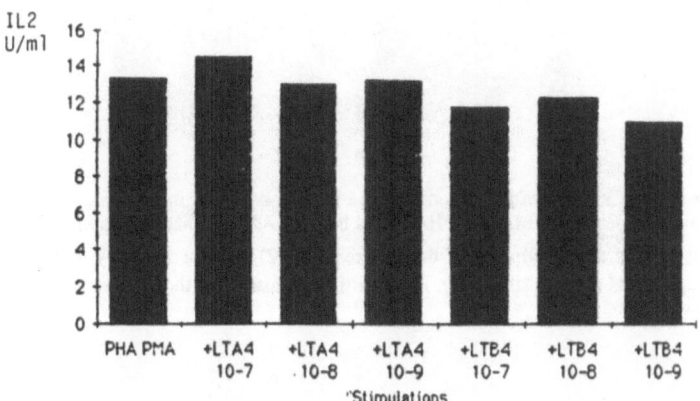

Figure 10: Effect of leukotrienes addition on IL2 synthesis. Leukotriene concentrations are expressed in μM.

266

Detection of LTB4

Detection of LTB4 in the cultures set up for IL2 production was undertaken to verify the possible involvment of leukotrienes in this function. Special culture conditions were required to demonstrate some level of LTB4 in Jurkat cell extracts. It was only in RPMI medium without fetal calf serum and PBS also without fetal calf serum that sizable amounts of LTB4 were detected, and more in PBS than in RPMI (Table IV). These conditions lead readily to cell death, and it was shown that LTB4 detection was parallel to (Fig. 12) and correlated with cell mortality (Fig. 13).

Thin layer chromatography was performed on cell extracts prepared from the two different cultures (RPMI versus PBS). For samples with a cell viability beyond 50%, 64 to 69 % of the radio- labelling was observed in phospho-lipids and 24 to27 % in triglycerids, with a slight proportion of 12- or 15-HETE. for samples with viability less than 50%, radioactivity of phospho-lipids decrease to 31 %, triglycerids decrease to 3.2-6.2 %, and HETEs remain comparable at 2.4 to 2.9. The major peak (51 to 61 %) appeared to be at a Rf value of 0.75, the same as arachidonic acid in this solvent system (illustrated in Gerber et al, 1988).

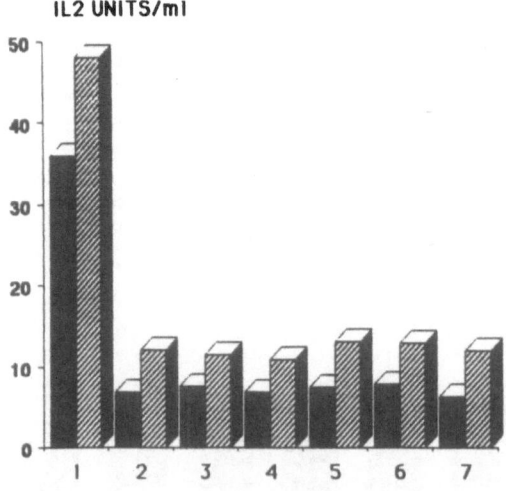

Figure 11: Absence of reverse effect of leukotrienes and 15 HETE on NDGA inhibition. Jurkat cells were cultured for 24 h at 37°C, in the presence of PMA and PHA (filled bars) or A23187 (hatched bars). In experiments 2-7, 10 μM NDGA was also added at the initiation of the culture alone (2), or concomitantly with $10^{-7}\mu$M LTA4 (3), LTB4 (4), LTC4 (5), LT-D4 (6), 15 HETE (7). IL2 was then measured in the supernatants. This experiment is representative of many others.

Table IV. RIA detection of LTB4 in Jurkat cells.

	Stimulation (PHA/PMA)				
	None	30 mn	60 mn	4 h	24 h
RPMI(cell extract)	104±21	102±20	63±11	56±13	107±20
(sup. extract)	183±50	n.d.	28±5	30±13	135±18
PBS (cell extract)	1750±250	2200±230	2560±440	n.d.	1665±194
(sup. extract)	93±6	n.d.	n.d.	n.d.	142±5

Cells were incubated for 24 hours at 37°C, either in RPMI without fetal calf serum or in PBS supplemented with 2mM MgCl$_2$ and 0,5mM CaCl$_2$. Stimulation was performed during this time with PHA (25 /ml) and PMA 10ng/ml. Supernates were extracted twice with 2 volumes (v:v) of ethyl-acetate. cell pellets were extracted in the same way with 2ml ethanol. Results are expressed in pg/10^7 cells.

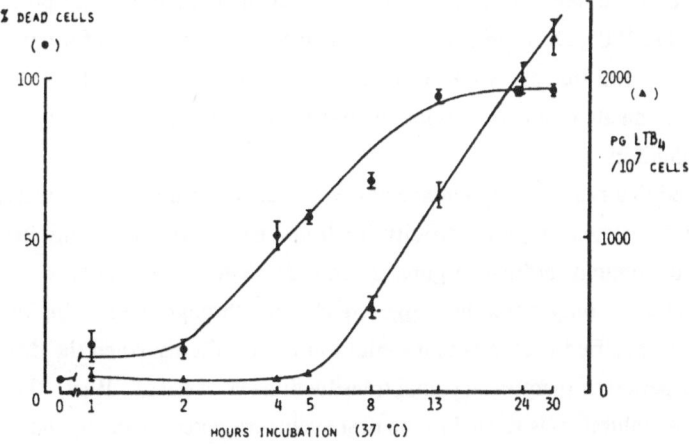

Figure 12. Kinetics of LTB4 detection in Jurkat cell extracts and cell mortality. Jurkat cells were incubated at 10^7/ml at 37°C . Mortality is evaluated by trypan blue exclusion and expressed as percentage of dead cells.

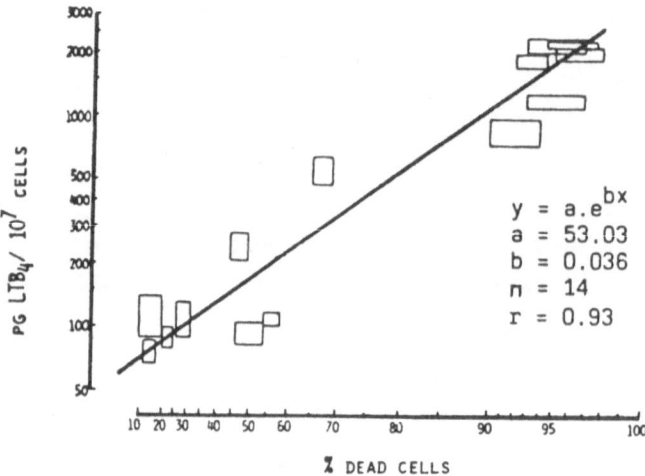

Figure 13. Correlation between intra-cellular LTB4 content and cell mortality.

Discussion

In previous reports, we have shown that several antioxidants inhibited IL2 production in Jurkat cells (Gerber et al., 1985, Dornand et al., 1987). This was also described for normal lymphocytes by several authors and ourselves (Kato et al., 1986; Rola-Pleszczynski, 1988; Dornand and Gerber, 1989). In an attempt to determine the mechanisms involved, we focused our interest on lipoxigenase inhibitors .

First, we showed that reactive oxygen species were generated upon T cell activation; then that several LO inhibitors with high specificity for lipoxygenase pathway inhibited IL2 synthesis with a dose dependent effect (Figure 2 and 3). The compounds were inhibitory at pharmacological doses only when the triggering occured through the binding of the ligand onto the receptor. We specified (Table I) that inhibition occurred only when the drugs were present during the time-period of membrane signal transduction (Weiss et al., 1987), thus indicating that the effect of these inhibitors is related to the first events triggered by the ligand-receptor binding. We showed (Fig. 4 and 5) that the effect of the LO inhibitors on the $[Ca^{2+}]i$ rise occured when the triggering resulted from the binding of the ligand onto the receptor and not when the signalling was by-passed by incubation of the cells with a calcium ionophore (Table II). Furthermore, we found that the effect of the LO inhibitors on the $[Ca^{2+}]i$ rise resulted from the effect of these compounds on the signal transduction pathway, upstream from IP3 formation (Fig

6). This finding agrees with a report on PHA-stimulated human T lymphocytes (Mire-Sluis et al., 1989).

The absence of effects of LO inhibitors on the three PKC-dependent events assayed here in conditions where the synthesis of second messengers was by-passed by incubation with PMA (Figures 7 and 8, Table III) is in line with their effect on IP3 synthesis and clearly indicates that they act at the PiP2-PLC-induced synthesis of diacyl-glycerol (DAG) and IP3 synthesis or upstream.

The effect of LO inhibitors on the tyrosine kinase pathway was then investigated, the tyrosine kinase being the first enzyme involved in protein phosphorylation to implement signal transduction. LO inhibitors did not inhibit IL2 production in Jurkat cells treated with $VO_4^=$ + PMA. $VO_4^=$ treatment, through inhibition of the dephosphorylation of proteins phosphorylated on tyrosine residues, is able to accumulate these phosphorylated proteins in the cell cytoplasm (Peyron et al., 1989a, 1989b), hence to keep activated the tyrosine kinase pathway. This situation results in an incomplete activation signal which, in the presence of a PKC activator, is sufficient to induce IL2 production. Since LO inhibitors did inhibit IL2 production by A23187(or PHA) + PMA-treated Jurkat cells but not that resulting from $VO_4^=$ + PMA treatment, this suggested that the drugs affect one or several events directly linked to tyrosine kinase activation. In normal lymphocytes, this effect can occur either at the phosphorylation induced by the association of the CD4 and the tyrosine kinase p56 lck, or at the level of the ζ chain, the ζAP-70, through the tyrosine kinase p59 fyn, which in turn activates the PLCγ1 for the production of the second messengers, IP3 and DAG (Fig. 14). The effect on CD 45 might also be considered. It is a tyrosine-protein phosphatase which has a role in the regulation of receptor linked signal transduction pathway, but also in cell mitoses (Isakov, 1993). The increase in IL2 production observed in the presence of high concentrations of NDGA (or of its congeners) can be easily explained: $VO_4^=$ activation is negatively affected by cAMP and by PGE2 (Peyron et al.,1989a, Felhmann,1989); at high concentrations, NDGA inhibited both the LO and CO pathways, and prevented PGE2 synthesis, as did indomethacin which is a CO inhibitor. Conversely, AA861, which is more specific to the LO pathway than NDGA, did not affect $VO_4^=$ induced synthesis even at high concentrations. Thus, we have some evidence that a LO metabolite(s) is(are) playing a role at the very early step of T cell activation. That its (their) effect is related to the blockade of lipoxigenase enzymes, and not only to their scavenging effect, is supported by the use of specific compounds, although, we do not have strong evidence that it is a metabolite of the 5-LO. Several authors have claimed that LO metabolites are involved in T lymphocyte functions (Gerber et al., 1985; Gualde et al.,1985; Farrar and Humes, 1985; Kato et al., 1986; Dornand et al., 1987; Delebasse et al.,1988; Sekkat et al., 1988; Rola-Pleszczynski, 1988; Dornand et al., 1989.), no specific product(s) has(ve) yet been characterized.

270

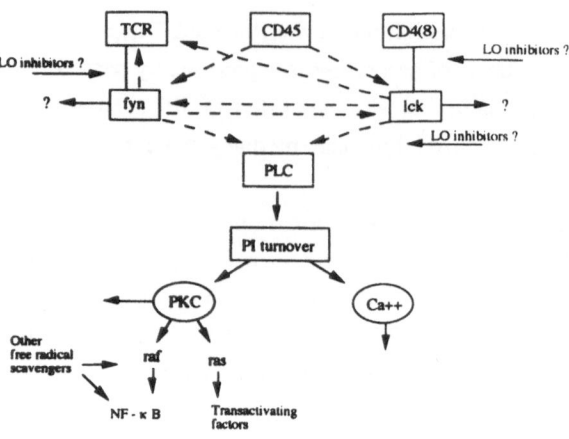

Fig 14. Putative targets of LO inhibitors.

While testing by RIA, the ability of T cells to metabolize arachidonic acid via LO pathways, we found that under special cell damage conditions, Jurkat cells were able to synthezise a product cross-reacting with an anti-LTB4 antibody . This finding which agrees with those of Odlander et al.,1988, but disagrees with those of Goldyne and Rea, 1987 and Poubelle et al., 1987 was neither confirmed by thin layer chromatography (Gerber et al.,1988) nor by HPLC measurements. Therefore, the cell capacity to synthesize leukotriene(s) was not undoubtly demonstrated. We failed to observe the formation of such compounds (or of their closed precursors) upon Jurkat cell activation. Leukotrienes and HETE failed to increase IL2 synthesis and to reverse the NDGA-induced inhibition of IL2 production (Fig. 10 and 11), thus, end-products of the 5-LO pathway are not likely to participate in signal transduction. As suggested by Parker et al., 1979, Goetzl 1981, and Mire-Sluis et al., 1989, other upstream metabolites may be responsible for the up-regulation of signal transduction. Good candidates could be 5-, 12- and 15-HPETE. It may be also other uncharacterized and/or unstable metabolite(s) from the arachidonic acid cascade. Moreover, there may be metabolites stemming from other substrates for the lipoxigenase pathway than arachidonic acid, namely linoleic acid or palmitic acid, as described in other systems (Walstra et al., 1987, Honn et al., 1988).

Besides the findings on signal transduction discussed above, other targets of anti-oxidant may be assumed in the cascade of events resulting in IL2 synthesis. Indeed, we have shown that on the one hand, higher concentrations of LO inhibitors impaired IL2 synthesis also when ligand receptor binding is by-passed, and on the other hand that other anti-oxidants like DMSO and Desferal are without effect on second messenger synthesis, although inhibiting IL2 production.

This indicates that they exert their inhibitory effect on events distal from the early steps of T cell activation, when the two signals $[Ca^{2+}]i$ rise and PKC activation synergize to induce and activate transactivating factors required for IL2 gene transcription (Crabtree, 1989). Other transactivating factors involved in IL-2 gene transcription include NF-κB and other octamer binding complex. NF-κB, have been shown to be modulated by a redox mechanism (Toledano and Leonard, 1991 and reviewed by P. A. Bauerle, P. Fergelot and N. Israël in this volume).But it might be also at the level of the NF-AT formation, which binds to a response element in the IL2 enhancer (Jain et al, 1992).

Thus the research on anti-oxidants in T cell activation and IL2 synthesis, resulting in the unraveling of the role of redox mechanism in this complex event, has been seminal in this domaine.

If we know now that T cell activation leads to apoptosis death in specific situations, (eg. in the thymus for the deletion of immature CD4 and CD8 lymphocytes and in AIDS for the CD4 lymphocytes), radiobiologists have known for a long time that the same apoptosis death occurs in lymphocyte after irradiation, which generates reactive oxygen species.

Acknowledgements.

This work has been financially supported by 'the Association de Recherche sur le Cancer and by the Institut National de la Santé et de la Recherche Médicale (INSERM, CRE 89-2009).

References

Alexander D.R. and Cantrell, D.A. (1989) Kinases and phosphatases in T cell activation. *Immunol. Today* 10, 200-204.

Benichou, G., Kanellopoulos, J.M., Mitenne, F., Galanaud, P. and Leca, G. (1989) T-cell chemiluminescence. A novel aspect of T-cell membrane activation studied with a Jurkat tumor cell line. Scand. *J. Immunol.* 30, 265-269.

Bertez, C., Miquel, M., Coquelet, C., Sincholle, D. and Bonne, C. (1984) Dual inhibition of cyclooxigenase and lipoxigenase by 2-acetylthiophene 2-Thiazolyhydrazone (CBS-1108) and effect on leukocyte migration in vivo. *Biochem. Pharmacol.* 33, 1757-1762.

Cantrell, D.A., Davies, A.A. and Crumpton, M.J. (1985) Activators of protein kinaseC down-regulate and phosphorylate the T3/T-cell antigen receptor complex of human T lymphocytes. *Proc. Natl. Acad. Sci. USA* 82, 8158-8162.

Chaudhri, G. and Clark, I.A. (1989) Reactive oxygen species facilitate the in vitro and in vivo lipopolysaccharide-induced release of tumor necrosis factor. *J.Immunol.* 143, 1290-1294.

Crantree, G. R. (1989). Contingent genetic regulatory events in T lymphocyte activation. *Science*,243,355-361.

Delebasse, S. and Gualde, N. (1988) Effect of arachidonic acid metabolites on thymocyte proliferation. *Ann. Inst. Pasteur/Immunol.* 139, 383-399.

Dornand, J., Sekkat, C., Mani, J.C. and Gerber, M. (1987) Lipoxigenase inhibitors suppress IL2 synthesis: Relationship with $[Ca^{2+}]$ rise and the events dependent on protein kinase C activation. *Immunol Lett.* 16, 101-106.

Dornand, J. and Gerber, M. (1989) inhibition of T-cell responses by anti-oxidants: the targets of lipo-oxygenase pathway inhibitors. *Immunol.* 68, 384-391.

El Moatassim, C., Tangui, M., Mani, J.C. and Dornand, J. (1989) The $(CA^{2+})i$ increase induced in thymocytes by extracellular ATP does not involved ATP hydrolysis and is not related to phosphoinositide metabolism. *FEBS Lett.*. 242, 391-396.

Farrar, W.L. and Humes, J.L. (1985) The role of arachidonic acid metabolism in the activity of interleukin 1 and 2. *J. Immunol.*135,1153.

Felhmann, M. (1989) L'activation des lymphocytes T: toutes les voies mènent à une protéine kinase. *MS Med.Sci.*,15-19.

Fidelus, R.K. (1988) The generation of oxygen radicals: A positive signal for lymphocyte activation. *Cell. Immunol.*113, 175-182.

Gelfand, E.W., Cheung, R.K. and Grinstein, S. (1986) Mitogen-induced changes in Ca^{2+} permeability are not mediated by voltage-gate K^+channels. *J. Biol. Chem.* 261, 11520-11523.

Gerber, M. (1984). Radiosensitivity of Murine T-lymphocyte cytotoxicity. *Radiation Res*, 100, 365-377.

Gerber, M., Ball, D., Michel, F. and Crastes de Paulet, A. (1985) Mechanism of enhancing effect of irradiation on production of IL 2. *Immunol.Lett.* 9, 279-283.

Gerber, M., Longhi, B., Pioch, Y., Michel, F. and Crastes de Pault, A. (1988) Role of lipoxigenase pathway products in IL2 synthesis. In: J.C. Mani and J. Dornand (Eds.), *Fundamental and Clinical Aspects of Lymphocyte Activation and Differentiation*. Walter de Gruyter, Berlin, pp. 600-605.

Gerber, M., Guichard, M., Pioch, Y., Dubois, J.B. (1989). The influence of Interleukine-2, feeder cells, and timing of irradiation on the radiosensitivity of human T lymphocytes assessed by the colony forming assay. *Radiation Res.* 120, 164-176.

Goetzl, EJ. (1981) Selective feedback inhibition of the 5-lipoxigenation of arachidonic acid in human T-lymphocytes. *Biochem. Biophys. Res. Commun.* 101, 344-350.

Goldyne, M.E. and Rea, L. (1987) Stimulated T and natural killer (NK) cell lines fail to synthesize leukotriene B4. *Prostaglandins* 34, 783-795.

Grinstein, S., Cohen, S., Goetz, J.D., Rothstein, A. and Gelfand, E.W. (1985) Characterisation of the activation of Na^+/H^+ exchange in lymphocytes by phorbolesters. Change in cytoplasmic pH dependence of the antiport. *Proc. Natl. Acad. Sci. U.S.A.* 82, 1429-1433.

Grinstein, S., Mack, E. and Mills, G.B. (1986) Osmotic activation of the Na^+/H^+ antiport in protein kinase C depleted lymphocytes. *Biochem. Biophys. Res. Comm.* 134, 8-13.

Grinstein, S., Smith, J.D., Rowatt, C. and Dixon, S.J. (1987) Mechanism of activation of lymphocyte Na^+H^+ exchange by Concanavalin A. *J. Biol. Chem.* 262, 15277-15284.

Gualde, N., Alturu, D. and Goodwin, J.S. (1985) Effects of lipoxigenase metabolites of arachidonic acid on proliferation of human T cells and T cell subsets. *J. Immunol.* 34, 1125-1129.

Honn, K.V., Grossi I.M., Fitzerald, L.A., Umbarger, C.A., Diglio, C.A. and Taylor, J.D. (1988) Lipoxigenase products regulate IRGpIIb/IIa receptor mediated adhesion of tumor cells to endothelial cells, subendothelial matrix and fibronectin. *Proc. Soc. Exp. Biol. Med.* 189, 130-137.

Imboden, J.B., Weiss, A. and Stobo, J.D. (1985) The antigen receptor on a human T cell line initiates activation by increasing cytoplasmic free calcium. *J. Immunol.* 134, 663-665.

Isakov, N. (1993) Tyrosine phosphorylation and dephosphorylation in T lymphocyte activation. *Molecular Immunol.* 30,197-210.

Jain, J., McCaffrey, P.G., Valge-Archer, V.E., and Rao, A. (1992). Nuclear factor of activated T cells contains Fos and Jun. *Nature* 356, 801-804.

June, C.H., Fletcher, J., Lledbetter, A. and Samelson, L.E. (1990) Increase in tyrosine phosphorylation are detectable before phospholipase C activation after T cell receptor stimulation. *J. Immunol.* 144, 1591-1599.

Koretzky, G.A., Wahi, M., Newton, M.E. and Weiss, A. (1989) Heterogeneity of protein kinase C isoenzyme gene expression in human T cell line. *J. Immunol.* 143, 1692-1695.

Kato,K., Yoshikaza, Y. and Murota, S. (1986) Contribution of lipoxigenase metabolites to IL2 production in the early phase of lymphocyte activation. *Prost. Leukotr. Med.* 22, 301-311.

Marth, J., Lewis, D., Wilson, C., Gearn, M., Krebs, E. and Perlmutter, R. (1987) Regulation of p56[lck] during T cell activation: functional implications for the scr-like proteine tyrosine kinases. *Embo J.* 6, 2727-2734.

Mire-Sluis, A.R., Cox, C.A., Hoffbrand, A.V. and Wickremasinghe E, R.G. (1989) Inhibitors of arachidonic acid lipoxygenase impair the stimulation of phospholipid hydrolysis by the T lymphocyte mitogen phytohaemagglutinin. *FEBS Lett.* 258, 84-88.

Nel, A.E., Schabort, I., Rheeder, A., Bouic, P. and Wooten, M.W. (1987) Inhibition of antibodies to CD3 surface antigen and phytohemagglutinin mediated T cellular responses by inhibiting Ca^{2+} dependent protein kinase activity with the aid of 1 (5isoquinolinyl sulfonyl) 2 methyl piperazine dihydrochloride. *J. Immunol.*139, 2330-2236.

Novogrodsky, A., Ravid, A., Rubin, A.L., StenzelL, K.H. (1982). Hydroxyl radical scavengers inhibit lymphocyte mitogenesis. *Proc. Natl Acad. Sci. USA.*, 1171-1174.

Odlander, B., Jakobsson, P.J., Rosen, A. and Claesson, H.E. (1988) Human B and T lymphocytes convert leukotriene A4 into leukotriene B4. *Biochem. Biophys. Res. Comm.*153, 203-208.

Parker, C.W., Stenson, W.F., Huber, M.G. and Kelly, J.P. (1979) Formation of tromboxane B and hydroxy arachidonic acids in purified human lymphocytes in the presence or in the absence of phytohemagglutinin. *J. Immunol.*122, 1572-1577.

Peyron, J.F., Aussel, C., Ferrua, B., Haring, H. and Felhmann, M. (1989a) Phosphorylation of two cytosolic proteins: an early event in T cell activation. *Biochem. J.* 258, 505-510.

Peyron, J.F., Ferrua, B. and Felhmann, M. (1989b) Activation of human T cells is associated with tyrosine phosphorylation of several proteins. *Cell. Signalling* 1, 313-322.

Poubellle, P.E., Borgeat, P. and Rola-Pleszczynski, M. (1987) Assessment of leukotriene B4 secretion in human lymphocytes by using high performance liquid chromatography and radioimmunoassay methods. *J. Immmunol.* 139, 1273-1277.

Rola-Pleszczynski, M. (1988) Leukotriene B4 in T cell activation . In A. Lewis (Ed), *Advances in Inflammation*. Raven Press, New York,NY, pp. 91-99.

Rossio, J.L., Thurman, G.B., Long, C., Vargosko, A. and Pinski, C. (1985). The BRMP IL2 reference reagent. *Lymphokine Res.* 5 (supp 1), S13.

Sekkat, C., Dornand, J. and Gerber, M. (1988) Oxidative phenomena are implicated in human T cell stimulation . *Immunology* 63, 431-438.

Thuiller, L., Metezeau, P., Selz, F., and Perignon, J.L.(1989) The activation of protein kinase C is not necessary for the monoclonal antibody-induced modulation of CD3 and CD4 antigens. (1990) *Eur J. Immunol.* 20,1197-1200.

Weiss, A., Imboden, J., Hardy, K., Manger, B., Terhorst, C. and Stobo J. (1986) The role of the T3/antigen receptor complex in Tcell activation. *Annu. Rev. Immunol.* 4, 593-619.

Weiss, A., Shields, R., Newton, M., Manger, B. and Imboden J. (1987) Ligand-receptor interactions required for commitment to the activation of interleukin 2 gene. *J. Immunol.* 138, 2169-2176.

Wrongemann, K., Weidmann, M.J., Peskar, B.A., Staudinger, H., Rietschel, E.T. and FISHER, H. (1978) Chemiluminescence and immune cell activation.I. Early activation of rat thymocytes can be monitored by chemiluminescence measurements. *Eur. J. Immunol.* 8,748-752.

Walstra, P., Verhagen, J., Vermeer, M.A.,Veldink, G.A. and Vlieegenthart J.F. (1987) Demonstration of 12-lipoxigenase activity in bovine polymorphonuclear leukocytes. *Biochim. Biophys. Acta* 921, 312-318.

Yoshimoto, T., Yokoyama, C., Ochi, K., Yamamoto, S., Maki, Y., Ashida, Y., Terao, S. and Shiraishi, M. (1982) 2,3,5-Trimethyl-6-(12hydroxy-5,10dodecadiynyl)-1,4 benzoquinone (AA861), a selective inhibitor of the 5-lipoxigenase reaction and the biosynthesis of slow-reacting substance of anaphilaxis. *Biochim. Biophys. Acta*, 713, 470-477.

Oxidative Stress, Cell Activation and Viral Infection
C. Pasquier et al. (eds)
© 1994 Birkhäuser Verlag Basel/Switzerland

Intracellular damages induced by singlet oxygen are signals for HIV-1 Reactivation

S. Legrand-Poels and J. Piette

Laboratory of Virology, Institute of Pathology B23, University of Liège, B-4000 Liège, Belgium

Summary

The infection of humans by human immunodeficiency virus type 1 (HIV-1) is characterized by a prolonged stage of clinical quiescence. This clinically asymptomatic period may be based, in part, on the development of cell populations within the body that maintain HIV-1 in a state of latency. Recent progresses are made in understanding how important cellular regulatory factors can activate quiscent proviral DNA. Among them, NF-kB is a multisubunit transcription factor that can rapidly, after induction, migrate to cell nucleus and activate the expression of the viral genes by binding on its DNA responsive elements situated in the Long Terminal Repeat (LTR) of the provirus. Hydrogen peroxide can promote viral reactivation in T-cells through the NF-kB pathway which involves a post-translational activation of NF-kB in the cytoplasm. Other reactive oxygen intermediate like singlet oxygen are capable to promote HIV-1 reactivation from cells harboring the provirus in a quiescent state. This activation occurs only when singlet oxygen is generated intracellularly like in the cytoplasm or in the nucleus; extracellular generation of singlet oxygen promotes an important cytotoxic effect but no HIV-1 reactivation. A signalling pathway initiated by an oxidative intracellular damage, involving either a post-transcriptional activation of regulatory proteins or chromatin unwinding can in fine lead to the LTR transactivation and HIV-1 reactivation.

Introduction

Considerable interest has recently been focused on the role that select host transcriptional factors may play in the initial activation of HIV-1 gene expression. The HIV-1 long terminal repeat (LTR) is known to contain discrete responsive-elements that serve as binding sites for various host transcription factors, and its transcriptional induction is critically dependent upon the presence of these elements (for review, see Greene, 1991; Cullen, 1991). Of particular importance are the binding sites for Sp1 and the TATA factor TFIID (Jones et al., 1986), as well as the inducible transcriptional factor NF-kB (Nabel and Baltimore, 1987). Moreover, the HIV-1 LTR contains other dispensable binding sites for nuclear transcription factors such as NFAT-1, USF, AP-1 and LBP (Garcia et al., 1987; Shaw et al., 1988).

The constitutively expressed cellular factors Sp1 and TFIID play an important role in mediating promoter function in many cellular and viral genes (for review see, Lewin, 1990). These binding interactions appear critical for the LTR promoter function because the deletions of the DNA-responsive elements result in defective HIV-1 proviruses (Jakobovits et al., 1988; Leonard et al., 1989).

The inducible transcription factor NF-kB serves as a pleotropic mediator of both tissue specific and inducible gene expression in a wide range of human cell types (Lenardo and Baltimore, 1989). Related kB enhancers are present in various cellular genes like interleukin-2 (IL-2), IL-2 receptor-α, interleukin-6 (IL-6), and tumor necrosis factor-α (TNF-α) and have further been shown to be functionally involved in their transcriptional induction by various T-cell stimulants (Lenardo and Baltimore, 1989). Although NF-kB binding to its cognate enhancer element is an inducible event in most cell types, the genes encoding the 50 kD and 65 kD subunits are constitutively expressed, but these two proteins are sequestred in the cytoplasm by an inhibitory protein termed IkB (Baeuerle and Baltimore, 1988). Cellular activation promotes disassembly of this cytoplasmic complex permitting a rapid translocation of the NF-kB p50/p65 complex to the nucleus where it acts (for review, see Baeuerle, 1991).

A stimulation to increased HIV-1 expression can be obtained by an array of immune stimuli such as T-cell mitogens, antigens, select cytokines and lipopolysaccharide (Clouse et al., 1989, Folks et al., 1987). Most of these compounds appear to act via activation of NF-kB. Treatment of T-cells with hydrogen peroxide (H_2O_2) induces also NF-kB DNA-binding activity and nuclear appearance of this factor (Schreck et al., 1991) followed by a reactivation of the latent proviral DNA (Legrand-Poels et al., 1991). The activation of NF-kB by treatment of T-cells with H_2O_2 appears to be a specific event because its occurs at low extracellular concentrations of H_2O_2 and other DNA-binding proteins do not seem to be affected (Schreck et al., 1991). Presumably the mechanism involves a passive diffusion of H2O2 through the cell membrane where it would indirectly trigger either a controlled proteolytic degradation of IkB or its phosphorylation providing therefore an irreversible NF-kB activation. This idea is consistent with observations that (i) IkB released in vivo during a PMA treatment of pre-B cells can apparently not be reused to inhibit the inactivated NF-kB and (ii) phosphorylation of IkB decreases its affinity for p65 causing its loss from the p50-p65 complex (for a review see, Baeuerle, 1991; Grimm and Baeuerle, 1993).

Because, H_2O_2 can promote HIV-1 reactivation from T-cells through a mechanism involving NF-kB, it is worth studying in details the eventual role of other reactive oxygen species in such a process, especially singlet oxygen ($^1O_2$) which is known to be produced by activated eosinophils (Kanovsky et al., 1988) and is a species reacting avidly with many biomolecules especially nucleic acids (Sies, 1986). In addition, this reactive oxygen species can be easily generated at different locations inside the cell allowing to investigate the importance of cellular compartments in the initiation of the signalling pathway. This second point is very important. Indeed, as it is mentionned above, the penetration of H_2O_2 in the cytoplasm is the primary event in the post-translational activation of NF-kB by a mechanism leading to the release of the inhibitory subunit and the migration of the p65/p50 complex to the nucleus. Thus, it is worth to know in details whether such a mechanism of HIV-1 reactivation can be triggered by an oxidative stress initiated outside the cytoplasm like in the nucleus or in the extracellular medium. Answering to this question would permit to characterize eventually other signalling pathways capable to promote LTR transactivation.

HIV-1 provirus activation by an oxidative stress mediated by $^1O_2$.

To study the possible role of cellular damages produced by $^1O_2$ as a signal HIV-1 reactivation, we have used cells harboring the provirus in a latent state together with photosensitization as a way to generate $^1O_2$. A Type II photosensitization is a reaction which involves a molecule absorbing light and a transfer of the absorbed energy towards molecular oxygen which in turn is activated and produces $^1O_2$ (Foote, 1988). This reactive oxygen species is a powerful oxidant capable to react with almost all biomolecules like cholesterol and phospholipids (Sies, 1986), aromatic amino-acids of proteins (Sies, 1986) and with nucleic acids (for review see Piette, 1991; Epe, 1991). One interesting point concerning photosensitization, is the possibility to initiate the reaction at various places inside the cell depending on the localization of the photosensitizer. In our work, we used three photosensitizers : (i) Rose Bengal (RB) bound to beads which generate $^1O_2$ outside the cells because the beads block the chromophore uptake, (ii) free RB which is water soluble, penetrates in the cytoplasm and in the nucleus and (iii) proflavine (PF) which intercalates inside the DNA, produces both $^1O_2$ and a type I reaction (Cadet et al., 1983) generating predominently 8-oxo-4-hydroxy-guanine in DNA (Epe et al., 1993) without any other alteration to other cellular compartments.

Two cell lines have been used as model in this work : the U1 monocytic and the ACH-2 lymphocytic cells lines currently employed as model systems to explore HIV-1 post-integration latency in cell culture (Folks et al., 1988, Clouse et al., 1989). In the base line unstimulated state, these cells express the multiply-spliced HIV-RNA and some singly-spliced HIV-1 RNA but extremely low level of the full-length unspliced RNA (Pomerantz et al., 1990). Upon stimulation, this pattern undergoes a switch to the synthesis of unspliced transcripts with a concomittant upregulation of total viral RNA transcription (Pomerantz et al., 1990). Thus, these two cell lines express a specific RNA pattern analogous to early stage and appear to be blocked from progressing to the late stage of productive infection unless they are stimulated. A similar situation is recorded in numerous asymptomatic patients, where by in vitro reverse transcription coupled to PCR, it is shown that the ratio between multiply spliced to unspliced HIV-1 RNA is dramatically higher than in patients with AIDS (Pomerantz et al., 1992).

Both U1 and ACH-2 cells lines have been reacted with $^1O_2$ produced by the three different conditions described above and the susceptibility of these two cell lines to $^1O_2$ generated in these conditions is investigated by measuring cell survival. No lethal effect is detected in the absence of light demonstrating that these photosensitizers by themselves have no cytotoxic effect on these cells. However, when visible light is used to irradiate the cells complexed with the photosensitizers, cell survival begins to decrease proportionnally with the irradiation time (Fig. 1). In all cases, ACH-2 cells turn out to be more sensitive to the treatment than U1 cells and especially important cytotoxic effects can be depicted when the photosensitization reaction is mediated by RB bound to beads generating $^1O_2$ from the extracellular medium. When the photosensitization reaction is mediated by free photosensitizers like RB (located in the cytoplasm and in the nucleus) or PF (nuclear localization), the survival is dependent on the photosensitizer concentration. Fig.1 shows an example of U1 cell survival after photosensitization during increasing period of time in the presence of two concentrations of PF.

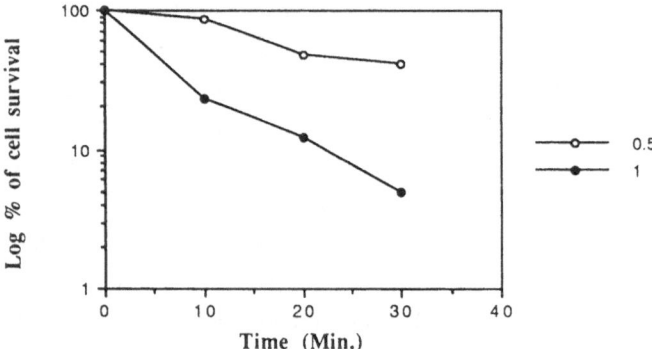

Figure 1 : U1 cell line survival after photosensitization mediated by two different concentrations of PF: -O- 0.5 mM and -●- 1 mM.

HIV-1 reactivation from both U1 and ACH-2 cell lines photosensitized in these three conditions can be determined by measuring RT activity in the cell supernatants respectively 24 and 48 h after the oxidative stress. RT activities measured 24 h after the photosensitization reactions are always very low whatever the irradiation time choosen. However, 48 h after photosensitization, significant RT activities can be detected in supernatants taken from cell photosensitized with free RB and with PF. In the case of RB bound to beads, no RT activity can be detected although cell viability is strongly affected by this kind of treatment. Thus, an extracellular generation of 1O2 mainly leads to important cytotoxic effect due to membrane damages but does not seem to be a signal capable to trigger HIV-1 reactivation. On the other hand, intracellular generation of 1O2 by either free RB or PF can constitute a reactivation signal. Fig. 2 shows these results on

U1 cells treated with free PF and increasing irradiation times. Two important features can be pointed out from RT measurements : (i) RT activities exhibit an optimal value as a function of the irradiation time and (ii) for similar survival conditions, RT activities are somewhat higher for ACH-2 than for U1 cells.

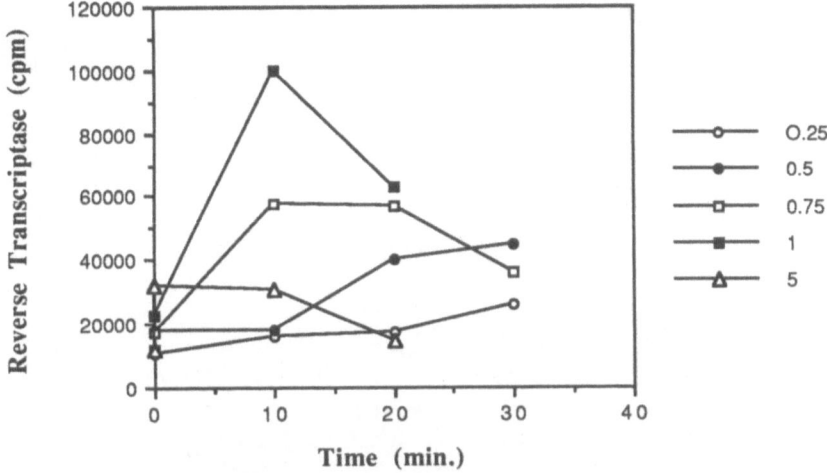

Figure 2 : Induction of a HIV-1 reactivation after photosensitization of the monocytic cell line U1 with various PF concentrations (from 0.25 to 5 mM) during various irradiation times (from 0 to 30 min.).

In order to characterize the molecular mechanisms underlying these reactivations induced by both a cytoplasmic and nuclear generation of $^1O_2$, U1 and ACH-2 cells photosensitized with free RB and PF are collected between 2 and 4 h after the reaction and fractionated. Nuclear salt extracts were then prepared and analyzed for the specific DNA-binding of NF-kB and AP-1 factors using electrophoretic mobility shift assays (EMSA). Binding activities of these two regulatory proteins are studied because they are both known as inducible regulatory factors capable to have positive effects on HIV-1 LTR activation (Cullen 1991). Controls are made with extracts from cells treated with PMA which is known as an inducer of these factors. Nuclear extracts performed from both U1 and ACH-2 cells treated with PMA yield activities that retarded, in native gels, the migration of 32P-labeled DNA probes encompassing the NF-kB and AP-1 motives. Competition with a 50-fold molar excess of unlabelled probe eliminates the formation of the radioactive protein-DNA complex. Using the monocytic U1 cell line as target for both free RB and PF photosensitization reactions, only weak DNA-binding activities towards both AP-1 and NF-kB probes are detected in the nuclear extracts of these cells. As shown on Fig. 3, the retarded band corresponding to AP-1 probe specific binding is not clearly increased after PF photosensitzation compared to non-photosensitized cells whereas important RT stimulation can be observed in the supernatant taken 48 h after the phototreatment. Similar results can be obtained with the two photosensitizers used and both AP-1 and NF-kB probes.

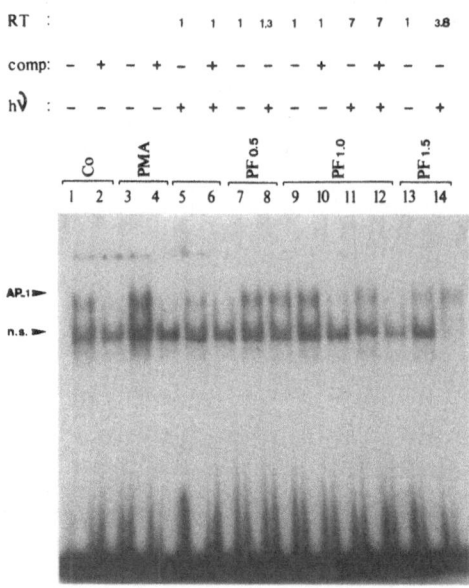

Figure 3: Effect of $^1O_2$ generated in the nuclei of U1 cells by PF photosensitization. 32P-labelled AP-1 probe was mixed with nuclear extracts prepared 2 h after photosensitization and run on native polyacrylamide gel. Co : untreated control cells, PMA: PMA-treated U1 cells. Binding specificity was assessed by competition with unlabelled AP-1 probe (comp), RT stimulation factors were determined for each experimental conditions and indicated (RT). hu + or - indicated the presence or the absence of light.

Similar experimental conditions were used to characterize the molecular mechanisms driving HIV-1 reaction in photosentized ACH-2 cells. In order to know the responsivness of ACH-2 cells towards an oxidative stress, these cells are, at first, treated with increasing concentrations of H2O2 (from 0 to 250 mM). Nuclear cell extracts are prepared and mixed with the labelled NF-kB probe and analyzed by the EMSA technique. Important specific DNA binding activities were observed even at low H2O2 concentrations confirming that in ACH-2 cells, NF-kB is the central intermediate in HIV-1 reactivation. These cells were then photosensitized in the presence of PF or free RB and nuclear extracts are prepared and analyzed for the specific DNA-binding of NF-kB. When ACH-2 cells are photosensitized with PF, only very slight NF-kB binding activity can be detected while RB photosensitization leads to better NF-kB induction in the cell nuclei.

Conclusion

From the data presented in this paper, it is obvious that an oxidative stress mediated by 1O2 can reactivate HIV-1 remaining latent in both monocytes and T-cells. However, important differences can be detected with regards to the site of 1O2 generation and the type of cells used as target. Indeed, when 1O2 is produced outside the cells, it produces a strong cytotoxic effect probably due to membrane oxidation but no HIV-1 reactivation can be observed. On the other hand, both intracytoplamic or intranuclear generation of 1O2 have important cytotoxic effects but promote HIV-1 reactivation in both U1 and ACH-2 cells. Cellular damages induced by free RB likely involve protein and nucleic acid oxidations whereas PF, which vastly intercalates in the chromosomal DNA (Lerman, 1961), generates upon illumination DNA oxidative damages such as 8-oxo-4-hydroxy-guanine, known to be produced when 1O2 reacts with DNA (Floyd et al., 1989). The use of PF photosensitization as a way to generate an intranuclear stress demonstrates that DNA damages, such as the well-known 8-oxo-4-hydroxy-guanine lesion, can be considered as a primary event in the HIV-1 reactivation. The second important point described in this paper is the striking difference observed between U1 and ACH-2 cells on the mechanisms involved in the reactivation. In the U1 cells, we are unable to observe a clear activation of both NF-kB and AP-1 regulatory factors while important RT activities are detected in cell supernatants demonstrating that HIV-1 reactivation occurred. It could be postulated that in monocytes both NF-kB and AP-1 could be constitutively expressed and DNA lesions such as those induced by 1O2 could be sufficient to induce chromatin unwinding (Valerie and Rosenberg, 1990) together with a concommitant activation of integrated HIV-1 genes by non-induced regulatory factors. In the T-cells, the situation appears to be clearer, because intranuclear damages seem to be detected by the cells as a signal with leads to the cytoplasmic activation of NF-kB and in turn to the LTR transactivation.

Acknowledgements

J.P. is Senior Research Associate from the Belgian National Fund for Scientific Research (Brussels, Belgium). S.L-P. is a predoctoral fellow granted by Pasteur-Merieux (France). This research program was supported by the Belgian NFSR and Pasteur-Merieux (France).

References

Baeuerle, P.A. (1991) The inducible transcription activator NF-kB: regulation by distinct protein subunits. *Biochim. Biophys. Acta* 1072:63-80..

Baeuerle, P.A., and Baltimore, D. (1988) IkB: a specific inhibitor of the NF-kB transcription factor. *Science* 242: 540-546.

282

Cadet, J., Decarroz, C., Wang, S.Y., and Midden, W.R. (1983) Mechanisms and products of photosensitized degradation of nucleic acids and related model compounds. *Israel J. Chem.* 23: 420-429.

Clouse, K.A., Powell, D., Washington, I., Poli, G., Strebel, K., Farrar, W., Barstad, D., Kovacs, J., Fauci, A.S., and Folks, T.M. (1989) Monokine regulation of HIV-1 expression in a chronically infected human T-cell clone, *J. Immunol.*142: 431-438.

Cullen, B.R. (1991) Regulation of gene expression in the human immunodeficiency virus type 1. *Adv. Virus Res.* 40: 1-17.

Epe, B. (1991) Genotoxicity of singlet oxygen, *Chem.-Biol. Interactions* 80: 239-260.

Epe, B., Pflaum, M., Häring M., Hegler, J. and Rüdiger, H. (1993) Use of repair endonucleases to characterize DNA damage induced by reactive oxygen species in cellular and cell-free systems. *Toxicology letters.* 67: 57-72.

Floyd, R.A., West, M.S., Eneff, K.L., and Schneider, J.E. (1989) Methylene blue plus light mediates 8-hydroxyguanine formation in DNA. *Arch. Biochem. Biophys.* 273: 106-111.

Folks, T.M., Justement, J., Kinter, A., Dinarello, C.A., and Fauci, A.S. (1987) Cytokine induced expression of HIV-1 in a chronically-infected promonocyte cell line. *Science* 238: 800-802.

Foote, C. (1988) Mechanistic characterization of photosensitized reaction. In : G. Moreno, R. Pottier and T.G. Truscott (eds.): Photosensitization: Molecular, cellular and medical aspects, *NATO Asi series*, Springer Verlag, Berlin, pp. 125-144.

Garcia, J.A., Wu, F.K., Mitsuyasu, R., and Gaynor, R.B. (1987) Interactions of cellular proteins involved in the transcriptional regulation of the human immunodeficiency virus. *EMBO J.* 6: 3761-3770.

Greene, W.C. (1991) The molecular biology of human immunodeficiency virus type 1 nfection. *New Engl. J. Med.* 324: 308-317.

Grimm, S., and Baeuerle, P.A. (1993) The inducible transcription factor NF-kB: structure-function relationship of its protein subunits. *Biochem. J.* 290, 297-308.

Jakobovits, A., Smith, D.H., Jakobovits, E.B., and Capon, D.J. (1988) A discrete element 3' of human immunodeficiency virus (HIV-1) and HIV-2 initiation sites mediates trannscriptional activation by an HIV trans-activator. *Mol. Cell .Biol.* 8: 2555-2561.

Jones, K.A., Kadonaga, J.T., Luciw, P.A., and Tijan, R. (1986) Activation of the AIDS retrovirus promoter by the cellular transcription factor, Sp1. *Science* 232: 755-759.

Kanovsky, J.R., Hoogland, H., Wever, R., and Weiss, S.J. (1988) Singlet oxygen production by human eosinophils. *J. Biol. Chem.* 263: 9692-9696.

Legrand-Poels, S., Vaira, D., Pincemail, J., Van de Vorst, A., and Piette, J. (1991) Activation of human immunodeficiency virus type 1 by oxidative stress. *AIDS Res.* 12: 1389-1397.

Leonard, J., Parott, C., Buckler-White, A.J., Turner, W., Ross, E.K., Martin, M.A., and Rabson, A.B. (1989) The NF-kB binding sites in the human immunodeficiency virus type 1 long terminal repeat are not required for virus infectivity. *J. Virol.* 63: 4919-4924.

Lenardo, M.J., and Baltimore, D. (1989) NF-kB: a pleiotropic mediator of inducible and tissue-specific gene control. *Cell* 58: 227-229.

Lerman, L.S. (1961) Structural considerations in the interactions with DNA. *J. Mol. Biol.* 3: 18-30.

Lewin, B. (1990) Commitment and activation at Pol II promoters: a tail of protein-protein interactions. *Cell* 61: 1161-1164.

Nabel, G., and Baltimore, D. (1987) An inducible transcription factor activates expression of human immunodeficiency virus in T cells. *Nature* 325: 711-713.

Piette, J. (1991) Biological consequences associated with DNA oxidation mediated by singlet oxygen. *J. Photochem. Photobiol. B.* 11: 241-260.

Pomerantz, R.J., Trono, D., Feinberg, M.B., and Baltimore, D. (1990) Cells non-productively infected with HIV-1 exhibit an aberrant pattern of viral RNA expression: a molecular model for latency. *Cell* 61: 1271-1276.

Pomerantz, R.J., Bagasra, O., and Baltimore, D. (1992) Cellular latency of immunodeficiency virus type 1. *Current Opinion Immunology* 4: 475-480.

Schreck, R., Rieber, P., and Baeuerle, P.A. (1991) Reactive oxygen intermediates as apparently widely used messengers in the activation of the NF-kB transcription factor and HIV-1. *EMBO J.* 10: 2247-2258.

Shaw, J.-P., Utz, P.J., Durand, D.B., Toole, J.J., Emmel, E.A., and Crabtree, G.R. (1988) Identification of a putative regulator of early T cell activation genes. *Science* 241: 202-205.

Sies, H. (1986) Biochemistry of oxidative stress. *Ang. Chem. Int. Ed. Engl.* 25: 1058-1072.

Valerie, K., and Rosenberg, M. (1990) Chromatin structure implicated in activation of gene expression by ultraviolet light. *The New Biologist* 2: 712.

Abnormal Redox Regulation in HIV Infection and other Immunodeficiency Diseases

W. Dröge, H.-P. Eck[*], S. Mihm[**], and D. Galter

*Division of Immunochemistry, Deutsches Krebsforschungszentrum
D-6900 Heidelberg, Germany
*Current address: Alter Weg 38, 2190 Cuxhafen 12
**Current address: Universitätsklinikum Göttingen, Abt. Gastroenterologie und Endokrinologie, D-3400 Göttingen,
Germany*

Summary

A large body of evidence indicates that AIDS may be the consequence of a virus-induced cysteine deficiency. HIV-infected persons at all stages of the disease have on the average markedly decreased plasma cystine and cysteine concentrations, decreased intracellular glutathione and elevated plasma glutamate levels. Elevated extracellular glutamate levels aggravate the cysteine deficiency since glutamate inhibits competitively the membrane transport of cystine. Lymphocyte functions *in vitro* are augmented even by moderate elevations of extracellular cysteine and inhibited by elevation of the extracellular glutamate concentrations. A significant correlation between individual $CD4^+$ T cell numbers and individual cystine and glutamate levels has also been found in a cohort of HIV-infected persons, in healthy human blood donors, and in chimpanzees. $CD8^+$ T cells showed no significant correlation. A rapid and significant decrease of plasma cysteine levels and increase of plasma glutamate was also found in rhesus macaques 2 weeks after infection with the closely related SIV_{mac}, but not in HIV-infected chimpanzees or SIV_{agm}-infected African green monkeys. (The latter two species do not develop AIDS-like symptoms.). Last not least, elevated plasma glutamate levels were found to be negatively correlated with lymphocyte functions also in cancer patients.
In view of the decreased levels of the *bona fide* antioxidants cysteine and glutathione one may expect to find manifestations of oxidative damage. Indeed, elevated levels of malondialdehyde have been demonstrated, but the contribution of oxidative damage to the immunopathology of HIV infection remains to be determined. A cysteine deficiency is also expected to compromise certain glutathione-dependent immunological functions, such as IL-2 dependent proliferation and activation of cytotoxic T cells. The activation of the transcription factor NFκB which controls the inducible transcription of several immunologically relevant genes, in contrast, was found to be negatively correlated with the extracellular cysteine supply. This indicates that the overactivation of several immunological functions in the early stages of the disease, including the overexpression of an interleukin-2 receptor α-chain cleavage product, TNFα and β_2-microglobulin may also be the consequence of the HIV-induced cysteine deficiency. The replication of HIV-1, i.e. another gene under control of NFκB binding sites, was shown to be inhibited by cysteine or N-acetyl-cysteine (NAC). In view of the established cysteine and glutathione deficiency in HIV-infected persons, we have proposed to consider N-acetyl-cysteine for the treatment of these patients. NAC is a well established and safe drug with well documented toxicology and pharmacokinetics.

Introduction

The acquired immunodeficiency syndrome (AIDS) is caused by the human immunodeficiency virus (HIV), a non-transforming retrovirus of the lentivirus family. A closely related simian immunodeficiency virus (SIV) causes AIDS-like symptoms in rhesus macaques. The hallmarks of HIV infection are the progressive depletion of the $CD4^+$ T cell population and the cellular dysfunction that is seen not only in $CD4^+$ T cells but also in other T cell subsets and in B cells. The high incidence of opportunistic infections in the late stages of the disease is generally explained by these functional defects. Numerous hypotheses on the mechanism of the HIV-induced immunopathology have been published during the last 10 years. A large body of clinical studies and complementary laboratory experiments supports the conclusion that the HIV-induced immunopathology may be largely the consequence of a virus-induced cysteine deficiency (Dröge et al. 1988a; Eck et al. 1989a; Eck et al. 1991; Eck et al. 1992; Dröge et al. 1992). There is also evidence that similar mechanisms account for the decreased immunological reactivity and the increased incidence of opportunistic infections in patients with (certain types of) advanced malignancies (Dröge et al. 1988b; Eck et al. 1989b; Eck et al. 1990; Eck et al. 1992).

Evidence for a dysregulation of plasma amino acid levels in diseases with different aetiologies

Abnormal plasma amino concentrations and decreased intracellular glutathione levels in HIV-infected persons and SIV-infected rhesus macaques
The first analysis of the amino acid concentrations in serum samples from HIV-infected patients of all stages of the disease revealed a significant and progressive decrease of serum cystine and methionine levels and a progressive increase of serum glutamate levels (Table I) (Dröge et al. 1988a). Subsequent studies with different groups of HIV-infected persons confirmed these results and revealed in addition a significant decrease of the acid soluble thiol levels (mainly cysteine) in the plasma and decreased intracellular glutathione levels (Table II) (Eck et al. 1989a; Eck et al. 1992). The glutathione deficiency was confirmed also by other laboratories (Buhl et al. 1989; Roederer et al. 1991; Staal et al. 1992). The studies revealed generally a high degree of variability of the plasma amino acid levels in HIV-infected persons in contrast to the relatively small interindividual and intraindividual variations of plasma glutamate, cystine and cysteine levels in healthy human subjects. Even asymptomatic HIV-infected persons were found to have in some cases more than 5-fold the normal level of plasma glutamate and/or less than 1/5 of the normal

Table I. Free amino-acid concentration in the serum of HIV-infected or healthy persons

Group	n	Average age (years) ± S.E.M.	CD4$^+$ cells (per mm^3) ± S.E.M.	Glutamate (μmol/l) ± S.E.M.	Cystine (μmol/l x 2) ± S.E.M.
A) Blood donors group					
	69	27.7		59.0 ± 2.3	59.5 ± 1.5
B) Anti-HIV antibody negative considered at risk					
	58	32.9		114.8 ± 12.0^{xxx}	49.4 ± 1.5^{xx}
C) Anti-HIV antibody positive without overt symptoms					
	32	32.7	684 ± 40	155.4 ± 13.6^{xxx}	54.7 ± 4.0^{xx}
D) LAS/ARC					
	23	30.5	523 ± 41	188.1 ± 18.7^{xxx}	44.2 ± 2.9^{xx}
E) AIDS					
	13	32.0	270 ± 71	178.5 ± 27.5^{xxx}	42.1 ± 4.0^{xxx}

^{xxx} P < .00001 ^{xx} P < .01
Taken from: Dröge et al. 1988a

level of plasma cysteine (acid soluble thiol), and in other cases relatively normal glutamate and cysteine levels. This is interpreted to mean that the retroviral infection impairs primarily the homeostatic control mechanisms and especially the glutamate clearance capacity that account for these fluctuations. Since the coincidence of abnormal plasma amino acid levels and HIV infection by itself does not prove a direct cause and effect relationship, it was important to establish that plasma glutamate levels increase and plasma cysteine (acid soluble thiol levels) decrease also in rhesus macaques within 2 weeks after infection with the closely related simian immunodeficiency virus (SIV$_{mac\ 251}$) (Eck et al. 1991). The studies on the SIV-infected rhesus macaques and the abnormal plasma amino acid levels in asymptomatic HIV-infected persons show clearly that this dysregulation is an early consequence of retroviral infection and not merely a late and indirect consequence of the advanced disease.

Chimpanzees and African green monkeys can be infected with HIV-1 and SIV$_{AGM}$, respectively, but do not come down with AIDS-like symptoms. These two species show also no significant changes of plasma glutamate or cystine levels after infection (Dröge et al. 1993).

288

Table II. Plasma amino acid and thiol concentrations and intracellular glutathione concentrations in HIV-1 infected persons and healthy blood donors

	Healthy blood donors	Untreated HIV$^+$ with LAS
n	29	18
Plasma glutamate mM ± S.E.M.	32.1 ± 1.3	91.1 ± 12.2^{xxx}
Plasma methionine mM ± S.E.M.	30.2 ± 0.8	24.9 ± 1.2^{xx}
Plasma cystine plus cysteine mM ± S.E.M.	74.7 ± 1.6	68.2 ± 3.7^{xx}
Plasma acid-soluble thiol mM ± S.E.M.	14.5 ± 0.7	5.3 ± 0.6^{xxx}
Glutathione in PBMC nmol/mg protein	24.3 ± 0.8	18.6 ± 0.5^{xx}
Glutathione in monocytes nmol/mg protein	33.6 ± 0.7 (n=26)[a]	24.8 ± 1.1^{xx} (n=14)[a]

^{xxx} P < .00001 ^{xx} P < .01;

[a]Some blood samples of these groups contained too few cells for this analysis
Taken from: Eck et al. 1989a

Elevated plasma glutamate levels in patients with advanced malignancies are inversely correlated with lymphocyte reactivity

It has been known for many years that patients with advanced malignancies have commonly elevated plasma glutamate levels. For example, elevated glutamate levels have been reported to occur in patients with gastrointestinal tumors (Fürst et al. 1981), bronchial carcinomas (Knauff and Leweling 1981), malignant lymphomas (Knauff and Leweling 1981), mammary carcinomas (Zenz et al. 1981; Roth et al. 1984), ovarian cancer (Zenz et al. 1981), Hodgkin's disease (Kluthe et al. 1981), and various other types of tumors (Beaton et al. 1951; White et al. 1952) as well as in tumor-bearing rats (Wu and Bauer 1960; White et al. 1954). Moreover elevated plasma glutamate levels have also been observed in clinical conditions other than malignancy (White et al. 1952).

Patients with advanced malignancies, especially those with lung cancer, suffer also frequently from opportunistic infections. Peripheral blood lymphocytes from patients with malignant diseases were shown to exhibit diminished proliferative responses to mitogens in comparison with healthy controls (Braun et al. 1980; Collins et al. 1980; Watkins 1973, Wanebo et al. 1975; Yron et al. 1986; Müller et al. 1984). An even more profound suppression has been observed in various *in vitro* assays of the reactivity of tumor-infiltrating lymphocytes (Tötterman et al. 1980; Vose et al. 1977; Miescher et al. 1986). A study on 134 persons including 39 patients with colorectal carcinoma and 64 patients with carcinoma of the lung and 31 apparently healthy persons revealed a highly significant inverse correlation between individual plasma glutamate concentrations and lymphocyte reactivity as determined by mitogenic stimulation (Dröge et al. 1988b). In the case of the patients with colorectal carcinoma, plasma glutamate levels were found to return to essentially normal levels within one week after potentially curative surgical operation. This was followed by a partial and relatively slow recovery of immunological reactivity (Eck et al. 1990).

Correlation between the survival time of lung cancer patients and their individual plasma glutamate and cystine levels

In contrast to patients with colorectal carcinoma, lung cancer patients can usually not be cured by surgical operation. We have therefore performed a study on patients with different types of bronchial carcinoma to test the hypothesis that abnormal plasma amino acid levels may also be correlated with disease progression and poor prognosis. The results of these studies revealed that patients with plasma glutamate levels above 120 µM have not only a markedly decreased lymphocyte reactivity to mitogens but also a substantially higher death rate than those with lower glutamate levels. This observation applied to a group of patients with small cell carcinoma of the lung and a group with non-small cell carcinoma of the lung (Eck et al. 1989b). The most clearcut correlation was observed in a group of patients with non-small cell carcinoma. When the group was divided into 4 subgroups as defined by the median glutamate and the median cystine levels at the time of hospital admission, it was found that all 6 of the 24 patients that survived for 9 months or more belonged to the group with low glutamate and high cystine levels. This subgroup had accordingly a significantly higher mean survival time than any of the other 3 subgroups. All patients of the other 3 subgroups died in less than 9 months with an average of 4-8 months (Eck et al. 1992). Future studies will have to show whether plasma amino acid levels may serve as a valuable prognostic parameter with a good prognostic value.

The influence of extracellular glutamate concentrations on intracellular cysteine and glutathione levels and on lymphocyte functions

Glutamate and cystine share the same membrane transport system (x_c^-), and elevated extracellular glutamate levels therefore inhibit competitively the membrane transport of cystine into the cell (Watanabe and Bannai 1987; Makowske and Christenson 1982; Takada and Bannai 1984; Bannai and Tateishi 1986). The importance of this membrane transport system for the metabolism and function of glutathione in mammalian cells has been pointed out (Bannai and Tateishi 1986). The effects of moderate variations of extracellular glutamate and cystine levels in the pathologically relevant range on macrophages have also been studied in cell cultures with approximately physiological extracellular amino acid concentrations (Eck and Dröge 1989). A 3-5-fold elevation of the extracellular glutamate concentrations was found to cause not only an increase of intracellular glutamate but also a substantial decrease of the intracellular cysteine and glutathione levels. These studies showed also that murine peritoneal macrophages, human peripheral blood lymphocytes and murine fibroblastoid cells consume cystine and release reduced cysteine into the extracellular space in cultures with approxiamtely physiological amino acid concentrations. This phenomenon is believed to be important for the stimulation of T-lymphocytes, since the antigen-presenting macrophages come into close contact with antigen-specific T cells and supply these cells thereby with additional amounts of cysteine and raise their intracellular glutathione level (Gmünder et al. 1990a). Importantly, the capacity of macrophages to release cysteine is markedly suppressed by a 3-5-fold increase of the extracellular glutamate concentration (Eck and Dröge 1989). We therefore proposed that the elevation of plasma glutamate levels in HIV-infected persons and in cancer patients is indeed a pathogenetic factor that aggravates the cysteine deficiency and may account at least partly for the immunopathology in these patients.

Moderate variations of the extracellular glutamate concentrations in the pathological range were also found to affect the functions of mitogenically stimulated T cell preparations and T cell lines in culture (Dröge et al. 1988a; Dröge et al. 1988b).

Studies on the consequences of the cysteine deficiency

Evidence for oxidative damage in HIV-infected patients

Glutathione and its precursor cysteine are antioxidants and components of an important cellular defense mechanism against oxidative stress. In view of the cysteine deficiency in HIV-infected patients, it was therefore not unexpected that these patients express also manifestations of oxidative stress (Sönnerborg and Jarstrand 1986; Sönnerborg et al. 1988). It remains to be determined, however, whether oxidative damage contributes significantly to the pathogenesis of AIDS. There is no evidence to suggest that the major hallmarks of the disease, i.e. the severe depletion of $CD4^+$ T cells and the cellular dysfunctions in T cells, B cells and macrophages are a consequence of oxidative damage. It is an intersting speculation that the exquisite sensitivity of the cytotoxic T cell response against the moderate depletion of intracellular glutathione levels (following paragraph) may result from oxidative processes. But this hypothesis remains to be tested.

In vitro studies on the limiting role of cysteine and intracellular glutathione levels in lymphocytes
The importance of thiols (Fanger et al. 1970; Cerottini et al. 1974) and especially of glutathione (Hamilos and Wedner 1985; Dröge et al. 1986; Gmünder et al. 1990a) to lymphocyte function has been known for many years. It was also shown that T cells have an extremely weak membrane transport activity for cystine (Ishii et al. 1987; Gmünder et al. 1991, Lim et al. 1992) which is the quantitatively most important cysteine source not only in standard cell culture media but also in the blood plasma. The membrane transport activity for cystine and glutamate (x_c^-) is more than 10-fold lower in these cells than the transport activities for cysteine, alanine or arginine. Cystine accounts for approximately 90% of the low molecular weight cysteine in the blood plasma, while reduced cysteine is present only at extremely low concentrations (15 µM) in comparison with other protein forming amino acids. As a consequence, T cells have a low baseline supply of cysteine even under healthy physiological conditions *in vivo* and under standard cell culture conditions.
The cysteine supply to T cells is further regulated by at least two additional mechanisms: Firstly, the cysteine transport activity is upregulated in stimulated T cells (Ishii et al. 1987), and secondly, macrophages can provide T cells with elevated extracellular concentrations of cysteine as discussed earlier (Gmünder et al. 1990a). Because of the weak transport activity for cystine, the intracellular GSH level of T cells is strongly influenced even by relatively moderate variations of the extracellular cysteine concentration under cell culture conditions with approximately physiological amino acid levels (Gmünder et al. 1990a; Eck et al. 1989a; Gmünder et al. 1991). Under such cell culture conditions, a cysteine deficiency impairs especially the IL-2 dependent proliferation and the development of large $CD8^+$ T cell blasts and cytotoxic T cell activity (Eck et al. 1989a; Gmünder et al. 1991; Gmünder and Dröge 1991). These functions require high intracellular GSH levels and are strongly inhibited by buthionine sulfoximine, a specific inhibitor of the biosynthesis of glutathione. The induction of IL-2 production, in contrast, was not inhibited even by severe depletion of intracellular glutathione (Gmünder et al. 1990b).
Taken together, there is a large body of evidence that important T cell functions are exquisitely sensitive even against moderate changes of extracellular cysteine concentrations and intracellular glutathione levels, but there is no evidence that $CD4^+$ T cells may be more sensitive against such changes than $CD8^+$ T cells. The effects that have been seen *in vitro* may therefore account for the cellular dysfunction in HIV-infected patients that is already demonstrable before the $CD4^+$ T cell

numbers start to decline and which also affects CD8$^+$ T cell subsets and B cells (Rosenberg and Fauci 1989; Giorgi and Detels 1989; Miedema et al. 1988). The other hallmark of HIV infection, namely the selective loss of CD4$^+$ T cells (Rosenberg and Fauci 1989; Fahey et al. 1984; Giorgi and Detels 1989), cannot be explained on the basis of these *in vitro* studies. In fact, the large CD8$^+$ T cell blasts and the activation of cytotoxic T cell activity were found to be even more sensitive against a moderate glutathione depletion than corresponding CD4$^+$ T cell blasts and IL-2 production (Gmünder and Dröge 1991; Gmünder et al. 1990b).

Evidence for a limiting role of cysteine and glutathione in vivo

Recent *in vivo* and *in vitro* studies have indicated that the early phase of blast transformation after T cell stimulation increases the nutritional requirements and causes a temporary decrease of intracellular glutathione levels unless these requirements are met by an increased supply of cysteine or an exogenous source of glutathione. Accordingly, excessive antigenic stimulation of T cells by superoptimal numbers of stimulator cells can cause a profound inhibition that is reversed by exogenous application of glutathione (Roth et al. 1993). This suggests the possibility that the cysteine and glutathione deficiency in HIV-infected patients is aggravated by an excessive stimulation of CD4$^+$ T cells, for example, by the retroviral GP120 envelope protein, which may ultimately lead to a selective depletion of this subset. Whether this phenomenon of an *induced* GSH depletion may contribute to the immunopathology of HIV infections remains to be established.

There are indications, however, that the cysteine supply is limiting for the immune system also in healthy human subjects and that the cysteine supply has a greater impact *in vivo* on the CD4$^+$ T cell population than CD8$^+$ T cells even in the absence of HIV infection (Eck et al. 1992). When groups of healthy human subjects were divided into 4 subgroups according to the median glutamate and median cystine levels, the subgroup with low glutamate and high cystine levels had a significantly higher CD4$^+$ T cell number than the other 3 subgroups, while the mean numbers of CD8$^+$ T cells were not different among the 4 subgroups. Another study with a different cohort has shown that the subgroup with low glutamate and high cystine levels produced, on the average, a significantly higher response to a T cell mitogen than the other 3 subgroups (Eck et al. 1992). In view of the influence of the extracellular glutamate level on the intracellular cysteine level, these findings suggest that the transport of cystine does play a limiting role for the immune system also *in vivo* even in healthy human individuals in the absence of HIV infection and that glutamate and cystine levels have a strong influence on the immune system in healthy human individuals. Studies with HIV-infected patients confirmed that the subgroup with a combination of high plasma cystine and low plasma glutamate levels has, on the average, higher CD4+ T cell counts than the other 3 subgroups and that CD8$^+$ T cell counts again were not different among the 4 subgroups (Eck et al.

1992). We therefore proposed the hypothesis that the elevated mean glutamate levels and decreased cystine and cysteine levels in HIV-infected persons may be responsible for the selective decrease of CD4$^+$ T cells. It should be noted, however, that CD4$^+$ T cell counts decline slowly and progressively with the stages of the disease, whereas glutamate, cystine and cysteine levels show strong interindividual and intraindividual variations at all stages of the disease (Eck et al. 1992). We were therefore led to propose that the damaging immunological consequences of occasional episodes with extremely high glutamate and/or low cysteine and cystine levels on the CD4$^+$ T cell population are essentially irreversible and may accumulate over many years.

Complementary studies with chimpanzees revealed that individual CD4$^+$ T cell numbers are also correlated with individual plasma glutamate and cystine levels in these primates (Dröge et al. 1993). However, chimpanzees show no changes of plasma cystine or glutamate levels and no AIDS-like symptoms after HIV infection as discussed earlier.

Effects of cysteine on the intracellular glutathione disulfide level

The cysteine and glutathione deficiency in HIV-infected patients suggested at face value an abnormal redox regulation in favor of prooxidant conditions. Whether the effects of glutathione on lymphocyte functions and the correlation between plasma amino acid levels and CD4$^+$ T cell counts are indeed based on antioxidant effects remains to be determined. There is evidence, however, that the *bona fide* "antioxidant" compound cysteine can facilitate also oxidative processes simply by serving as a precursor for the oxidizing compound glutathione disulfide (GSSG). And there is a possibility that this prooxidant effect may account for certain aspects of the immunopathology of HIV infection and especially for the overactivation that has its visible manifestation in the generalized lymphadenopathy syndrome. Other manifestations are the abnormally high and variable concentrations of TNFα (Lahdevirta et al. 1988; Scott-Algara et al. 1991), IL-2 receptor α-chain cleavage products (Kloster et al. 1987; Prince et al. 1988) and β$_2$-microglobulin (Fahey et al. 1990; Osmond et al. 1991). The inducible transcription of the corresponding genes involves in all three cases the transcription factor NFκB (Ullmann et al. 1990). The elevated TNFα levels are generally believed to contribute to weight loss (cachexia) of these patients.

Recent experiments from our laboratory have shown that increasing extracellular cysteine concentrations in the physiologically relevant micromolar range lead not only to an increase of intracellular glutathione levels but cause also an even stronger increase of intracellular glutathione disulfide levels (Mihm et al. 1993). Glutathione disulfide, in turn, was shown to inhibit strongly the DNA binding activity of the transcription factor NFκB. The overexpression of TNFα, interleukin-2 receptor α-chain cleavage products and β$_2$-microglobulin in HIV-infected persons suggests that this prooxidant effect of cysteine may operate also *in vivo* under physiological and pathological conditions. Since the HIV-proviral DNA contains also two binding sites for the transcription factor NFκB, it seems likely that the virus-induced cysteine deficiency facilitates also the viral replication in these patients. As expected from the effect of cysteine on the NFκB activity, we found that cysteine and N-acetyl-cysteine inhibit the replication of HIV in a dose-dependent fashion (Mihm et al. 1991).

It is important to point out that there is no "redox state of a person" or a "redox state of a cell". Each cell contains rather a number of different redox couples the redox states of which are not necessarily linked to each other. The redox states of the redox couples glutathione and thioredoxine, for example, appear to be essentially independent. In experiments with the human T lineage cell Molt-4, we found that incubation for 1 h with 100 μM hydrogen peroxide does not lead to an increase of intracellular glutathione disulfide levels unless the glutathione reductase activity of the cell is suppressed by corresponding inhibitors (Mihm et al. 1993). This is an important prerequisite for the stimulation of T lineage cells since they require activation of NFκB to induce their receptor for the T cell growth factor interleukin-2 and since the activation of NFκB involves reactive oxygen intermediates (Schreck et al. 1991; Schreck et al. 1992). As mentioned earlier, GSSG levels must remain low to permit binding of the activated NFκB to the corresponding oligonucleotide sequences in the DNA. In order to reconcile these requirements, it is important that elevated levels of reactive oxygen intermediates do not automatically increase intracellular GSSG levels. Our studies suggest that the intracellular cysteine levels may play a more important role in the regulation of intracellular GSSG levels than reactive oxygen intermediates (Mihm et al. 1993).

Conclusion

Taken together there is now a large body of evidence suggesting that the immunopathology of HIV infection may be the consequence of a virus-induced cysteine deficiency. This cysteine deficiency is aggravated by a virus-induced increase of plasma glutamate levels which leads to a competitive inhibition of the membrane transport of cystine. Elevated glutamate levels have also been found in patients with advanced malignancies and were in this case also correlated with an impairment of immunological functions.

The cysteine deficiency appears to have several immunologically relevant consequences. But it is not in all cases clear to what extent these may contribute to the immunopathology of HIV infection. HIV-infected patients show definitively manifestations of oxidative damage that are indicative for impaired antioxidant defense mechanisms. There is also evidence from *in vitro* studies that a cysteine deficiency and the resulting decrease of intracellular glutathione levels inhibit the IL-2 dependent proliferation and may therefore contribute to the cellular dysfunction that has been observed in HIV-infected persons already at early stages of the disease. Whether these effects are based on oxidative processes is presently not known. While *in vitro* studies provided no indication that $CD4^+$ T cells have a greater requirement for extracellular cysteine or intracellular glutathione than the $CD8^+$ T cell counts, *in vivo* studies revealed a striking correlation between plasma amino acid levels and individual $CD4^+$ T cell counts but not $CD8^+$ counts. This correlation was found even in healthy human individuals in the absence of retroviral infection. Whether glutathione is causally involved in this phenomenon is not known. It is important to note, however, that the *bona fide* antioxidant cysteine serves not only as a precursor for reduced glutathione but also for glutathione disulfide. The elevation of extracellular cysteine concentrations in the physiological range causes an even stronger increase of GSSG than GSH at least in some cells. Some of the effects of cysteine in cell cultures and some aspects of the HIV-induced pathology may be explained by this "prooxidant" effect. The prooxidant effect of cysteine was shown to be involved in the regulation of the DNA binding activity of the transcription factor NFκB. The overexpression of the NFκB-dependent cytokine TNFα is widely believed to contribute to the progressive loss of muscle mass (cachexia) in these patients.

In view of the important role of cysteine in the immune system and in view of the various consequences of a cysteine deficiency for lymphocyte functions, we have proposed to consider cysteine or cysteine derivatives such as N-acetyl-cysteine (NAC) for the treatment of HIV-infected persons (Dröge et al. 1992; Dröge 1989).

Acknowledgement

The dedicated assistance of Mrs. I. Fryson in the preparation of this manuscript is gratefully acknowledged.

References

Bannai S and Tateishi N (1986) Role of membrane transport in metabolism and function of glutathione in mammals. *J Membrane Biol* 89:1-8.

Beaton JR, McGanity WJ, and McHenry EW (1951) Plasma glutamic acid levels in malignancy. *Can Med Assoc J* 65:219-221.

Braun DP, Cobleigh MA, and Harris JE (1980) Multiple concurrent immunoregulatory effects in cancer patients with depressed PHA induced lymphocyte DNA synthesis. *Clin Immunol Immunopathol* 17:89-101.

Buhl R, Holroyd K, Mastrangeli A, Cantin AM, Jaffe HA, Wells FB, Saltini C, Crystal RG (1989) Systemic glutathione deficiency in symptom-free HIV-seropositive individuals. The *Lancet* Dec. 2:1294-1298.

Cerottini J-C, Engers HD, MacDonald HR, and Brunner KT (1974) Generation of cytotoxic T lymphocytes *in vitro*. I. Response of normal and immune mouse spleen cells in mixed leucocyte cultures. *J Exp Med* 140:703-717.

Collins PB, Johnson AH, and Moriarty M (1980) T lymphocytes in human cancer. I. Mitogen-responsiveness of lymphocytes in cancer patients. *Ir J Med Sci* 149:301-303.

Dröge W, Pottmeyer-Gerber C, Schmidt H, and Nick S (1986) Glutathione augments the activation of cytotoxic T lymphocytes *in vivo*. *Immunobiol* 172:151-156.

Dröge W, Eck H-P, Näher H, Pekar U, and Daniel V (1988a) Abnormal amino acid concentrations in the blood of patients with acquired immunodeficiency syndrome (AIDS) may contribute to the immunological defect. *Biol Chem Hoppe-Seyler* 369:143-148.

Dröge W, Eck H-P, Betzler M, Schlag P, Drings P, and Ebert W (1988b) Plasma glutamate concentration and lymphocyte activity. *J Cancer Res Clin Oncol* 114:124-128.

Dröge W (1989) Metabolic disorders in HIV infection. *Project News - AIDS-Forschungsförderung des Bundesministeriums für Forschung und Technologie* 2:4-5.

Dröge W, Eck H-P, Gmünder H, and Mihm S (1991) Modulation of lymphocyte functions and immune responses by cysteine and cysteine derivatives. *Am J Med* 91 (suppl 3C):140S-144S.

Dröge W, Eck H-P, and Mihm S (1992) HIV-induced cysteine deficiency and T-cell dysfunction - a rationale for treatment with N-acetyl-cysteine. *Immunol Today* 13:211-214.

Dröge W, Krishna K, Stahl-Hennig C, Hartung S, Plesker R, Rouse S, Peterhans E, Kinscherf R, Fischbach T, Eck H-P (1993) Plasma amino acid dysregulation after lentiviral infection. Submitted for publication.

Eck H-P and Dröge W (1989) Influence of the extracellular glutamate concentration on the intracellular cyst(e)ine concentration in macrophages and on the capacity to release cysteine. *Biol Chem Hoppe-Seyler* 370:109-113.

Eck H-P, Gmünder H, Hartmann M, Petzoldt D, Daniel V, and Dröge W (1989a) Low concentrations of acid-soluble thiol (cysteine) in the blood plasma of HIV-1-infected patients. *Biol Chem Hoppe-Seyler* 370:101-108.

Eck H-P, Drings P, and Dröge W (1989b) Plasma glutamate levels, lymphocyte reactivity and death rate in patients with bronchial carcinoma. *J Cancer Res Clin Oncol* 115:571-574.

Eck H-P, Betzler M, Schlag P, and Dröge W (1990) Partial recovery of lymphocyte activity in patients with colorectal carcinoma after curative surgical treatment and return of plasma glutamate concentrations to normal levels. *J Cancer Res Clin Oncol* 116:648-650.

Eck H-P, Stahl-Hennig C, Hunsmann G, and Dröge W (1991) Metabolic disorder as early consequence of simian immunodeficiency virus infection in rhesus macaques. *Lancet* 338:346-347.

Eck H-P, Mertens T, Rosokat H, Fätkenheuer G, Pohl C, Schrappe M, Daniel V, Näher H, Petzoldt D, Drings P, and Dröge W (1992) T4[+] cell numbers are correlated with plasma glutamate and cystine levels: association of hyperglutamataemia with immunodeficiency in diseases with different aetiologies. *Int Immunol* 4:7-13.

Fahey JL, Prince H, Weaver M, Groopman J, Vissher B, Schwartz K, and Detels R (1984). Quantitative changes in T helper or T suppressor cytotoxic lymphocyte subsets that distinguish acquired immune deificiency syndrome from other immune subset disorders. *Am J Med* 76:95-.

Fahey JL, Taylor JMG, Detels R, Hofmann B, Melmed R, Nishanian P, and Giorgi JV (1990) The prognostic value of cellular and serologic markers in infection with human immunodeficiency virus type I. *New Engl J Med* 322:166-172.

Fanger MW, Hart DA, Wells JV, and Nisonoff A (1970) Enhancement by reducing agents of the transformation of human and rabbit peripheral lymphocytes. *J Immunol* 105:1043-1045.

Fürst P. Bergström J, Hellström B, Vinnars E, Herfarth Ch, Klippel C, Merkel N, Schultis K, Elwyn D, Hardy M, Kinney J (1981) Amino acid metabolism in cancer. In: Kluthe R, Löhr GW (eds) *Nutrition and metabolism in cancer*. Thieme, Stuttgart New York, pp 75-89.

Giorgi JV and Detels R (1989) T cell subset alterations in HIV-infected homosexual men: NIAID multicenter AIDS cohort study. *Clin. Immunol. Immunopathol.* 52:10

Gmünder H, Eck H-P, Benninghoff B, Roth S, and Dröge W (1990a) Macrophages regulate intracellular glutathione levels of lymphocytes. Evidence for an immunoregulatory role of cysteine. *Cell Immunol* 129:32-46.

Gmünder H, Roth S, Eck H-P, Gallas H, Mihm S, and Dröge W (1990b) Interleukin-2 mRNA expression, lymphokine production and DNA synthesis in glutathione-depleted T cells. *Cell Immunol* 130:520-528.

Gmünder H and Dröge W (1991) Differential effects of glutathione depletion on T cell subsets. *Cell Immunol* 138:229-237.

Gmünder H, Eck H-P, and Dröge W (1991) Low membrane transport activity for cystine in resting and mitogenically stimulated human lymphocyte preparations and human T cell clones. *Eur J Biochem* 201:113-117.

Hamilos DL, and Wedner HJ (1985) The role of glutathione in lymphocyte activation. I. Comparison of inhibitory effects of buthionine sulfoximine and 2-cyclohexene-1-one by nuclear size transformation. *J Immunol* 135:2740-2747.

Ishii T, Bannai S, and Sugita Y (1981) Mechanism of growth stimulation of L1210 cells by 2-mercaptoethanol *in vitro*. Role of the mixed disulfide of 2-mercaptoethanol and cysteine. *J Biol Chem* 256:12387-12392.

Kloster BE, John PA, Miller LE, Rubin LA, Nelson DL, Blair DC, and Tomar RH (1987) Soluble interleukin 2 receptors are elevated in patients with AIDS or at risk of developing AIDS. *Clin Immunol Immunopathol* 45:440-446.

Kluthe R, Adam G, Billmann U, Leins R, and Wannenmacher M (1981) Serum amino acids and proteins in Hodgkin's disease. In: Kluthe R, Löhr GW (eds) *Nutrition and metabolism in cancer*. Thieme, Stuttgart New York, pp 95-100.

Knauff HG and Leweling H (1981) Amino acid metabolism and supplementation in cancer. In: Kluthe R, Löhr GW (eds) *Nutrition and metabolism in cancer*. Thieme, Stuttgart New York, pp 101-110.

Lahdevirta J, Maury CPJ, Teppo A-M, and Repo H (1988) Elevated levels of circulating cachectin/tumor necrosis factor in patients with acquired immunodeficiency syndrome. *Am J Med* 85:289-291.

Lim J-S, Eck H-P, Gmünder H, and Dröge W (1992) Expression of increased immunogenicity by thiol-releasing tumor variants. *Cell Immunol* 140:345-356.

Makowske M and Christensen HN (1982) Contrasts in transport systems for anionic amino acids in hepatocytes and a hepatoma cell line HTC. *J Biol Chem* 257:5663-5670.

Miedema F, Petit AJC, Terpstra FG, Schattenkerk JKME, de Wolf F, Al BJM, Roos M, Lange JMA, Danner SA, Grandsmit J, and Schellekens PTA (1988) Immunological abnormalities in human immunodeficiency virus (HIV)-infected asymptomatic homosexual men. *J Clin Invest* 82:1908.

Miescher S, Whiteside TL, Carrel S, and von Fliedner V (1986) Functional properties of tumor-infiltrating and blood lymphocytes in patients with solid tumors: effects of tumor cells and their supernatants on proliferative responses of lymphocytes. *J Immunol* 136:1899-1907.

Mihm S, Ennen J, Pessara U, Kurth R, and Dröge W (1991) Inhibition of HIV-1 replication and NFκB activity by cysteine and cysteine derivatives. *AIDS* 5:497-503.

Mihm S, Galter D, Bockstette M, and Dröge W (1993) Inhibition of glutathione reductase, elevation of glutathione disulfide levels and inhibition of NFκB activity in intact cells by cysteine. Submitted for publication.

Müller DS, Manger B, Zawatzky R, Kirchner H, and Kalden JR (1984) Mitogen-induced γ-interferon production in peripheral blood lymphocytes from patients with colorectal tumors. *Immunobiol* 166:494-499.

Osmond DH, Shiboski S, Bacchetti P, Winger EE, and Moss AR (1991) Immune activation markers and AIDS prognosis. *AIDS* 5:505-511.

Prince HE, Kleinman S, and Williams AE (1988) Soluble IL-2 receptor levels in serum from blood donors seropositive for HIV. J Immunol 140:1139-1141.

Roederer M, Staal FJT, Osada H, Herzenberg LA, and Herzenberg LA (1991) CD4 and CD8 T cells with high intracellular glutathione levels are selectively lost as the HIV infection progresses. *Int Immunol* 3:933-937.

Rosenberg ZF and Fauci AS (1989) The immunopathogenesis of HIV infection. *Adv Immunol* 47:377-.

Roth E, Lenzhofer R, Ollenschläger G, and Funovics J (1984) Influence of parenteral nutrition on plasma amino acid levels of mammary carcinoma patients. *Nutrition* 8:408-411.

Roth S and Dröge W (1987) Regulation of T-cell growth factor (TCGF) production by hydrogen peroxide. *Cell Immunol* 108:417-424.

Roth S, Bockstette M, and Dröge W (1993) Glutathione reverses the inhibition of T cell responses by superoptimal numbers of "nonprofessional" antigen presenting cells. Submitted for publication.

Schreck R, Rieber P, and Baeuerle PA (1991) Reactive oxygen intermediates as apparently widely used messengers in the activation of the NFκB transcription factor and HIV-1. *EMBO J* 10:2247-2258.

Schreck R, Meier B, Männel DN, Dröge W, and Baeuerle PA (1992) Dithiocarbamates as potent inhibitors of NFκB activation in intact cells. *J Exp Med* 175:1181-1194.

Scott-Algara D, Vuillier F, Marasescu M, de Saint Martin J, and Dighiero G (1991) Serum levels of IL-2, IL-1α, TNFα, and soluble receptor of IL-2 in HIV-1-infected patients. *AIDS Res Hum Retrovir* 7:381-386.

Sönnerberg A and Jarstrand C (1986) Nitroblue tetrazolium (NBT) reduction by neutrophilic granulocytes in patients with HTLV-III infection. *Scand J Infect Dis* 18:101-103.

Sönnerberg A, Carlin G, Akerlund B, and Jarstrand C (1988) Increased production of malondialdehyde in patients with HIV infection. *Scand J Infect Dis* 20:287-290.

Staal FJT, Ela SW, Roederer M, Anderson MT, Herzenberg LA, and Herzenberg LA (1992) Glutathione deficiency and human immunodeficiency virus infection. *The Lancet* 339:909-912.

Takada A and Bannai S (1984) Transport of cystine in isolated rat hepatocytes in primary culture. *J Biol chem* 259:2441-2445.

Tötterman TH, Parthenais E, Häyry P, Timonen T, Saksela E (1980) Cytological and functional analysis of inflammatory infiltrates in human malignant tumors. *Cell Immunol* 55:219-226.

Ullmann KS, Northrop JP, Verweij CL, and Crabtree GR (1990) Transmission of signals from the T lymphocyte antigen receptor to the genes 18 responsible for cell proliferation and immune function: the missing link. *Annu Rev Immunol* 8:421-452.

Vose BM, Vanky F, Klein E (1977) Human tumor-lymphocyte interaction *in vitro*. V. Comparison of the reactivity of tumor-infiltrating blood and lymph-node lymphocytes with autologous tumor cells. *Int J Cancer* 20:895-902.

Wanebo HJ, Jun MY, Strong EW, and Oettgen H (1975) T-cell deficiency in patients with squamous cell cancer of the head and neck. *Am J Surg* 130:445-448.

Watanabe H and Bannai S (1987) Induction of cystine transport activity in mouse peritoneal macrophages. *J Exp Med* 165:628-640.

Watkins SM (1973) The effects of surgery in lymphocyte transformation in patients with cancer. *Clin Exp Immunol* 14:69-76.

White JM, Beaton JR, and McHenry EW (1952) Observations on plasma glutamic acid. *J Clin Lab Med* 40:703-706.

White JM, Ozawa G, Ross GAL, McHenry EW (1954) An effect of neoplasms on glutamic acid metabolism in the host. *Cancer Res* 14:508-512.

Wu Ch and Bauer JM (1960) A study of free amino acids and of glutamine synthesis in tumor-bearing rats. *Cancer Res* 20:848-852.

Yron I, Schickler M, Fisch B, Pinkas H, Ovadia J, and Witz IP (1986) The immune system during the pre-cancer and the early cancer period. IL-2 production by PBL from postmenopausal women with and without endometrial carcinoma. *Int J Cancer* 38:331-338.

Zenz M, Hilfrich J, and Neuhaus R (1981) Gynecologic cancer and amino acid metabolism. In: Kluthe R, Löhr GW (eds) *Nutrition and metabolism in cancer*. Thieme, Stuttgart New York, pp 90-94.

Oxidative Stress, Cell Activation and Viral Infection
C. Pasquier et al. (eds)
© 1994 Birkhäuser Verlag Basel/Switzerland

Influence of redox status of lymphocytes and monocytes on HIV expression and immune functions. Evaluation *in vitro* of antioxidant molecules as potential anti-HIV therapy

N. Israël*, M.-A. Gougerot-Pocidalo**, F. Aillet* , and J.-L. Virelizier*

*Unité d'Immunologie Virale, Institut Pasteur, 75724 Paris Cedex 15 France. ** Unité INSERM U 294, Hopital Bichat, Paris, France.

Summary

We used BHA, a phenolic, lipid-soluble, chain-breaking antioxidant to show that peroxyl radical scavenging blocks NF-κB activation and HIV1 enhancer activity in PMA- or TNF-stimulated lymphoblastoid T (J.Jhan) and monocytic (U937) cell lines. The anti-oxidative effect of BHA was accompanied by an increase in thiol, but not glutathione, content in stimulated and unstimulated T cells, whereas TNF stimulation itself barely modified the cellular thiol level. Oxidative stress obtained by the addition of H_2O_2 to the culture medium of J.Jhan or U937 cells could not by itself induce NF-κB activation. These observations suggest that TNF and PMA do not lead to NF-κB activation through induction of changes in the cell redox status. Rather, TNF and PMA can exert a full activation effect only if cells are in a basal redox equilibrium tending towards oxidation since prior modification towards reduction by BHA treatment prevents their activation effects.
The effects of BHA or NAC, a known glutathione precursor, were investigated also on the regulation of HIV1 expression in latently infected U1 cells and in the productively and chronically infected U937 cells. Both antioxidants inhibited TNF- or PMA-induced NF-κB activity in U1 cells in parallel with a partial decrease in induction of HIV replication. Both prevented the sustained NF-κB activity permanently induced by the virus in HIV chronically infected U937 but intriguingly did not modify HIV replication. This may be a limitation to potential antiviral effects of antioxidant therapies. Another limitation may be that antiviral (at least partially) concentrations of NAC or BHA inhibited IL2-induced human PBMC proliferation and also secretion of TNF in PMA-stimulated U937 cells.

Introduction

Reactivation of latent HIV genome is intimately linked to activation of infected cells (Virelizier, 1990). HIV responds to transcriptional stimuli similar to those leading to induction of a series of cellular genes during T cell activation (Bauerle, 1991). Probably the pivotal event is the activation of NF-κB transcription factor. This ubiquitous factor, composed of two subunits (p50 and p65) is retained in an inactive form in the cytoplasm by interaction with inhibitory molecules designated IκBs (Bauerle 1991; Blank et al., 1992; Nolan 1992; Rice et al., 1992). After stimulation by phorbol myristate acetate (PMA) (Nabel et al., 1987) or by a number of cell surface molecules such as receptors to antigens (Hazan, et al., 1990) or to cytokines (Duh, et al., 1989; Israël et al., 1989; Osborn et al., 1989), this factor is translocated into the nucleus. There, it interacts with specific cis-acting sequences present in promoter/enhancer region of cellular genes encoding proteins involved in immunological or inflammatory response and also in the regulatory region (LTR) of HIV (Nabel et al., 1987).

Various antioxidant molecules have been shown to block cellular activation events. Indeed, hydroxyl and peroxyl radical scavengers inihibit lymphocyte mitogenesis (Novogrosky et al., 1982). A series of free radical scavengers were shown to inhibit *in vitro* proliferation of murine T cells initiated by either alloantigens or PMA and calcium ionophores (Chaudhri et al., 1986). Inhibition of murine proliferation was accompanied by decreased IL2 secretion and induction of IL2 receptor (Dornand et al., 1989).

We have tested the hypothesis that phenolic, lipid-soluble, chain-breaking antioxidants that act at the membrane level, such as nordihydroquairetic (NDGA) acid, α-tocopherol (Vit E), and particularly butylated hydroxyanisole (BHA) would decrease PMA- or TNF-induced HIV LTR activation of this NF-κB-dependent promoter. Using both lymphoblastoid T (J.Jhan) and monocytic (U937) cell lines, we showed that these free radicals scavengers inhibit HIV LTR transactivation by blocking NF-κB activation. The mechanism of this inhibitory effect was studied by comparing the cellular thiol content in the various situations of stimulation by TNF or PMA or of inhibition by BHA. This study was completed by experiments using oxidative stress in the two cell systems used. These various experiments lead us to the conclusion that the inhibitory effect of BHA on HIV transactivation was the consequence of an imbalance of the initial redox equilibrium of cells towards reduction leading to a refractory state of cells to TNF or PMA signals (Israël et al.,1992). We showed also that this initial redox status of the cells was not modified by TNF and probably not by PMA.

Besides BHA, (Israël et al.,1992) various antioxidants as dithiocarbamates (Schreck et al., 1992) and N-acetylcysteine (NAC) or others glutathione precursors (Roederer et al., 1990; Staal et al., 1990; Mihn et al.,1991; Kalebic et al., 1991) have been shown to decrease HIV transcription by blocking NF-κB activation induced by PMA or TNF. NAC has been shown also to inhibit virus replication in normal circulating lymphocytes infected *in vitro* with HIV1 and stimulated with mitogens (Roederer et al. 1990), or decrease virus replication in infected cell lines (Mihn et al.,1991; Kalebic et al., 1991). These effects on HIV transcription and replication have suggested a potential therapeutic use of antioxidants in treatment of HIV infected patients.

Since these molecules were shown to be active also on cellular activation events, this suggestion deserved a careful examination in parallel of their effects on HIV replication and on immunological functions.

First, anti-HIV therapies might aim, at the same time, at inhibition of induced replication in latent cells and constitutive replication in cells which multiply actively the virus since *in vivo* the integrated provirus may either remain latent like in normal circulating lymphocytes, or be actively transcribed like in lymphoïd organs (Pantaleo et al., 1991; Pantaleo et al., 1993; Embretson et al. 1993) or in tissue macrophages (Vazeux et al., 1987 ; Embretson et al., 1993).

With these notions in mind, we have investigated whether the two prototypic antioxidants, namely NAC which is a glutathione precursor but with some free radical scavenging properties, and BHA, influence HIV replication in two human monocytic cell systems of the same origin, but exhibiting different levels of HIV replication. We chose the U937 cell line, because it permits to analyse the effects of antioxidants both at early stages of HIV activation from its quiescent state in the U1 subclone of infected U937 cells which was previously characterized, as a model of HIV latency (Folks et al., 1988), and in U937 chronically and actively replicating the virus. In parallel we tested the same concentrations of antioxidants on human PBMC proliferation as a major criterion of immune competence. TNF secretion by PMA stimulated U937 was also evaluated. We show that the concentrations of NAC and BHA which inhibit NF-κB activation were only partially antiviral on U1 cells and totally inefficient on U937 chronically infected cells, but inhibit profoundly IL2-induced PBMC proliferation and TNF secretion. These results incite to prudence in the design of anti-AIDS therapies based on antioxidants.

Influence of redox status of lymphocytes and monocytes on HIV transcription.

Treatment of T or monocytic cells with phenolic peroxyl radical scavengers lead to inhibition of TNF or PMA-induced HIV transcription by blocking NF-κB activation.

In this study we used BHA, Vit E, and NDGA, which are free radical scavengers and can interrupt the propagation of free radicals among unsaturated fatty acids and thereby inhibit lipid peroxidation of cell membranes. These three antioxidant molecules were found to decrease the luciferase activity directed by HIV LTR in lymphoblastoid T cells stably transfected with an HIV LTR-luciferase plasmid and stimulated by TNF or PMA. BHA was consistently proved to be the most efficient in the cell system used. Then we transiently transfected, into T (J.Jhan) or monocytic (U937) cells, the 3Enh-TK-Luc construct which contains three copies of a synthetic oligonucleotide corresponding to the two NF-κB sites of HIV1 enhancer cloned upstream of the basal thymidine kinase promoter encoding the luciferase gene. TNF and PMA increased the NF-κB-dependent luciferase activity and BHA inhibits this induction. This experiment led to the conclusion that the target of BHA effect in the HIV-LTR was the NF-κB sequences. Using bandshifts assays we confirmed that BHA conteracted induction of nuclear NF-κB activity induced by TNF or PMA in both J. Jhan and U937 cells. In both cell lines, BHA was more efficient in blocking PMA than TNF-induced HIV enhancer activity.

TNF or PMA do not modify the redox status of T or monocytic cells but rather the redox state of these cells modulate the efficiency of membrane stimuli to induce NF-κB activation.

Table I : BHA led to an increase of total thiol, but not glutathione, content in an HIV-LTR constitutive T cell clone. TNF barely modified thiol and glutathione contents.

BHA addition μM	TNF addition	Thiol content nm/10⁶ cells	Luciferase activity RLU/10⁶ cells	Glutathione content nmol/10⁶ cells
0	-	14,0	92	3,90
100	+	12,2	2266	4,90
200	+	13,7	2137	4,40
300	+	18,4	1841	3,50
400	+	19,1	1166	5,20
500	+	24,7	214	3,75

Cells from an HIV-LTR constitutive T cell clone were incubated for 6 h with 500 IU/ml TNF (+) or without TNF addition (-) in the presence of various concentrations of BHA (100 to 500 μM). Luciferase activity was expressed as relative light units (RLU) per 10^6 cells. Total thiol content was measured as described in Gougerot-Pocidalo et al., 1985; Glutathione analysis was carried out by a modification of the method described by Tietze 1969. Each value is the mean of two duplicates in this experiment representative of three.

Table I shows that TNF treatment did not modify much the thiol content of lymphoblastoid T cells stably transfected with the LTR-Luciferase vector, eventhough it led to an intense LTR-transactivation as measured by luciferase activity. In contrast, addition of BHA actively modified the redox status towards reduction in the cell system used, as indicated by an increase in total thiol content. It seems likely that BHA "spares" the cellular thiol content, and this was accompanied by a decrease in TNF-induced LTR-transactivation in a BHA dose dependent manner. Thus TNF does not seem to dysbalance the cellular redox status towards oxidation (no decrease in thiol content), but the signal generated by this lymphokine was rendered inefficient when the redox status of cells was progressively imbalanced towards reduction (as shown by increased thiol content) by BHA.

In contrast, glutathione, which represents about 30% of the total cell thiol content in lymphoblastoid cells, was not modified in BHA-treated cells. However, thiol forms of proteins are greatly dependent on the glutathione system (Kosover et al.,1989; Meister et al., 1989) and we suggest that BHA might increase the glutathione turnover leading to an accumulation of proteins in their thiol forms, since cooperative and synergestic interactions between different antioxidant systems have been postulated (Barclay et al., 1988). This might permit a tight homeostasis of glutathione content.

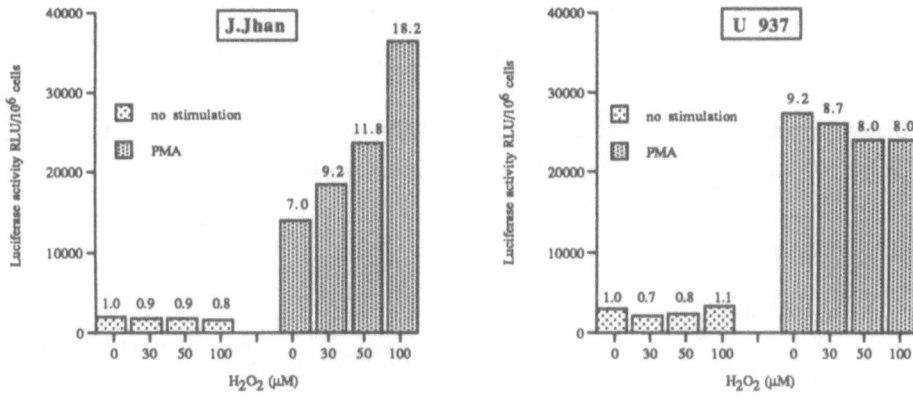

Figure 1. T (J.Jhan) or monocytic (U937) Cells were transfected in quadruplicate in a microtransfection assay using DEAE dextran. Cells (3×10^5) were transfected with 50 ng of the 3Enh-TK-Luciferase plasmid. Cells were then exposed to various concentrations of H_2O_2 alone, or in association with PMA for 6 h, from 24 to 30 h after transfection. Luciferase activity was measured according to standard procedure (Schwartz et al., 1990). Results were expressed in relative light units (RLU) per 10^6 viable cells. Numbers above columns refer to fold induction compared to the untreated control.

Since TNF does not transactivate the HIV-LTR through modification of the redox state of T or monocytic cells, it is not necessary to postulate that ROI are necessary to mediate NF-kB activation. Indeed, we made the observation that oxidative stress (such as H_2O_2 or hyperoxic treatments) is not effective by itself in significantly increasing the NF-κB dependent activation in T (J.Jhan) or monocytic (U937) cells (Figure 1). Thus ROI do not appear to mediate NF-κB activation in the cell lines (and subclones) here tested. These negative results were not due to experimental conditions since H_2O_2 was able to increase HIV LTR activity in a particular Jurkat subclone JR: cells, as shown by Schreck et al., 1991; and confirmed by us (Israël et al., 1992). H_2O_2 alone does not induce NF-κB activation above its constitutive level in most other hematopoietic cell lines tested, as we and others (Molitor et al. 1991; Stylianou et al. 1992) have shown. However in J.Jhan cells (but not in U937 cells), H_2O_2 clearly synergizes with PMA stimulation for HIV LTR induction. This synergistic effect further suggests that an imbalance towards oxidation facilitates (but does not mediate) NF-kB activation.

How may the cellular redox status influence NF-kB activation ?
We postulate that modification of the cell redox status influences the functionning of membrane-associated enzyme(s) implicated in the signalling pathway induced by the immunomodulators. Indeed oxydative stress, due to UV irradiation, initiated at the plasma membrane an activation of Src tyrosine kinase (Devary et al., 1992). PMA by itself was not able to induce the same phenomenon. This is further evidence that PMA does not act by inducing ROI. In contrast BHA

shown before to act at the plasma membrane through the scavenging of ROI, might inactivate these enzymes, thus impairing the transmission of the signals generated by PMA or TNF.

Influence of antioxidant NAC and BHA on HIV replication and on immunological functions. Evaluation of antioxidant as potential anti-HIV therapy.

The results of this section are summarized in Table II.

Antioxidants	NF-κB binding activity *				HIV-Replication **				IL-2 induced *** PBMC proliferation
	U1			U937	U1			U937	
	C	PMA	TNF	C	C	PMA	TNF	C	
NONE	-	+	+	+	100	9000	2200	100	100.0
NAC 10 mM	-	+/-	+/-	+/-					31.8
20 mM	-	+/-	+/-	-				107	17.3
30 mM	-	-	-	-	93	5000	1200		4.1
BHA 100 μM	-	+/-	+/-	+/-				105	33.7
200 μM	-	+/-	+/-	-				114	8.0
300 μM	-	-	-	-	139	2800	1900	124	5.0

Table II.: NAC or BHA counteracted PMA- or TNF-induced (U1 cells) or HIV-induced (chronically infected U937 cells) NF-κB activity .This resulted in a partial inhibition of HIV replication in U1 cells but no effect was observed in chronically infected U937 cells. Antiviral concentrations of NAC or BHA inhibited IL2-induced human PBMC proliferation.
* Nuclear NF-κB activity was determined by bandshift assay (Israël et al., 1992): in U1 cells, untreated: C or stimulated for 6 h by TNF or PMA in presence of NAC or BHA at the indicated concentrations; in U937 used from 50 to 80 days after HIV infection. The oligonucleotide used as the probe contained the HIV enhancer erncompassing two NF-κB sequences.
** HIV replication was evaluated by measuring p24 gag antigen production in the supernatant of U1 and chronically infected U 937 cells in the indicated culture conditions, using an ELISA technique (NEN Research Products). The results are expressed as a percent of p24 production determined in the supernatant of untreated cells: C.
*** Human peripheral blood mononuclear cells (PBMC) were isolated from heparinized blood of healthy adult donors on a ficoll-Isopaque gradient (J. Bio). Cells, 10^5, were incubated in triplicate in 200 ml culture medium with 30 ng/ml recombinant human (rh IL-2) (Cetus) in presence of NAC or BHA at the indicated concentrations. Cultures were incubated for 72 h at 37°C in 95% air and 5% CO2. During the last 6 hr of incubation, cells were pulsed with 1μCi of ^{3}H-thymidine (^{3}H-TdR, 2 Ci/mmol), (CEA, Saclay, France). The results are expressed as percent of counts per minute (cpm) obtained in absence of antioxidant treatment.

Influence of antioxidant NAC or BHA on NF-κB *activity and subsequent effect on HIV replication level in U1 cells or U937 chronically infected cells.*

We showed that BHA and NAC both counteract TNF- or PMA-induced NF-κB activation in the U1 cell line and suppress the constitutive NF-κB activity permanently induced by the virus itself in the nucleus of HIV chronically infected U937 cells (Bachelerie et al., 1989). Inhibition of NF-κB activity by TNF or PMA was accompanied by a partial inhibition of induction of HIV replication in U1 cells as measured by p24 antigen concentration. Furthermore, suppression of NF-κB activity in chronically infected U937 cells (used 50 to 80 days post infection) did not lead to any detectable decrease in HIV replication level (Table II). The reason why the suppression of NF-κB was unsufficient to provoke a decrease in viral replication is very likely that BHA (Israël et al., 1992) or NAC (not shown) cannot counteract the intense transactivating effect of HIV Tat, as reported previously. These *in vitro* results suggest that one of the limitation of antioxidant therapy *in vivo* might be that tissue macrophages (Vazeux et al., 1987 ; Embretson et al., 1993) multiplies actively the virus. Although antioxidant molecules cannot suppress HIV particle production, it is not excluded that they might prevent the virus propagation since NAC was shown by others (Roederer et al., 1990) to prevent acute infection of activated PBL.

Influence of NAC or BHA on immunological responses.

The use of antioxidant in AIDS therapy deserves a carefull examination since the concentrations which suppress NF-κB-activity and partially inhibit HIV replication in latent cells might prevent the NF-κB dependent events implicated in immunological or inflammatory responses. We tested these range of concentrations on two criterions of immunocompetence : IL2-induced PBMC proliferation and TNF secretion in PMA-stimulated U937. As shown in Table II, BHA or NAC inhibited the IL2-induced proliferation of human PBL in a concentration dependent manner. The efficient concentrations (NAC 30mM and BHA 300 μM) to inhibit partially TNF- or PMA-induced replication in U1 cells inhibited almost completely the PBMC proliferation.(4.1, and 5% of the control).We also reported previously that the secretion of TNF by U937, stimulated with PMA, was abolished by similar concentrations of BHA (Israël et al., 1992). The breakdown of TNF secretion might not prevent *in vivo* the activation of infected cells since we showed *in vitro* that chronically infected U937 cells cannot be stimulated by TNF or PMA. This was not due to the disappearance of the TNF receptors, since they were normally detected by immunofluorescence on the cell membrane of infected cells (data not shown). This might result of a general defect in the signalling pathway due to the replication of the virus itself since the two different activators PMA and TNF were both unable to further enhance HIV transcription or replication in chronically infected monocytes.

Conclusion

In vitro evaluation of antioxidants as potential anti-HIV therapy :
Antioxidants such as NAC or glutathione derivatives (Roederer et al., 1990; Staal et al., 1990; Mihn et al.,1991; Kalebic et al., 1991) have been proposed for treatment of AIDS patients but according to our findings there are several limitations : First, antioxidants were shown *in vitro* to have incomplete (U1) or undetectable (infected U937) effect on HIV replication in monocytes eventhough they inhibit NF-κB activation, in our define experimental conditions. *In vivo* the limitation would be that a number of cells (Pantaleo et al., 1991; Pantaleo et al., 1993; Embretson et al. 1993; Vazeux et al., 1987) which multiplies actively the virus and which probably might not respond to antioxidant therapy. Second, it is not excluded that these antioxidants might prevent the virus propagation since NAC was shown by others (Roederer et al., 1990) to prevent acute infection of PBL. However the concentrations needed to prevent NF-κB activation would be difficult to obtain *in vivo*, and if obtained, they might have deleterious effect on immune functions such as TNF production or PBL proliferation.

Furthermore Low concentrations of antioxidants might have opposite effects to that observed with high concentrations, resulting in an NF-κB activation. These bimodal effects were shown on HIV transcription, using concentrations lower than 20 μM for vitamin E (Israël et al., 1992) and 1μM for BHA (data not shown), and by others with dithiocarbamates derivatives (Schreck et al., 1992). The effect of these molecules on the PBMC proliferation is also bimodal, depending on the concentrations used. For instance stimulation was observed with low concentrations of thiols (Gougerot-Pocidalo et al., 1985; Dröge et al. 1992). The bimodal effect of the antioxidant molecules is another limitation of the use of these molecules in AIDS therapy.

Altogether, these considerations warrant prudence in the design of antioxidant-based anti-HIV therapies.

References

Bachelerie, F., Alcami, J., Arenzana-Seisdedos, F., and Virelizier, J.L. (1991) HIV enhancer activity perpetuated by NF-κB induction on infection of monocytes *Nature* 350: 709-712.

Barclay, L. R. C. (1988) The cooperative anti-oxidant role of glutathione with a lipid-soluble and a water-soluble anti-oxidant during peroxidation of liposomes initiated in the aqueous phase and in lipid phase. *J. Biol. Chem.* 263 : 16138-16142.

Bauerle, P.A. (1991) The inducible transcription activator NF-κB: regulation by distinct protein subunits. *Biochec. Biophys. Acta.* 1072: 63-80.

Blank, V., Kourilsky, P. and Israël, A. (1992) The NF-κB family : rel/dorsal homology and ankyrin-like repeats. *Trends in Biochemical Sciences.* 17: 135-140.

Chaudhri, G., Clark, I.A., Hunt, N.H., Cowden, W.B. and Ceredig, R. (1986) Effect of antioxidants on primary alloantigen-induced T cell activation and proliferation. *J. Immunol* 137: 2646-2652.

Devary, Y., Gottlieb R.A., Smeal T., and Karin M. (1992) The mammalian ultraviolet response is triggered by activation of Src tyrosine kinase. *Cell* 71: 1081-1091.

Dornand, J., and Gerber, M. (1989) Inhibition of murine T-cell responses by anti-oxidants : the targets of lipo-oxygenase pathway inhibitors. *Immunology* 68: 384-391.

Dröge, W., Eck, H. P. , and Mihm, S. (1992) Hiv-induced cysteine deficiency and T-cell dysfunction- a rationale for treatment with N-acetylcysteine. *Immunol. Today.* 13: 211-214.

Duh, E. J., Maury, W. J., Folks, T. M., Fauci, A. S., and Rabson, A. B. (1989) Tumor necrosis factor-a activates human immunodeficiency virus through induction of nuclear factor binding to the NF-κB sites in the long terminal repeat. *Proc. Natl. Acad. Sci.* USA 86: 5974-5978.

Embretson, J., Zupanic, M., Ribas, J.L., Burke, A., Racz, P., Tenner-Racz, K., and Haase A.T. (1993) Massive covert infection of helper T lymphocytes and macrophages by HIV during the incubation period of AIDS. *Nature.* 362: 359-362.

Folks, T. M., Justement, J., Kinter, A., Schnittman, S. M., Orenstein, J., Poli, G., and Fauci, A. S. (1988) Characterization of a promonocyte clone chronically infected with HIV and inducible by 13-phorbol-12-myristate acetate. *J. Immunol.* 140: 1117-1122.

Gougerot-Pocidalo, M.-A., Fay, M., Roche, Y., Lacombe, P., and Marquetty, C. (1985) Immune oxydative injury induced in mice exposed to normobaric O2 : effects of thiol compounds on the splenic cell sulfhydryl content and Con A proliferative response. *J. Immunol.* 135: 2045-2052.

Hazan, U., Thomas, D., Alcami, J., Bachelerie, F., Israël, N., Yssel, H., Virelizier, J.-L. and Arenzana-Seisdedos, F. (1990) Stimulation of a human T-cell clone with anti-CD3 or tumor necrosis factor induces NF-κB translocation but not human immunodeficiency virus 1 enhancer-dependent transcription. *Proc. Natl. Acad. Sci. USA* 87: 7861-7865.

Israël, N., Hazan, U., Alcami, J., Munnier, A., Arenzana-Seisdedos, F., Bachelerie, F., Israël, A., and Virelizier.J.-L. (1989) Tumor necrosis factor stimulates transcription of HIV-1 in human T lymphocytes, independently and synergistically with mitogens. *J. Immunol.* 143: 3956-3960.

Israël, N., Gougerot-Pocidalo, M.-A., Aillet, F., and Virelizier, J.-L. (1992) Redox status of cells influences constitutive or induced NF-κB translocation and HIV long terminal repeat activity in human T and monocytic cell lines. *J. Immunol.* 149: 3386-3393.

Kalebic, T., Kinter, A., Poli, G., Anderson, M. E., Meister, A., and Fauci, A. S. (1991) Suppression of human immunodeficiency virus expression in chronically infected monocytic cells by glutathione, glutathione ester, and N-acetyl cysteine. *Proc. Natl. Acad. Sci.* USA 88: 986-990.

Kosower, E. M. (1989) Glutathione : chemical, biochemical and medical aspects. vol III part A. D. Dolphin, R. Poulzon, O. Avramovic, eds. Wiley-interscience, p.103.

Meister, A. (1989) Glutathione : chemical, biochemical, and medical aspects. vol III part A. D. Dolphin, R. Poulzon, O. Avramovic, eds. Wiley-interscience, p.1.

Mihn, S., Ennen,J., Pessara, U., Kurth, R., and Dröge, W., (1991) Inhibition of HIV-1 replication and NF-κB activity by cysteine and cysteine derivatives. *AIDS.* 5: 497-503.

Molitor, J. A., Ballard, D. W., and Greene, W. C. (1991) κB-specific DNA binding proteins are differentially inhibited by enhancer mutations and biological oxidation. *The new biologist* 3: 987-996.

Nabel, G., and Baltimore, D. (1987) An inducible transcription factor activates expression of human immunodeficiency virus in T cells. *Nature* 326: 711-713.

Nolan, G. P., and Baltimore, D. (1992) The inhibitory ankyrin and activator Rel proteins. *Current Opinions in Genetics and Development.* 2: 211-220.

Novogrosky, A., Ravid, A., Rubin, A. L., and Stenzel, K. H. (1982) Hydroxyl radical scavengers inhibit lymphocyte mitogenesis. *Proc. Natl. Acad. Sci.* USA 79: 1117-1122.

Osborn, L., Kunkel, S., and Nabel, G. J. (1989) Tumor necrosis factor and interleukin 1 stimulate the human immunodeficiency virus enhancer by activation of the nuclear κB. *Proc. Natl. Acad. Sci.* USA 86: 2336-2340.

Rice, N. R., MacKichan, M. L., and Israël, A. (1992) The precursor of NF-κB p50 has IκB-like functions. *Cell* 71: 243-253.

Roederer, M., Staal, F. J. T., Raju, P. A., Ela, S. W., Herzenberg, L. A. and Herzenberg, L. A. (1990) Cytokine-stimulated human immunodeficiency virus replication is inhibited by N-acetyl-L-cysteine. *Proc. Natl. Acad. Sci.* USA. 87: 4884-4888.

Schreck, R., Rieber, P., and P. A. Baeuerle. (1991) Reactive oxygen intermediates as apparently widely used messengers in the activation of the NF-κB transcription factor and HIV-1. *EMBO J.* 10 : 2247-2257.

Schreck, R., Meier, B., Mannel, D. N., Dröge, W., and Bauerle, P. (1992) Dithiocarbamates as potents inhibitors of nuclear kappa B activation in intact cells. *J. Exp. Med.* 175: 1181-1194.

Staal, F. J. T., Roederer, M., Herzenberg, L. A., and Herzenberg, L. A. (1990) Intracellular thiols regulate activation of nuclear factor kB and transcription of human immunodeficiency virus. *Proc. Natl. Acad. Sci. USA* 87: 9943.-9947

Stylianou, E., O'Neill, L. A. J., Rawlinson, L., Edbrooke, M. R., Woo, P., and Saklatvala, J. (1992) Interleukin 1 induces NF-κB through its type I but not its type II receptor in Lymphocytes. *J. Biol. Chem.* 267: 15836-15841

Schwartz, O., Virelizier, J.-L., Montagnier, L., and Hazan, U. (1990) A microtransfection method using the luciferase-encoding reporter gene for the assay of human immunodeficiency virus LTR promoter activity. *Gene 88 : 197.*

Tietze, F. (1969) Enzymatic method for quantitative determination of nanograms amounts of total and oxidized glutathione. *Anal. Biochem.* 27 : 502-522.

Vazeux, R., Brousse, N., Jarry, A., Henin, D., Marche, C., Vedrenne, C., Mikol, J., Wolff, M., Michon, C., Rozenbaum, W., Bureau, J. F, Montagnier, L., and Brahic, M. (1987) AIDS subacute encephalitis: identification of HIV infected cells. *Am. J. Pathol.* 126: 403-410

Virelizier, J.-L. (1990) Cellular activation and human immunodeficiency virus infection. *Curr. Opin. Immunol.* 2: 409-413.

Oxidative Stress, Cell Activation and Viral Infection
C. Pasquier et al. (eds)
© 1994 Birkhäuser Verlag Basel/Switzerland

Place for an antioxidant therapy in human immunodeficiency virus (HIV) infection

S. Baruchel[1,2] G. Bounous[2], P. Gold[2]

1 McGill University, Dept. of Pediatrics; McGill AIDS Centre, Montreal, Qc, H3H 1P3, Canada
2 McGill University, Dept of Medicine; McGill AIDS Centre, Montreal, Qc, H3G 1A4, Canada

Summary

Oxidative stress, a known activator of HIV replication in vitro, has a potential role as a cofactor of HIV disease progression. Arguments supporting the role of oxidative stress as a cofactor in HIV activation are summarized in this review. The role of intracellular antioxidants such as glutathione (GSH), and drugs and nutriceutical agents promoting GSH synthesis, are discussed. The review also includes the early results of nutritional interventions based on a diet enriched with IMMUNOCAL[TM], a whey protein concentrate prepared in a proprietary manner.

Introduction

In human immunodeficiency virus (HIV) mediated infection, after primary infection and viral dissemination, most patients have a period of "clinical latency" that may last for years (Pantaleo et al., 1993). Factors that stimulate HIV to replicate and to determine the period of latency are poorly understood in vivo. Results from close observations have shown HIV-infected individuals to have a decreased level of acid-soluble thiol; in particular, cysteine and GSH in plasma and leucocytes (Buhl et al., 1989; Eck et al., 1991) suggesting that oxidative stress may play an important role in the progression to full blown acquired immunodeficiency syndrome (AIDS) (Baruchel and Wainberg, 1992; Halliwell and Cross, 1991).

Rationale for an antioxidant therapy in HIV infection

Oxidative stress occurs when the balance between free radical generation and antioxidant defense is upset. In such cases, active oxygen species and free radicals are so reactive and short-lived that their levels are difficult to measure directly. For these reasons, most methods measure only the product of oxidative stress. GSH is a tripeptide and a major intracellular antioxidant which accounts for over 90% of the intracellular non-protein thiols. One mechanism of action of GSH is through removal of intracellular H_2O_2 by providing substrate for GSH peroxidase, the major H_2O_2 removing enzyme.

Indirect evidence indicates that HIV infection is associated with an increased consumption of antioxidants. The concentration of intracellular GSH in the peripheral blood mononuclear cells and lymphocytes of asymptomatic HIV-seropositive patients was found to be lower than in the healthy control group (Roederer et al., 1990; Smith et al., 1990; Staal et al., 1992). In addition, GSH levels were profoundly depressed in patients with AIDS and AIDS-related complex.

HIV-infected patients commonly excrete higher than average quantities of malondialdehyde into their urine, reflecting an increased level of lipid peroxidation. Rhesus monkeys that were acutely infected with the simian immunodeficiency virus had lower plasma thiol levels within two weeks of infection than uninfected animals. These early metabolic changes coincide with a rise in the level of urinary neopterin, a non-specific marker of macrophage activation (Eck et al., 1991).

A high plasma glutamate is also associated with a low level of thiol, leading to a reduction of intracellular GSH and impairment in T cell function (Eck et al., 1989). The consumption of intracellular non-protein thiol in these situations may play a role in increased oxidative stress. HIV-infected individuals show a 30-40% decrease in GSH in both CD4+ and CD8+ T cells. This decrease is due primarily to the specific removal from the circulation of a class of T cells with high GSH content (Roederer et al., 1990).

Oxidative stress and hydroxyradical formation can lead to increased lipid peroxidation and modification in both membrane fluidity and receptor conformation (Carson et al., 1986). In the absence of a proper antioxidant system, the DNA repair capacity of the cells may be altered and functional lymphocytes may be destroyed in situ (apoptosis) or suffer from impaired function (Ameisen and Capron, 1991; Arends and Wyllie, 1991). Although lower magnitude oxidative bursts are associated with lymphocyte activation and are regulated by intracellular GSH (Hamilos and Wedner, 1985; Hamilos et al., 1989), depletion of GSH is associated with immunosuppression and down regulation of IL-2 receptors (Suthantiran and May, 1990). Moreover, HIV-infected patients are also known to manifest increased radiation sensitivity when GSH is depleted (Hughes-Davies et al., 1991; Valis, 1991).

Oxidative stress stimulates HIV replication in vitro

HIV gene expression can be activated in vitro by oxidative stress. H_2O_2 can induce the expression of HIV in human T cell lines by activating transcription of nuclear factor kappa-B (NF-kappaB) (Duh, 1989). NF-kappaB is inactive in the cytosol when complexed to a second regulatory molecule termed IkappaB. Activation with a variety of oxidative stimuli results in the dissociation of the complex between NF-kappaB and IkappaB and subsequent translocation of NF-kappaB to the nucleus. NF-kappaB binding activity in the nucleus is modulated by oxidoreduction in vitro. Activation of NF-kappaB nuclear binding activity by oxidative stress is specific and occurs at low concentration (Paeck et al., 1991; Toledano and Leonard, 1991). A variety of antioxidants, including GSH, GSH ester, pentoxyphylline, desferrioxamine and N-acetylcysteine can block activation of NF-kappaB (Baruchel et al., 1993; Fazely et al., 1991; Kalebic et al., 1991).

Mechanism of oxidative stress in HIV infection
Cytokines are important constituents in the regulation of the immune response. Tumor necrosis factor a (TNF-α) can up-regulate HIV expression in chronically infected T cells and monocytes (Griffin et al., 1990; Mellors et al., 1991; Poli et al., 1990). In this regard, TNF-α acts synergistically with either interleukin-6 or granulocyte-macrophage colony stimulating factor (GM-CSF) (Poli et al., 1990). TNF-α triggers virus expression by the induction of transcription of an activating factor which binds to NF-κB in the promoter region of the HIV long terminal repeat. This, in turn, results in increased transcription of HIV RNA and the eventual production of viral progeny (Paeck et al., 1991).

TNF-α is produced by activated macrophages and may contribute to disease progression by activation of HIV replication through the action of NF-κB. There are many controversies concerning the role of TNF-α in the activation of HIV infection in vivo. TNF-α levels have not been found to be constantly elevated in AIDS patients sera (Lahdevirta et al., 1988).

Intracellular thiol levels regulate lymphocyte functions
A number of studies have shown that depletion of GSH inhibits T cell proliferation. In fact, there is a critical need for adequate intracellular GSH levels for T cell proliferation (Fidelus and Tsan, 1986; Fidelus et al., 1987). Reduction of GSH by 10-40% in T cells completely inhibits T cell activation. While depletion of GSH leads to inhibition of some T-cell functions, supplementation can augment others functions, both in vitro and in vivo. GSH added exogenousely augments lymphocyte proliferation in response to lectin. 2-L-Oxothiazolidine-4-carboxylate, a cysteine precursor which increases GSH levels, acts synergistically with concanavalin A to stimulate T cells. An increased of previously-lowered GSH levels in mice augments activation of cytolytic T cells, demonstrating the importance of GSH in vivo (Fidelus et al., 1987).

Oxidative stress and wasting syndrome
HIV-infected patients have a higher rate of resting energy expenditure associated with an increased fat oxidation rate and an increased concentration of IL-6. This cytokine is closely interrelated with TNF-α and IL-1 which appear to have metabolic effects related to tissue wasting probably through the generation of hydroxyradicals. In summary, oxidative stress may

be implicated in the pathophysiology of the wasting syndrome experienced by AIDS patients (Hommes et al., 1991).

Use of antioxidant in HIV infection ?

Several authors have found that various thiols alter HIV expression in vitro. Kalebic et al. (1991) have found that the thiols GSH, GSH monoester, and N-acetylcysteine block HIV expression in vitro. They have suggested that these thiol agents may have therapeutic value in HIV-infected patients. This group used the U1 cell line, chronically infected with the human immunodeficiency virus. Expression of HIV is minimal in these cells unless activated by stimuli such as PMA, TNF-α, or IL-6. Some of these cytokines may stimulate HIV activity, in part, by stimulating production of reactive oxygen species (ROS) (Yamauchi, 1990), as discussed above.

The HIV-infected U1 cells were pre-treated with various concentration of thiols, then the activators PMA, TNF-α or IL-6 were added and the cells incubated for various periods of time. HIV activity was estimated by measuring the activity of the viral enzyme reverse transcriptase, HIV messenger RNA and synthesis of the major HIV viral proteins. Reverse transcriptase activity in the cell culture supernatant without thiol treatment increased ten to thirty-fold when treated with either PMA or TNF-a and three to five fold when treated with IL-6. A single pre-treatment with GSH at 1, 5 or 15 mM suppressed reverse transcriptase activity markedly in a dose dependant manner regardless of whether PMA, TNF-α or IL-6 was used as the activator. The duration of exposure to the thiols was important. Pre-treatment with thiols for six hours was more effective than pre-treatment for three hours or the simultaneous addition of thiols with the viral inducers. At thiol concentration of 15 mM, all three thiols inhibited the induction of the total HIV protein synthesis induced by PMA, TNF-α or IL-6. Other reducing agents that are not directly used for GSH synthesis were also found to affect HIV replication in acute or chronic in vitro system.

Harakeh et al. (1990) studied the action of ascorbic acid on HIV activity in a chronically HIV-infected T-lymphocytic cell line and found that the presence of nontoxic concentrations of ascorbic acid. in the cell culture medium reduced the level of extracellular reverse transcriptase activity by 99% and the expression of p24 antigen by 90%. Moreover, the presence of ascorbic acid caused both a time and dose-dependent decrease in HIV activity in acutely infected CD4+ T-lymphocytes. The molecular mechanism by which ascorbic acid suppresses HIV activity is unknown, but the close interrelation between ascorbic acid and GSH metabolism is intriguing.

N-Acetylcysteine (NAC) at 10mM/L caused less than a two-fold inhibition of RT and conferred a synergistic effect (approximatively eight-fold inhibition) when added in conjunction with ascorbic acid. Long term experiments have shown continuous exposure to ascorbate was necessary for HIV suppression.

Pentoxyphylline, another antioxidant agent, has shown some anti-HIV activity in chronically and acutely-infected cell systems (Fazely et al., 1991). More re cently, a lipoic acid has shown some activity in the inhibition of NF-kappaB induced by TNF-α or by PMA. This inhibitory action of a lipoic acid was found to be very potent and only 4 mM was needed for a complete inhibition, whereas 20mM was required for N-acetyl cysteineSuzuki et al., 1992). Moreover, in vitro and in vivo applications of a lipoic acid have been associated with an increase in GSH concentration (Busse et al., 1992).

In our laboratory, we have tested 2-L-oxothiazolidine4 carboxylate (OTC) on the U1 cell line system chronically infected with HIV-1 and stimulated by PMA. Inhibition of HIV activity was measured as reflected by reverse transcriptase (RT) activity and P24 antigen expression. OTC, a GSH prodrug, provided a cysteine delivery system. At 2 mM concentration, OTC inhibits HIV expression with 80% inhibition of RT activity and partial inhibition of P24 antigen expression.

We also tested other types of antioxidants, specifically desferrioxamine on the acutely HIV-infected cell system MT4. We have shown that DFO at an in vitro concentration similar to the one achieved in clinical situations blocked HIV expression probably by inhibition of NF-kappaB (Baruchel et al., 1993; Paeck et al., 1991). Many antioxidants were found to be of potential use in this regard. Some compounds which do not replenish GSH may have beneficial effects by inhibiting ROS production and sparing GSH to some extent.

Antioxidant in vivo
Various agents such as diethyldithiocarbamate (DDTC), lipoic acid, 2-L-oxothiazolidine-4-carboxylate have shown some GSH-promoting activity in non-infected animals, however except for DDTC none of these compounds have been studied in HIV-infected animals. We have studied OTC in non-HIV infected animals, and found a GSH-promoting activity in various tissues including bone marrow.

IMMUNOCAL™

IMMUNOCAL™ is a whey protein concentrate prepared in a proprietary fashion so as to preserve the most thermosensitive molecule of whey, such as serum albumin, in their undenatured form (serum albumin contains 6 glutamyl-cysteine groups per molecule). This group of milk proteins, when prepared appropriately, can produce a glutamyl-cysteine delivery system. IMMUNOCAL™ was kindly provided through courtesy of Immunotec Research Corporation Ltd., Montreal, Qc..

Bounous et al have shown that the humoral immune response of mice fed with 20g of IMMUNOCAL™/100g of diet was higher than mice fed formula diets of similar nutritional efficiency containing 20g/100g diet of any other types of commercially available semipurified food protein, such as casein. Bounous et al have further shown that the immunoenhancing activity of IMMUNOCAL™ concentrate is related to increased production of splenic GSH during the oxygen-requiring antigen driven clonal expansion of lymphocytes (Bounous et al., 1988; 1993).

Recent experiments in Japan have shown spleen cells of BALB/c male mice fed a 25g IMMUNOCAL™ concentrate/100g diet for 4 weeks had an increased immune response to SRBC in vitro and a higher content of L3T4+ cells than mice fed on isocaloric diet with 25g pure casein/100d diet. Similarly, the spleen L3T4+/Lyt-2+ ratio was 1,36 ± 0.07 in IMMUNOCAL™ fed mice and 0.55 ± 0.07 in the casein control group (p<0.001) (Hirai et al., 1990). This background represents the rationale for using IMMUNOCAL™ as a food supplementation in HIV-infected individuals.

Antioxidants in HIV infected individuals

Only a few reports are available on the use of antioxidants in HIV infected individuals.

Diethyldithiocarbamate

The only antioxidant drug widely studied in clinical trials has been diethyldithiocarbamate (DDTC). Despite encouraging early preliminary reports in two randomized placebo control studies and one randomized non-placebo study in AIDS patients, a larger study in asymptomatic individuals has failed to prove any benefit from the use of DDTC (Hersh et al., 1991; Lang et al., 1988; Picolet et al., 1993; Reisinger et al., 1990). The heterogenicity of the patients in the former study gives rise to the question of the benefits of antioxidant therapy in all HIV- infected patients. It also reinforces the argument for clear entry criteria into any study focusing on antioxidant metabolism and the need for careful monitoring of the oxidative stress status of patients before conducting any clinical trial using antioxidants.

Glutathione and Glutathione Pro-drugs

Recently published studies on the bio-availability of orally administrated GSH in human have been disappointing. These studies have shown no significant increase in plasma cysteine or GSH after oral administration of 3g of GSH in healthy volunteers. This may be due to hydrolysis of GSH by intestinal and hepatic gamma-glutamyl transferase (Yamauchi, 1990). However, oral administration of 25 mg OTC/kg in healthy volunteers and in HIV-infected individuals alone, or in association with intramuscular injection of 800 mg GSH per day, was successful in increasing the total GSH in the blood (Witschin et al., 1992).

N-acetylcysteine (NAC), another GSH-promoting agent, is currently under investigation. Four hours after a single dose of NAC given orally to HIV-infected patients, the concentration of cysteine in plasma and mononuclear cells increases and GSH concentration is moderately higher than before, or 2 hours after NAC administration. A sustained increased in intracellular cysteine may be necessary to normalize intracellular glutathione. This may be accomplished by repeat administration of NAC (Giorgi et al., 1992; Mihm et al., 1991; Ruffmann and Wendel, 1991).

Desferrioxamine (DFO)

A retrospective study looking at the median daily dose of desferrioxamine in a cohort of 69 HIV-1 infected thalassemic patients and the rate of progression to stage IV CDC classification over 6 years concluded that patients who received 40 mg/Kg daily of DFO as a chelating agent for iron overload had an 11% risk of progression as compared to 40% for those who received less than 40mg/kg/day (p <0.001). When the dose was taken as a continuous variable it was found that the rate of progression was significantly slower in thalassemic patients receiving a higher dose (p<0.003). (Costagliola/Baruchel personal communication).

IMMUNOCAL™

Based on the animal experiments, a pilot study was undertaken at McGill University in order to evaluate the effect of a diet enriched with IMMUNOCAL™ in four HIV-infected asymptomatic individuals and one AIDS patient during a period of 3 to 5 months. IMMUNOCAL™ was dissolved and taken in a cold flavored drink in quantities progressively increased from 8.4 to 39 g per day. In 4 HIV asymptomatic patients who regularly took the product, no side effects were noted and the patient's body weight increased progressively during the study by an amount varying from 2 to 7 kg. As expected, the blood mononuclear cells' GSH content was below normal value in all patients at the onset of the study. In 3 of the asymptomatic HIV-infected patients, the GSH level increased over a 3 month period, and in one case returned to normal values (70% increased in concentration). Three comparable patients on their usual standard diet over the same period showed some weight loss and no change in their blood GSH mononuclear cell content. After 5 months of treatment, the patient with AIDS showed weight stabilisation and had his intracellular GSH reach normal values. These preliminary data indicate that, whenever patients maintained their overall energy and protein intake, they either stabilized or increased their body weight as well increasing their GSH in lymphocytes. This confirms the potential for IMMUNOCAL™ as a glutamyl-cysteine delivery system (Bounous et al., 1989).

This new nutritional approach to HIV-positive individuals or patients suffering from full blown AIDS must be considered a nutriceutical form of therapy and is currently under investigation in a population of children suffering from AIDS and wasting syndrome in order to confirm and extend our preliminary data and to attempt to correlate the GSH modulating activity of IMMUNOCAL™ with both quantitative and qualitative immunological parameters.

Conclusion

There is certainly enough rationale and in vitro data to consider antioxidant therapy in association with antiretroviral therapy in HIV infection. Caution should probably be exercised before using any antioxidant in order to avoid any toxic side effects. An important prerequisite is good pharmacological monitoring of oxidative stress as an important part of any such trial. The parameters of such activity must be clearly defined if such therapy is to gain general utility. In addition, studies of viral burden will be necessary in the patients under investigation.

Furthermore, it must be noted that whereas lowering levels of reactive oxygen species may have beneficial biological effects, it is possible that excessive antioxidant protection could have some deleterious effects such as possible paradoxical immunosuppression. A good candidate for clinical monitoring of oxidative stress is GSH. Good biochemical techniques are available for measuring GSH in mononuclear cells. Flow-cytometric evaluation of GSH in T cell subsets is also available, but needs some standardisation and good correlation with the biochemical assay.

Pro-drugs of GSH such as N-acetyl-cysteine and 2-L-oxothiazolidine are currently in phase 1 investigations but results are not yet available. Nutritional interventions through a modified dairy product is an attractive approach to a cysteine delivery system and GSH elevation, and studies are currently ongoing in order to confirm our preliminary results.

References

Ameisen, J.C., and Capron, A. (1991) Cell dysfunction and depletion in AIDS: the programmed cell death hypothesis. *Immunol. Today*. 12: 102-105.

Arends, M., and Wyllie, A. (1991) Apoptosis mechanism and role in pathology. *Int. Rev. Exp. Pathol*. 32: 223-254.

Baruchel, S., Gao, Q., and Wainberg, M.A. Desferrioxamine and HIV. *Lancet*. 337: 1356.

Baruchel, S., and Wainberg, M.A. (1992) The role of oxidative stress in disease progression in individuals infected by the human immunodeficiency virus. J. Leucocyte Biol. 52: 111-114.

Baruchel, S., Wang, T., Farah, R., and Batist, G. (1993) In vivo selective modulation of tissue glutathione in a rat mammary carcinoma model. *Biochem. Pharmacol*. Submitted.

Bounous, G., Kong Shaun, P.A.C., and Gold, P. (1988) The immunoenhancing property of dietary whey protein concentrate. *Clin. Invest. Med*. 11: 271-278.

Bounous, G., Baruchel, S., Falutz, J., and Gold, P. (1993) Biological activity of dietary whey protein in HIV seropositive individuals. *Clin. Invest. Med*. In press.

Bounous, G., Batist, G., and Gold, P. (1989) Immunoenhancing property of dietary whey protein in mice: role of glutathione. *Clin. Invest. Med*. 12: 154-155.

Buhl, R., Holroyld, K.J., Mastrangeli, A., Cantin, A.M., Jaffe, H.A., Wells, F.A., Saltini, C., Crystal, R.G. (1989) Systemic glutathione deficiency in symptom free HIV positive individuals. *Lancet*. 2: 1294.

Busse, E., Zimmer, G., Schopohl, B., and Kornhuber, B. (1992) Influence of a lipoïc acid on intracellular glutathione in vitro and in vivo. *Drug Res*. 42: 829-831.

Carson, D.A., Seto, S., and Masson, D.B. (1986) Lymphocytes dysfunction after DNA damage by toxic oxygen species. *J. Exp. Med*. 63: 746-751.

Duh, E.J. (1989) Tumor necrosis factor alpha activates human immunodeficiency virus type 1 through induction of nuclear factor binding to the NF-κB sites in the long terminal repeat. *Proc. Natl. Acad. Sci. USA*. 86: 5974-5978.

Eck, H.P., Grunder, H., Hartmann, M., Petzoldt, D., Daniel, V., and Dröge, W. (1989) Low concentration of acid soluble thiol (cysteine) in the blood plasma of HIV-1 infected patients. *Biol. Chem*. 370: 101. Hoppe Seyler.

Eck, H.P., Stahl-Henning, C., Hussmann, G., and Droge, W. (1991) Metabolic disorder as earlyconsequence of simian immunodeficiency virus infection in rhesus macaque. *Lancet*. 1: 346.

Fazely, F., Dezube, B.J., Allan-Ryan, J., Pardee, A.B., and Ruprecht, R.M. (1991) Pentoxiphylline (Trental) decreases the replication of the human immunodeficiency virus type 1 in human peripheral blood mononuclear cells and cultured T-cells. *Blood*. 77: 1553-1556.

Fidelus, R.K., Ginouves, P., Lawrence, D., and Tsun, M.F. (1987) Modulation of intracellularglutathione concentrations alters lymphocyte activity and proliferation. *Exp. Cell Res*. 170: 269.

Fidelus, R.K., and Tsan, M.F. (1986) Enhancement of intracellular glutathione promotes lymphocyte activation by mitogen. *Cell Immunol*. 97: 155.

Giorgi, G., Micheli, L., Fiaschi, A.I., Cerretani, D., Romeo, R., Dalpra, P., and Bo, L. (1992) L-2-Oxothiazolidine-4-Carboxylic acid and glutathione in human immunodeficiency virus infection.*Current Therapeu. Res*. - Clinical and Experimental. 52: 461-467.

Griffin, G.E., Leung, K., Folks, T.M., Kunkel, S., and Nabel, G.J. (1990) Activation of HIV gene expression during monocyte differentiation by induction of NF-KappaB. *Nature*. 339: 70-73.

Halliwell, B., and Cross, C.E. (1991) Reactive oxygen species, antioxidants, and acquired immunodeficiency syndrome: Sense or speculation? *Arch. Intern. Med*. 151: 29-31.

Hamilos, D.L., and Wedner, H.J. (1985) Comparison of inhibitory effect of buthionine sulfoximine and 2-cyclohexene-l-one by nuclear size transformation. *J. Immunol*. 135: 2740.

320

Hamilos, D.L., Zelarney, P., and Mascali, J.J. (1989) Lymphocyte proliferation in glutathione-depleted lymphocytes: direct relationship between glutathione availability and the proliferative response. *Immunopharmacology*. 18: 223.

Harakeh, S., Jariwalla, R.J., and Pauling, L. (1990) Suppression of human immunodeficiency virus replication by ascorbate in chronically and acutely infected cells. *Proc. Natl. Acad. Sci. USA*. 87: 7245-7249.

Hersh, E.M., Brewton, G., Abrams, D., Bartlett, J., Galpin, J., Parkash, G., Gorter, R., Gottlieb, M., Jonikas, J.J., Landesman, S., Levine, A., Marcel, A., Petersen, E.A., Whiteside, M., Zahradnik, J., Negron, C., Boutitie, F., Caraux, J., Dupuy, J-M., and Salmi, L.R. (1991) Dithiocarb sodium (diethyldithiocarbamate) therapy in patients with symptomatic HIV infection and AIDS. A randomized double blind, placebo-controlled multicenter study. *JAMA*. 265: 1538-1544.

Hirai, Y., Nakay, S., Kiduishi, H., and Kawai, K. (1990) Report: Evaluation of the immunological enhancement activities of IMMUNOCAL. *Otsuka Pharmaceutical Co. Ltd.* Cellular Biology Institute, Osaka, Japan.

Hommes, M.T., Romijn, J.A., Endert, E., and Sauerwein, H.P. (1991) Resting energy expenditure and substrate oxidation in human immunodeficiency virus infected asymptomatic men HIV affects host metabolism in the early asymptomatic stage. *Am. J. Clin. Nutr.* 54: 311-315.

Hughes-Davies, L., Young, T., and Spittle, M. (1991) Radiosensitivity in AIDS patients. *Lancet*. 337: 1516.

Kalebic, T., Kinter, A., Poli, G., Anderson, M.E., Meister, A., and Fauci, A.S. (1991) Suppression of human immunodeficiency virus expression in chronically infected monocytic cells by glutathione, glutathione ester, and N-acetylcysteine. *proc. Natl. Acad. Sci. USA*. 88: 986-990.

Lahdevirta, J., Maury, C.P.J., Teppo, A.M. and Repo, H. (1988) Elevated level of circulating cachectin, tumor necrosis alpha in patients with acquired immunodeficiency syndrome. *Am. J. Med* 85: 289-191.

Lang, J.M., Touraine, J.L., Trepo, C., Choutet, P., Kirstetter, M., Falkenrodt, A., Herviou, L., Livrozet, J.M., Retornaz, G., Touraine, F., et al. (1988) Randomized double-blind placebo-controlled study of dithiocarb sodium ("Imuthiol") in human immunodeficiency virus infection. *Lancet*. 2: 702-706.

Mellors, J.W., Griffith, B.P., Ortiz, M.A., Landry, M.L., and Ryan, J.L. (1991) Tumor necrosis factor alpha/cachectin enhances human immunodeficiency virus type 1 replication in primary macrophages. *J. Infect. Dis.* 163: 78-82.

Mihm, S., Ennen, J., Pessara, U., Kurth, R., and Droge, W. (1991) Inhibition of HIV-1 replication and NF-kappa B activity by cysteine and cysteine derivatives. *AIDS*. 5: 497-503.

Paeck, R., Rubin, P., and Bauerle, P.A. (1991) Reactive oxygen intermediates as apparently widely used messengers in the activation of the NF-κB transcription factor and HIV. *EMBO J*. 10: 2247-2258.

Pantaleo, G., Graciozi, C., and Fauci, A.S. (1993) The immunopathogenesis of human immunodeficiency virus infection. *NEJM* 338: 327-335.

Picolet, H., Lang, J.M., Touraine, J.L., Livrozet, J.M., Saintmarc, T., Kirstetter, M., Trepo, C., Retornaz, G., Chossegros, P., Barrier, J., Raffi, F., Touze, J.E., Hovette, P., Kouchner, G., Lecompte, T., Hauteville, D., Jaubert, D., Danis, J.F., Philippe, G., Herson, S., Coutelier, A., Jasmin, C., Ecstein, E., Guillet, G., Plantin, P., Gentilini, M., Katlama, C., Kernbaum, S., Antiphon, P., Pinay, P., Cartier, W., Michelet, C., Reynes, J., Siffert, M., Blanc, A.P. Routy, J.P., Rozenbaum, W., Rouveix, B., Vilde, J.L., Leport, C., Perronne, C., Salmonceron, D., Vachon, F., Rozenbaum, W., Andrieu, J., Bideault, H., Bonnefond, R., Bousquet, R., and Bressot, H., et al. (1993). Multicenter, randomized, placebo-controlled study of ditiocarb (Imuthiol) in human immunodeficiency virus infected asymptomatic and minimally symptomatic patients - by the HIV 87 Study Group. *AIDS Res. Human Retroviruses*. 9: 83-89.

Poli, G., Bressler, P., Kinter, A., Duh, E., Trimmer, W.C., Rabson, A., Justement, J.S., Stanley, S., and Fauci, A.S. (1990) Interleukin 6 induces human immunodeficiency virus expression in monocytic cells alone and in synergy with tumor necrosis factor alpha by transcriptional and mechanisms. *J. Exp. Med*. 172: 151-158.

Poli, G., Kinter, A., Justement, J.S., Kerhl, J.H., Bressler, P., Stanley, S., and Fauci, A.S. (1990) Tumor necrosis factor alpha functions in an autocrine manner in the induction of humanimmunodeficiency virus expression. *Proc. Natl. Acad. Sci. USA*. 87: 782-785.

Reisinger, E.C., Kern, P., Ernst, M., Bock, P., Flad, H.D., and Dietrich, M. (1990) Inhibition of HIV progression by dithiocarb. *Lancet*. 335: 679-682.

Roederer, M., Staal, F.J., Osada, H., and Herzenberg, L.A. (1990) CD4 and CD8 T cells with high intracellular glutathione levels are selectively lost as the HIV infection progresses. *Int. Immunol*. 199: 933-937.

Ruffmann, R., and Wendel, A. (1991) GSH rescue by N-acetylcysteine. *Klin-Wochenschr*. 69: 857-662.

Smith, C.V., Hansen, T.N., Hanson, K., and Shearer, W.T. (1990) Glutathione concentration in plasma and blood are markedly decreased in HIV infected children. *Sixth Int. Conf. AIDS*. II: 368. Sonnenborg, A., Carlin, G., Akerlund, B., and Jarstrand, C. (1988) Increased production of malonedialdehyde in patients with HIV patients. *Scan. J. Infect. Dis*. 20: 287-290.

Staal, F.J.T., Roederer, M., Israelski, D.M., Buop, J., Mole, L.A., McShane, D., Deresinski, S.C., Ross, W., Sussman, H., Rago, P.A., Anderson, M.T., Moore, W., Elsa, W., Herzenberg, L.A., and Herzenberg, L.A. (1992) Intracellular glutathione levels in T cells subset decrease in HIV infected individuals. *AIDS Res. Human Retroviruses*. 2: 311.

Suthantiran, M., and May, M. (1990) Glutathione regulates activation dependent DNA synthesis in highly purified normal T lymphocytes stimulation via CD2 and CD3 antigens. *Proc. Natl. Acad.Sci. USA*. 87: 3343-3347.

Suzuki, Y.J., Aggarwal, B.B., and Packer, L. (1992) a lipoïc acid is a potent inhibitor of NFKB activation in human T cells. *Biochém. Biophys. Res. Com*. 189: 1709-1715.

Toledano, M., and Leonard, W. (1991) Modulation of transcriptor factor NF-KappaB by oxidation-reduction in vitro. *Proc. Natl. Acad. Sci. USA*. 88: 4328-4332.

Valis, K.A. (1991) Glutathione deficiency and radiosensitivity in AIDS patients. *Lancet*. 337: 918 -919.

Witschi, A., Reddy, S., Stofer, B., and Lauterburg, B.H. (1992) The systemic availability of oral glutathione. *Euro. J. Clin. Pharmacol*. 43: 667-669.

Yamauchi, N. (1990) Suppressive effect of intracellular glutathione on hydroxyradicals production induced by tumor necrosis factor. *Int. J. Cancer*. 46: 884-888.

Oxidative Stress, Cell Activation and Viral Infection
C. Pasquier et al. (eds)
© 1994 Birkhäuser Verlag Basel/Switzerland

Prevention of early cell death in peripheral blood lymphocytes of HIV infected individuals by an anti-oxidant: N-Acetyl-Cysteine

R. Olivier*, O. Lopez*, M. Mollereau*, T. Dragic°, D. Guetard*, L. Montagnier*

*Unité d'Oncologie Virale , Institut Pasteur, 25-28 Rue du Dr Roux, 75724 Paris cedex 15, France
*Hôpital de la Pitié -Salpétrière, 83 boulevard de l'Hôpital , 75013 Paris, France
°Inserm U332, Institut Cochin de Génétique moléculaire, 22 rue Mechain, 75014 Paris, France

Summary

Peripheral blood lymphocytes (PBLs) from HIV seropositive patients, cultured in growth factor free medium for 3 days show an important loss of viability and an apoptosis phenomenon in T cell subsets compared to healthy donor's PBLs. Different reports describe a decrease of intracellular Glutathione (GSH) in T lymphocyte subsets correlated to the stage of AIDS disease. N-Acetyl-cysteine (NAC) has been described to increase Glutathione level and restore cellular redox pathways. 15 patients were treated with 600 to 1200 mg of NAC per day over a year. Viability and apoptosis profiles were followed over a 10 day period. In a flow cytometry (FCM) test using Acridine Orange, we noted a cell population with low DNA staining intensity in PBLs from HIV seropositive patients. Agarose gel electrophoresis revealed DNA fragmentation correlated to this phenomenon, suggesting apoptosis. With NAC treatment we observed a systematic and stable amelioration of viability results from 2 months to 6 months of NAC treatment. At day 3 no cell death or apoptosis is detected in PBLs from healthy donors and patients treated over 6 months. At day 6 PBLs from healthy donors have up to 30% dead and apoptotic cells (60 to 100% for non or recently treated patients). There is no cell death before day 8 to 10 in PBLs from patients treated over 6 months. N-Acetyl-Cystein administration to HIV seropositive patients seems to prevent early cell death and to lead an "overprotection" of patients' PBLs.

Introduction

Since 1986, we have been following PBLs viability in HIV infected individuals. This study evaluated cell death in PBLs as a function of CD4+ cell decrease in different stages of the disease (Montagnier et al., 1989; 1991; 1992). Many reports indicate that HIV infected individuals including asymptomatic patients, were found to express decreased plasma cysteine and intracellular GSH levels (Dröge, 1988; 1989; Buhl, 1989; Eck, 1989; Jacobsen, 1990, Robinson, 1992; Roederer, 1993). This suggests that HIV infection is associated with highly reactive oxygen species and increased consumption of antioxidants. GSH is a constituent of a cellular and plasmatic redox regulation system involving enzymes (glutathione peroxydase, catalase, superoxide dismutase), a-tocopherol (Vit E), Vit A, Zinc and Selenium (Meister, 1983; 1988).

Decrease of GSH has also been correlated to a molecular process leading to cell damage and cell death (Hinshaw, 1986; Schraufstatter, 1988). Roederer (1991) and Staal (1992) confirmed the decrease of GSH levels in both CD4 and CD8 lymphocyte subsets in HIV seropositive patients and Eck (1992) correlated cysteine depletion to CD4+ lymphocyte number. It was therefore interesting to study the restoration of a global viability by repleating redox pathways with NAC, a well known non-toxic drug (Bonanomi, 1980; Holdiness, 1991) as a precursor of GSH (Ruffman, 1991). Many reports recommend the administration of cysteine derivatives to HIV infected individuals (Droge, 1992; 1993; Roederer, 1992; 1993)

Patients with less than 400 CD4 cells, were treated with AZT (Zidovudine: 400 to 600 mg per day), Rovamycine (6.10^6 I.U. per day), and NAC (Fluimucil, Zambon). A saturating administration of NAC (600 to 1800 mg per day) has been used. The half-life of NAC is 6 hours with an oral bioavailability of 9% (De Caro, 1989). The dosage is regulated between 600 and 1800 mg per day according to the digestive tolerance of each patient. Biological and clinical features were checked as well as *in vitro* cell viability. For this study we selected homosexual seropositive individuals with no major opportunistic infections, with or without lymphadenopathy and no oxidative stress factors like diabetes or alcoholism, excluding drug users. We have been applying a cell viability test since 1986 to more than 400 patients (Montagnier, 1989). Since in no case was an increase in viability status, a placebo control was not necessary in this study.

Material and methods

Cells
Peripheral blood was drawn from homosexual seropositive individuals on lithium heparin, conserved at room temperature and prepared on the same day for cell culture. Control blood was drawn the same day from seronegative healthy individuals.

Techniques
In order to quantify cell death in PBLs, handling a large number of samples, we set a flow cytometry assay on a Facscan (Becton-Dickinson), adapting a common technique in fluorescence microscopy with two nucleic acid staining molecules Acridine Orange (AO) and Ethidium Bromide (EB) (Becton Dickinson):
PBLs isolated on Ficoll were washed twice by gentle centrifugation in RPMI 1640 (Whittaker) to eliminate eventual toxicity or stimulation by seric factors, and cultured ($2x10^6$cells / ml) in growth factor free medium (RPMI 1640 with 10% foetal calf serum (FCS) (Whittaker) in 5% CO_2 at 37°C. Gentamicin (Boehringer Mannheim) was selected among commonly used

antibiotics as the least toxic in our culture conditions for normal lymphocytes. It was used at a lower concentration than usual: 10 µg/ml. On day 3, 500 µl of culture were gently centrifuged once at room temperature. Supernatant was removed and the pellet was resuspended in 200 µl of staining buffer (Facsflow, Becton Dickinson). 2 µl of staining solution (stock solution: AO: 1,5 µg/ml; EB: 5 µg/ml in PBS) were added to the suspension and incubated for 1 mn at room temperature. 200 µl of staining buffer were added to the suspension in order to dilute the remaining AO in the buffer and to lower staining background. The suspension was then analyzed with a flow cytometer (Facscan, Becton Dickinson) within 2 mn of the staining reaction.

EB is used to stain dead cells, emitting a red orange fluorescence with a 488 nm laser exitation. Under the same excitation, AO emits a yellow-green fluorescence when bound to a double stranded nucleic acid (DNA) and a red-orange fluorescence with a single stranded nucleic acid (RNA). With a low concentration of AO, only DNA emits enough fluorescence intensity to be detected with a cell sorter.

Using this assay, we observed not only the dead and viable cell populations usually described in fluorescence microscopy, but also a third population, presenting low DNA staining occuring only in HIV infected individuals (figure 1).

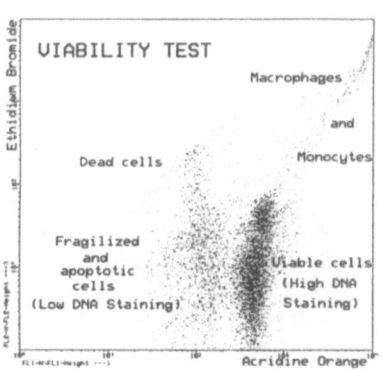

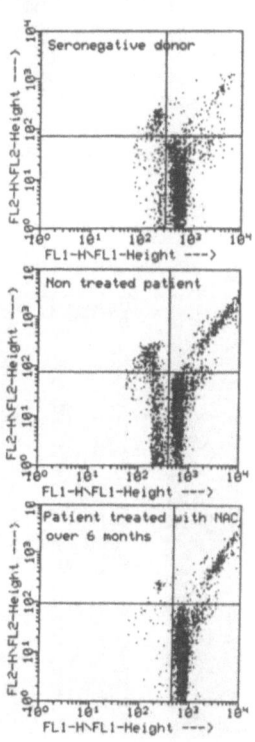

Figure 1: Flow cytometry analysis of cell viability.

326

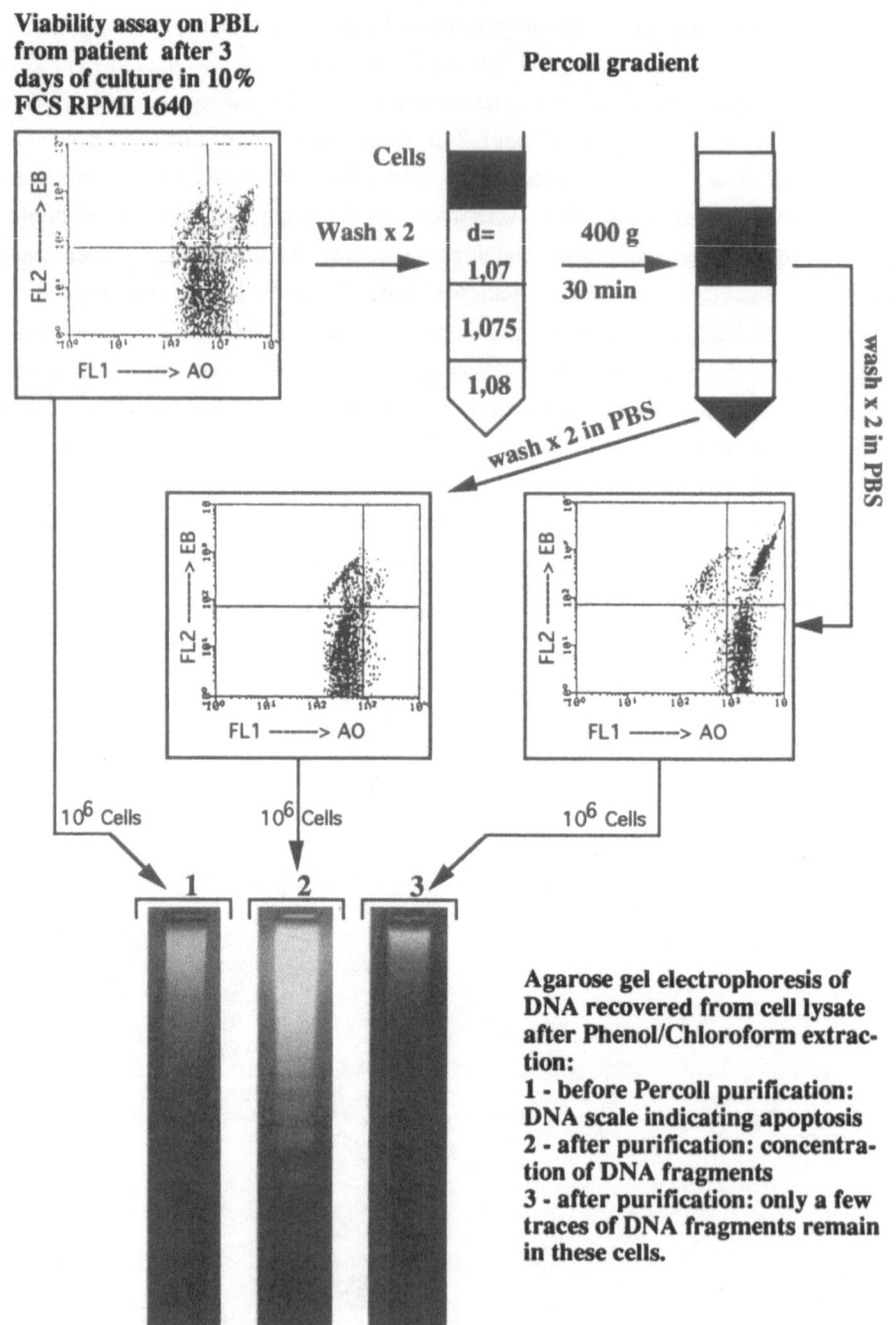

Viability assay on PBL from patient after 3 days of culture in 10% FCS RPMI 1640

Percoll gradient

Wash x 2

Cells

d=
1,07

1,075

1,08

400 g

30 min

wash x 2 in PBS

wash x 2 in PBS

10⁶ Cells

10⁶ Cells

10⁶ Cells

1   2   3

Agarose gel electrophoresis of DNA recovered from cell lysate after Phenol/Chloroform extraction:
1 - before Percoll purification: DNA scale indicating apoptosis
2 - after purification: concentration of DNA fragments
3 - after purification: only a few traces of DNA fragments remain in these cells.

Figure 2: Correlation of DNA damage in an apoptotic process with the low DNA staining observed by flow cytometry in the cell viability test.

Since it was important to characterize the additional population, we used a Percoll gradient technique published by Seamus Martin (1990), to separate dead, fragilized and apoptotic cells from viable cells. We reanalyzed the cell populations after separation to verify this purity. DNA from sorted cell populations was analyzed on agarose gel. We observed a DNA scale in the global population and in the dead and low stained cells (figure 2). This correlation between subdiploid DNA content and the apoptotic process has since been confirmed by cell sorting by Mary Collins and Rodriguez-Tarduchy (1990, 1992).

Results

In a preliminary study, we made the following observations. After 3 days of culture, lymphocytes from seronegative donors had less than 20 % dead and apoptotic cells while lymphocytes from patients non-treated or treated with NAC under 6 months had a range of 10 to 60% of dead and apoptotic cells, lymphocytes from patients treated over 6 months are significantly viable (figure 3).

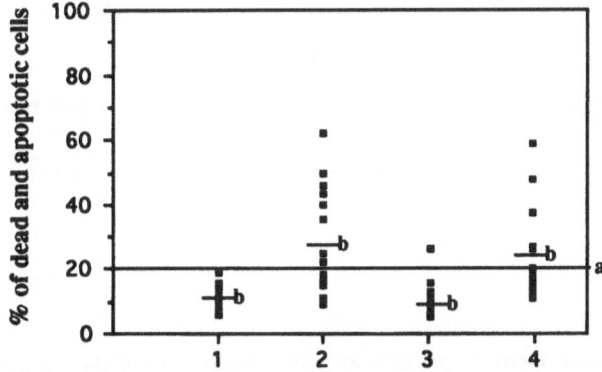

Figure 3: PBLs were cultured in 10% FCS RPMI 1640 without any additional growth factor. An FCM viability was performed on day 3 of culture. Comparison of cell death for (1) Seronegative donors, (2) Non treated patients and patients recently treated (t<6months) with NAC (last value), (3) Patients treated over 6 months (last value). (4) Initial value before treatment for the patients cited in (3). n =15 for each series. (a) highest value for seronegative donors. (b) average value for each series.

In a kinetics study over a 10 day period (figure 4), we observed a more interesting difference between treated patients and control individuals. An uniform progression of cell death was observed with seronegative donors. This progression was more rapid with HIV infected individuals. For NAC treated patients, progression of cell death is lowered. After 6 days of culture, the number of dead and apoptotic cells was still under the 20% threshold. After 10 days of culture, when all the cells from non treated patients are dead, the range of dead and apoptotic cells is 20 to 60%. For the seronegative donors, on day 10 the range is 30 to 100% of dead cells. In comparison to seronegative donors, patients seem to be over-protected against cell fragilization after a few months of treatment.

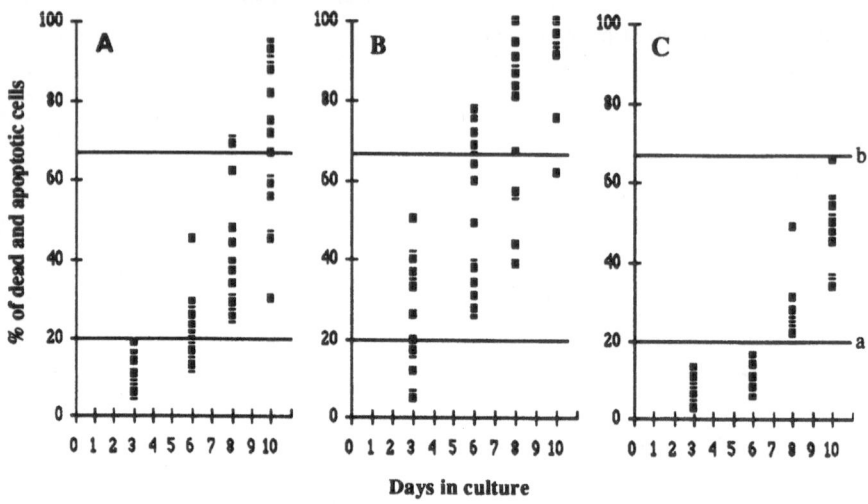

Figure 4: PBLs from (A) healthy seronegative donors, (B) patients not treated or recently treated with NAC (< 6months), (C) patients treated with NAC at least for 6 months. (a) reference threshold on lymphocyte status at day 3 of culture for normal PBLs, (b) highest value observed on day 10 with lymphocytes from patients treated with NAC over 6 months. No cell debris was detected in samples from these patients before day 8. (n=15 for each serie).

Discussion

Different observations were made in parallel to these results. The CD4 lymphocyte count in treated patients compared to non treated was followed. CD4+ lymphocyte decrease was arrested for most of them. This stabilization was followed by a slow but stable increase of T4 cells for 3 patients.

Patients with a positive HIV antigenemia before treatment kept their antigenemia. No decrease or increase of viral production accompagnied the amelioration of cell viability, the stabilisation or increase of T4 numbers.

In a triple staining flow cytometry technique using AO, EB and monoclonal antibodies against CD4, CD8, CD19 antigens we observed that Both T4, T8 and B cells are involved in the apoptotic process (data not shown). During NAC treatment, early cell death is prevented in CD4,

CD8 and CD19 lymphocytes. These results indicate that HIV infection may not have a direct effect on the CD4 lymphocyte depletion. We also observed that early cell death occurs right after seroconversion. Preliminary results about oxidative stress measurement (Zinc, Selenium, Malondialdehyde, Glutathione Peroxydase activity...) seem to correlate to these observations (data not shown).

Cysteine and cysteine derivatives like glutathione have been widely described as important factors in free radical scavenging systems (Meister; Dröge) and in T cell mediated immune responses (Dröge, 1991; Hamilos, 1989; Gmunder, 1990). Glutathione particularly protects cells against the toxic effects of oxidation, by participating directly in the elimination of reactive oxygen species and the preservation of reduced forms of other molecules like vitamin C or vitamin E, also involved in the redox regulation system (Larsson *et al.*, 1983; Dolphin *et al.*, 1989; Taniguchi *et al.*, 1989; Meister and Anderson, 1983, Meister, 1992).

In HIV infected individuals, different factors may contribute to immunodeficiency. Depletion of T4 lymphocytes was the initial observation of viral cytopathogenesis. Since only small percentage of CD4 lymphocytes are infected in patients, this cannot explain the wide ranging and radical depletion of CD4 lymphocytes observed. On the other hand, GSH depletion and cell fragility and mortality affect most of the lymphocyte subsets and can hardly be explained by a direct action of HIV on the cells. Different models of pathogenicity have been proposed. Among them, an antigenic stimulation either by viral constitutive molecules or by co-factors like mycoplasmas are considered (Montagnier, 1990; 1993; Dröge, 1992). The consequence of this stimulation could be a production of inflammatory cytokines able to produce a toxic effect on lymphocytes. Elevated levels of TNF-α and Interferon-γ have been described by Lahdevirta (1988) and Fuchs (1989) in AIDS patients' sera. Stimulation could also lead an over-production of reactive oxygen species by macrophages or neutrophils (Bertoni, 1985; Allen, 1986). Reactive oxygen species are well known to inhibit cell proliferation (Dröge, 1991) and to induce DNA damage by activation of nucleases (Hinshaw, 1986, Schraufstatter and coll., 1988) and can explain the results we observed in AIDS patients and the effect of NAC treatment.

Conclusion

NAC has been widely described in *in vitro* studies to inhibit HIV production (Dröge, 1992; Kalebic, 1991; Roederer, 1990; 1992; Israël, 1992; Staal, 1990; 1992; 1993; Schreck, 1991) but, the concentration used is far too high to be attained *in vivo* (in a mM range). In *in vitro* studies which we did to restore cell viability, using cells from untreated patients, the highest concentration used was in the µM range (data not shown) which compares to the highest value

330

detected in patient plasma after administration of NAC (Frank, 1984; Olsson, 1988). Above the mM concentration range, we observed a rapid induced cell toxicity. Nevertheless, considering the absence of an increase of antigenemia in HIV infected patients who had a negative antigenemia before NAC treatment, we can hypothesize that the stabilization of intracellular and plasmatic redox status could prevent HIV replication. This could suggest a long term anti-viral effect of NAC. Early therapy to prevent early cell death and also HIV replication should be applied.

Acknowlegment

We thank Prs Lea and Len Herzenberg for initiating this study and their helpful advices in the therapeutic use of N-acetyl-cysteine.

References

Bertoni, G., Keist, R., Groscurth, P., Wyler, R., Nicolet, J., and Peterhans, E. (1985) A chemiluminescent assay for mycoplasmas in cell cultures. *J. Immunol. Methods.* 78: 123-133.
Bonanomi, L., Gazzaniga, A. (1980) Toxicological, pharmacokinetic and metabolic studies on acetylcysteine. *Eur. J. Resp. Dis.* 61(S11): 45-51.
Buhl, R., Holroyld, K.J., Mastrangeli, A., Cantin, A.M., Jaffe, H.A., Wells, F.A., Saltini, C., Crystal, R.G. (1989) Systemic glutathione deficiency in symptom free HIV positive individuals. *Lancet.* 2: 1294.
Collins, M.K.L., Marvel, J., Malde, P., Lopez-Rivas, A. (1992) Interleukin 3 protects murine Bone Marrow cells from Apoptosis Induced by DNA Damaging Agents. *J. Exp. Med.* 176:1043-1051.
De Caro, L., Ghizzi, A., Costa, R., Longo, A. Ventresca, G.P., Lodola, E. (1989) Pharmacokinetics and Bioavailability of Oral Acetylcysteine in Healthy Volunteers. *Arzneim.-Forsch./Drug Res.* 39(1) 3, 382-386.
Dolphin, D., Poulson, R., and Avramovic, O. (Eds.) (1989) *Glutathione: Chemical, Biochemical, and Medical Aspects, Parts A and B, Coenzyme and Cofactors Series,* Vol. III. John Wiley, New York.
Dröge, W., Pottmeyer-Gerber, C., Schmidt, H., and Nick, S. (1986) Glutathione augments the activation of cytotoxic T lymphocytes *in vivo. Immunobiol* 172:151-156.
Dröge, W., Eck, H-P., Näher, H., Pekar, U., and Daniel, V. (1988) Abnormal amino acid concentrations in the blood of patients with acquired immunodeficiency syndrome (AIDS) may contribute to the immunological defect. *Biol Chem Hoppe-Seyler* 369:143-148.
Dröge, W., (1989) Metabolic disorders in HIV infection. *Project News - AIDS-Forschungsförderung des Bundesministeriums für Forschung und Technologie* 2:4-5.
Dröge, W., Eck, H-P., Gmünder, H., and Mihm, S. (1991) Modulation of lymphocyte functions and immune responses by cysteine and cysteine derivatives. *Am J Med* 91 (suppl 3C):140S-144S.
Dröge, W., Eck, H-P., and Mihm, S. (1992) HIV-induced cysteine deficiency and T-cell dysfunction - a rationale for treatment with N-acetyl-cysteine. *Immunol Today* 13:211-214.
Dröge, W. (1993) Cysteine and glutathione deficiency in AIDS patients: a rational for the treatment with N-Acetyl-cysteine.*Pharmacology* 46(2):61-5.
Eck, H-P., Gmünder, H., Hartmann, M., Petzoldt, D., Daniel, V., and Dröge, W. (1989) Low concentrations of acid-soluble thiol (cysteine) in the blood plasma of HIV-1-infected patients. *Biol Chem Hoppe-Seyler* 370:101-108.

Eck, H-P., Mertens, T., Rosokat, H., Fätkenheuer, G., Pohl, C., Schrappe, M., Daniel, V., Näher, H., Petzoldt, D., Drings, P., and Dröge, W. (1992) T4+ cell numbers are correlated with plasma glutamate and cystine levels: association of hyperglutamataemia with immunodeficiency in diseases with different aetiologies. *Int Immunol* 4:7-13.

Frank, H., Thiel, D., Langer, K. (1984) Determination of N-Acetyl-l-cysteine in biological fluids. *J. Chromato*. 309: 261-267.

Fuchs, D., Hausen, A., Reibnegger, G, Werner, E.R., Werner-Felmayer, G., Dierich, M.P., Watcher, H. (1989) Interferon-gamma concentrations are increased in sera from individuals infected with human immunodeficiency virus type 1. J. Acq. Imm. Defic. Synd., 2-158.

Gmunder H., Roth S., Eck H.P., Gallas H, Mihm S., Dröge W. (1990) *Cell. Immunol.* 130(2):520-8.

Hamilos D.L., Zelarney P., Mascali J.J. (1989) Lymphocyte proliferation in glutathione-depleted lymphocytes: direct relationship between glutathione availability and the proliferative response. *Immunopharmacology*. 18(4):557-8.

Hinshaw, D.B., Sklar, L.A., Bolh, B., Shraufstatter, I.;U., Hyslop, P.A., Rossi, M.W., Spragg, R.G., Cochrane C.G.(1986) Cytosskeletal and morphologic impact of cellular oxidant injury. *Am. J. Path.*, 123:454-464.

Holdiness, M.R. (1991) Clinical Pharmacokinetics of N-Acetylcysteine. *Clin. Pharmacokinet.* 20(2): 123-134.

Israël, N., Gougerot-Pocidalo, M.-A., Aillet, F., and Virelizier, J.-L. (1992) Redox status of cells influences constitutive or induced NF-kB translocation and HIV long terminal repeat activity in human T and monocytic cell lines. *J. Immunol.* 149: 3386-3393.

Jacobsen, D;W;, Green, R., Hebert, W., Longworth, D.L., and Crystal, R.G. (1990) Decreased serum glutathione with normal homocysteine level and cysteine level in patients with AIDS, *Clin. Res.* 38, 556.

Kalebic, T., Kinter, A., Poli, G., Anderson, M. E., Meister, A., and Fauci, A. S. (1991) Suppression of human immunodeficiency virus expression in chronically infected monocytic cells by glutathione, glutathione ester, and N-acetyl cysteine. *Proc. Natl. Acad. Sci.* USA 88: 986-990.

Lahdevirta, J., Maury, C.P.J., Teppo, A-M., Repo, H. (1988) Elevated levels of circulating cachectin/tumor necrosis factor in patients with acquired immunodeficiency syndrome. *Am. J. Med.* 85:289.

Larsson, A., Orrenius, S., Holmgren, A., and Mannervik, B. (Eds.) (1983) *Functions of Glutathione. Biochemical, Physiological, Toxicological, and Clinical Aspects*. Raven Press, New York.

Martin S.J., Lennon, S.V., Bonham, A.M., Cotter, T.G. (1990) Induction of apoptosis (Programmed Cell Death) in human Leukemic HL-60 cells by inhibition of RNA or protein synthesis. *J. Immunol.* 145, 1859-1867.

Meister, A., and Anderson, M.E. (1983). Glutathione. *Annu. Rev. Biochem.* 52: 711-760.

Meister, A. (1983) Selective modification of glutathione metabolism. *Science* 220: 471-477.

Meister, A. (1988) Glutathione metabolism and its selective modification. *J. Biol. Chem.* 263: 17205-17208.

Meister, A. (1991) Glutathione deficiency produced by inhibition of its synthesis and its reversal; Applications in research and therapy. *Pharmacol. Ther.* 51: 155-194.

Meister, A. (1992) On the antioxidant effects of ascorbic acid and glutathione. *Biochem. Pharmacol.* 44: 1905-1915.

Mihm, S., Ennen,J., Pessara, U., Kurth, R., and Dröge, W., (1991) Inhibition of HIV-1 replication and NF-kB activity by cysteine and cysteine derivatives. *AIDS*. 5: 497-503.

Montagnier, L., Guetard, D., Rame, V., Olivier, R.and Adams, M. (1989) Virological and immunological factors of AIDS pathogenesis, *Retroviruses of Human AIDS and related animal diseases,Colloque des cents gardes, Paris*, 4, 11-17

Montagnier, L., Gougeon, Tschopp, R., Guetard, D., Lecoeur, H., Rame, V., Garcia, S., Lopez, O., Olivier, R., (1991) New insights on the mechanisms of CD4+ lymphocytes depletion in

332

aids, *Retroviruses of Human AIDS and related animal diseases, Colloque des cents gardes, Paris,* 6, 9-17

Montagnier, L., Gougeon, M.L., Olivier, R., Garcia, S., Guetard, D., Dauguet, C, Bechet, J-M., Adams, M., (1992) Factors and Mechanisms of AIDS Pathogenesis, *Science Challenging AIDS, Basel, Karger,* 51-70.

Montagnier, L., Blanchard, A. (1993) Mycoplasmas as Cofactors in Infection Due to the Human Immunodeficiency Virus. *Clin. Inf. Dis.* 17(S1): S309-S315.

Olsson, B., Johansson, M., Gabrielsson, J., Bolme, P. (1988) Pharmacokinetics and Bioavailability of Reduced and Oxidized N-Acetylcysteine. *Eur. J. Clin. Pharmacol.* 34:77-82.

Robinson M.K., Hong R.W., Wilmore D.W. (1992) Glutathione deficiency and HIV infection. *Lancet* 339(8809):1603-4.

Rodriguez-Tarduchy, G., Collins, M., Lopez-Rivas, A. (1990) Regulation of Apoptosis in Interleukin-3-dependent hemopoietic cells by interleukin-3 and calcium ionophores. *EMBO J.* 9 (9) 2997-3002.

Roederer, M., Staal, F. J. T., Raju, P. A., Ela, S. W., Herzenberg, L. A. and Herzenberg, L. A. (1990) Cytokine-stimulated human immunodeficiency virus replication is inhibited by N-acetyl-L-cysteine. *Proc. Natl. Acad. Sci.* USA. 87: 4884-4888.

Roederer, M., Staal, F.J.T., Osada, H., Herzenberg, L.A., and Herzenberg, L.A. (1991) CD4 and CD8 T cells with high intracellular glutathione levels are selectively lost as the HIV infection progresses. *Int Immunol* 3:933-937.

Roederer, M., Ela, S. W., Staal, F. J. T., Herzenberg, L. A. and Herzenberg, L. A. (1992) N-Acetyl-Cysteine: a new approach to anti-HIV therapy. *AIDS Res Hum Retroviruses* 8(2):209-17.

Roederer, M., Staal, F. J. T., Anderson M., Rabin R., Raju, P. A., Herzenberg, L. A. and Herzenberg, L. A. (1993) Disregulation of leukocyte glutathione in AIDS. *Ann N Y Acad. Sci.* Mar 20:: 677 (1):113-25.

Roederer, M., Staal, F. J. T., Ela, S. W., Herzenberg, L. A. and Herzenberg, L. A. (1993) N-Acetyl-Cysteine: potential for AIDS therapy. *Pharmacology.* 46.

Ruffmann R, Wendel A. (1991) GSH rescue by N-Acetylcysteine. *Kin Wochenschr* 69(18):857-62.

Schraufstatter, I., Hyslop, P.A., Jackson, J., and Cochrane, C.G. (1988) Oxidant induced DNA damage of target cells. *J. Clin. Invest.* 82, 1040-1050.

Schreck, R., Rieber, P., and P. A. Baeuerle. (1991) Reactive oxygen intermediates as apparently widely used messengers in the activation of the NF-kB transcription factor and HIV-1. *EMBO J.* 10 : 2247-2257.

Staal, F. J. T., Roederer, M., Herzenberg, L. A., and Herzenberg, L. A. (1990) Intracellular thiols regulate activation of nuclear factor kB and transcription of human immunodeficiency virus. *Proc. Natl. Acad. Sci. USA* 87: 9943.-9947

Staal F.J.T., , Roederer M., Anderson M.T., Herzenberg L.A., and Herzenberg L.A. (1992) Glutathione deficiency and human immunodeficiency virus infection. *The Lancet* 339:909-912.

Staal, F.J.T., Roederer, M., Israelski, D.M., Buop, J., Mole, L.A., McShane, D., Deresinski, S.C., Ross, W., Sussman, H., Rago, P.A., Anderson, M.T., Moore, W., Elsa, W., Herzenberg, L.A., and Herzenberg, L.A. (1992) Intracellular glutathione levels in T cells subset decrease in HIV infected individuals. *AIDS Res. Human Retroviruses.* 2: 311.

Staal, F. J. T., Roederer, M., Raju, P. A., Anderson M., Ela S.W., Herzenberg, L. A. and Herzenberg, L. A. (1993) Antioxidants inhibit stimulation of HIV Transcription *AIDS Res. Human Retroviruses.* 9: 299-306.

Taniguchi, N., Higashi, T., Sakamoto, Y., and Meister, A. (Eds.) (1989) *Glutathione Centennial, Molecular Perspectives and Clinical Implications.* Academic Press, New York.

Alcanes measurements in human immunodeficiency virus infection

E. Postaire, L. Massias, O. Lopez[1], M. Mollereau [1] and G. Hazebroucq

Département de Pharmacie Clinique - Faculté de Pharmacie Paris XI - Chatenay-Malabry - France
[1] CNTS - Groupe Hospitalier Pitié-Salpétrière - Paris - France

Summary

Toxic reactive oxygen species (ROS) generated by mononuclear phagocytes and neutrophils by means of the reduced nicotinamide adenine dinucleotide phosphate (NADPH) oxidase system can produce, by chain reaction, a wide array of molecular alterations including lipid peroxidation. Pentane measurements in alveolar air breath, which can be considered as a specific marker of lipoperoxidation, are performed in Humans affected by immunodeficiency virus (HIV) which is characterized in its course by a progressive CD4+ lymphocytes depletion including reduced production of regulatory cytokines and oxidative burst. Pentane measurements are carried out after collection of the alveolar air breath on graphite cartridges followed by thermal desorption and gas chromatographic separation of alkanes. The 7 first patients show a mean value of 5.73 nmol/l of pentane.

Introduction

HIV infection affects various part of the immune system, including the CD4 lymphocytes and mononuclear phagocytes, and causes a progressive immunodeficiency. This renders the patient susceptible to various opportunistic infections and neoplasm. Reactive Oxygen Species (ROS) are important for the intracellular killing of microorganisms by mononuclear phagocytes and neutrophils (Segal, 1989). Although, data are discrepant, several studies suggest that the generation of ROS is impair in mononuclear phagocytes, and possibly also in neutrophils from HIV-infected patients. This may lead to deficient killing of intracellular microorganisms predisposing the HIV-infection patient to certain opportunistic infections. Recently, *in vitro* studies have shown that ROS activate the intracellular transcription nuclear factor (NF-κB) which stimulates HIV replication and that, during the phase expression, the splicing pattern of viral mRNA results in the preferential translation of regulatory proteins, including Tat which is essential for HIV replication (Schreck et al., 1991).

However, the role of ROS in regulation of HIV replication in vitro is unknown at present. the role of ROS in HIV infection is difficult to assess, both at the cellular and clinical level. And it seems now very important to know the exact level of the oxidative burst and to assess the consequences on the stimulation of HIV replication and cellular dammages which are responsible of the clinical deterioration.

Toxic ROS generated include superoxide anion, hydrogen peroxide, singlet oxygen and the highly reactive hydroxyl radicals which can produce by chain reactions a wide array of molecular alterations including membranal lipid peroxidation (Crastes de Paulet, 1987) (with generation of degradation products), protein and nucleic acid damage, ultimatly resulting in cancerogenesis or cell death.

Detailed studies of lipid peroxidation reactions began with recognition that unsatured lipids take up oxygen to form hydroperoxides with conjugated dienes that can decompose to form numerous other products, including carbonyl compounds (malonaldehyde and other thiobarbituric acid-reacting substances) and trace volatiles via ß-scission. Carbonyl products can react with sensitive biological molecules to form fluorescent chromophores, including those with carbonyl-amine Schiff base structures (Pryor and Godber, 1991). Measurement of trace volatile hydrocarbons is the only fully noninvasive method developped to date to measure a lipid peroxidation product in living animals (Wade and Van Rij, 1988; Van Gossum and Decuyper, 1989; Burk and Ludden, 1989). Riely et al. (1974) first reported the measurement of ethane expired by carbon tetrachloride treated mice. Ethane derives from $\omega 3$ fatty acids, but studies in our laboratory focused on measured of the $\omega 6$ -hydroperoxy product, pentane.

The chemical transformations involved in the conversion of $\omega 6$ fatty acids, i.e. arachidonic acid, to pentane are illustrated in the Figure 1. Radical chain peroxidation of polyunsaturated fatty acids produces a rather uniform distribution of isomeric hydroperoxides based on abstraction of bis-allylic H atoms and preservation of the conjugated diene in subsequent reactions of the pentadienyl radicals, but the possible contributions of chemical specificity or of enzymatically generated hydroperoxydes to the production of alkanes under some reactions should not be unawared contributions of chemical specificity or of enzymatically generated hydroperoxydes to the production of alcanes under some reactions should not be unawared. Reaction of the fatty acid hydroperoxides with chemically reactive chelates of ferrous iron to produce the corresponding alkoxyl radical. These alkoxyl radical also can undergo ß-scission reactions, and the alkyl radical may next abstract a hydrogen atom from an adjacent donor molecule to produce the alkane, but other reactions of these radicals are possible. In additions, pentane could be metabolized by microsomal cytochromes to 2- and 3-pentanols. The history of breath analysis can be traced back over 100 years, when, in 1874, Anstie reported the elimination of alcohol *via* human breath. It is claimed that in one hand, collection and analysis of breath is a nontraumatic, easily repeated, and useful non-invasive alternative to blood analysis. In the other hand non-invasive analysis are becoming increasingly important, especially for patients who require daily monitoring of therapeutical and biological parameters. And at last, a standardized, reproducible breath sample is critically important for quantitative breath analysis (Manolis, 1983; Wilson, 1986).

Arachidonic acid

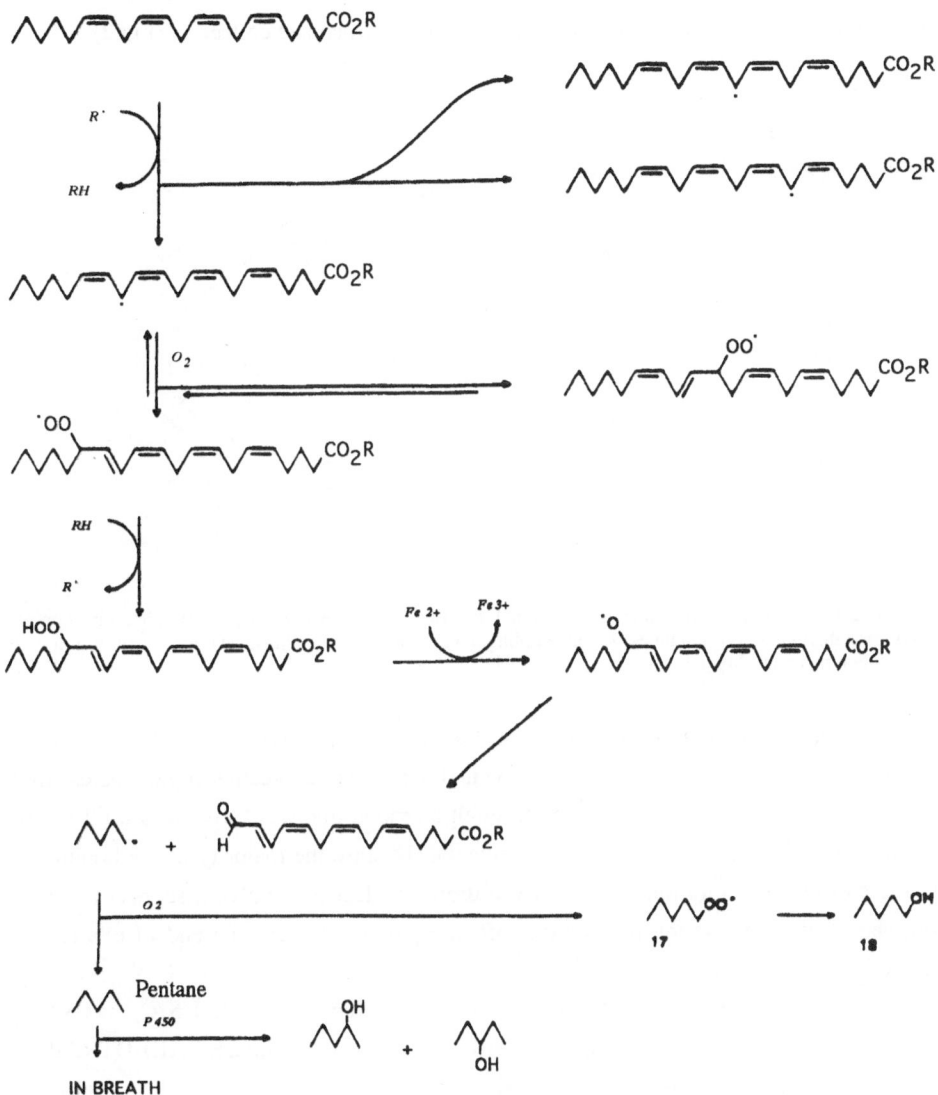

Fig. 1. Peroxidation of arachidonic acid

336

Material and Methods

We have developped and validated a portable apparatus for pentane collection (Fig.2).

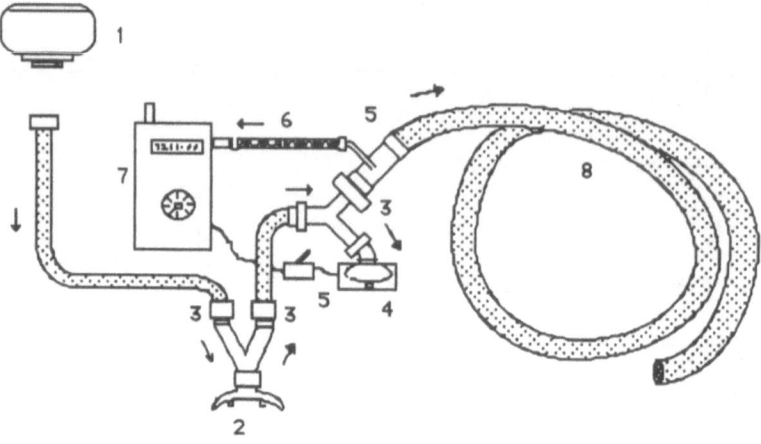

Fig. 2. 1, Organic vapor respirator cartridge; 2, mouth bit; 3, one-way valve; 4, electronic gate; 5, on-off contact; 6, Carbotrap® thermal desorption tubes; 7, Air sampling pump (continuously adjustable 0 to 200 ml/min flow rate); 8, Tube containing exhaled breath.

Collection of large volumes of breath over an extented period causes weariness, so we trap and concentrate only a part of the breath. The individual providing the breath sample inhales ambient air through a one-way valve and exhales through a second one way valve into a tube. Breath drawn back from the tube is predominately alveolar, because the majority of the breath comes from the end of the expiration. At each expiration, an electronic valve disconect a sampling pump, and at each inspiration the valve is off, the pump is on and the end of expired air is collected in adsorbent tubes.

Thermal desorption adsorbent tube, Carbotrap® (Supelco, Belfonte, USA), has valuable proporties for trapping, then releasing many airbone organic compounds. CARBOTRAP 200® (SUPELCO) contains graphitized carbon blaks with properties for trapping then releasing C_2-C_{14} hydrocarbures. Desorption efficiency is approximately 100% for trapped organic compounds.

Thermal Desorption Unit (SUPELCO) : Deliver undiluted airborne organic compounds onto the GC analytical column by heating the adsorbent tube from 35°C to 350°C in 16 sec.

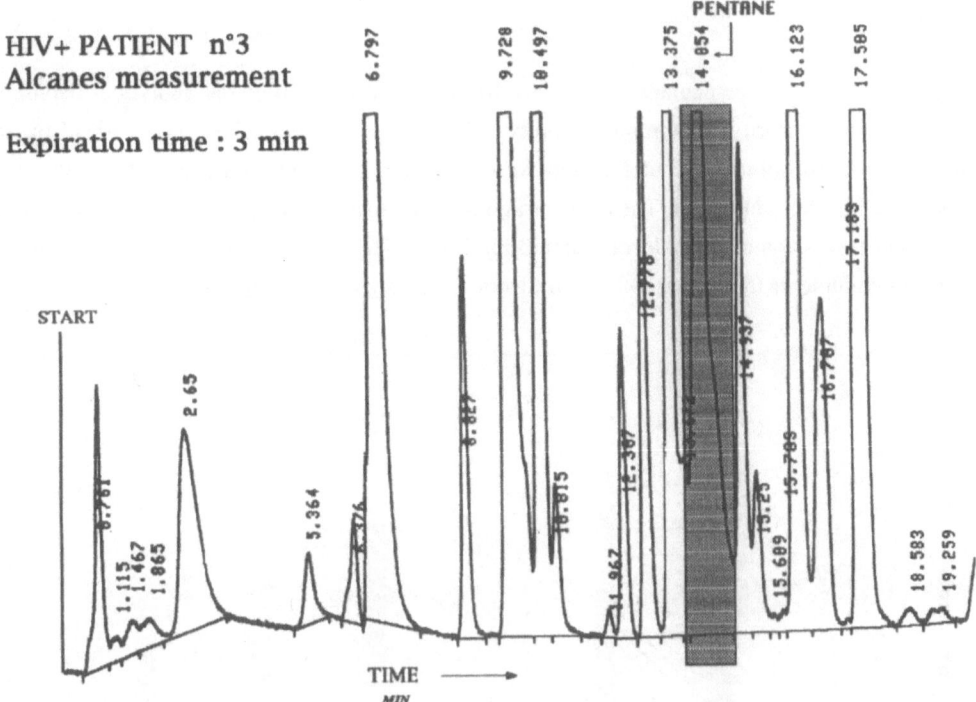

Fig. 3. Chromatogram

Gas Chromatograph : The gases were analyzed using a GIRDEL serie 30 gas chromatograph with a 60/80 carbopack B 6'x 1/8" SS column oven temperature from 30°C (2 min) to 260°C (16°C/min), detector (F.I.D) temperature 260°C . High purity Helium is the carrier gas at a column flow rate of 20 ml/min. The retention time of pentane was 14.2+/-0.2 min (Fig.3).

338

Results

The detection limit for pentane was 2.7+/-10-11 mg. Validation of the analytical technique indicates good linearity, good reproducibility and recovery (98% +/- 1.5%). Recoveries from the adsorption/desorption cycle of the trap were validated by comparing trapped injections of pentane with direct injection. The concentrations of the hydrocarbons were calculated by using the linear regression of multi-level standard injections. We collected samples of alveolar breath from ten volunteers (5 women and 5 men), from 23 to 35 years old (Fig. 4).

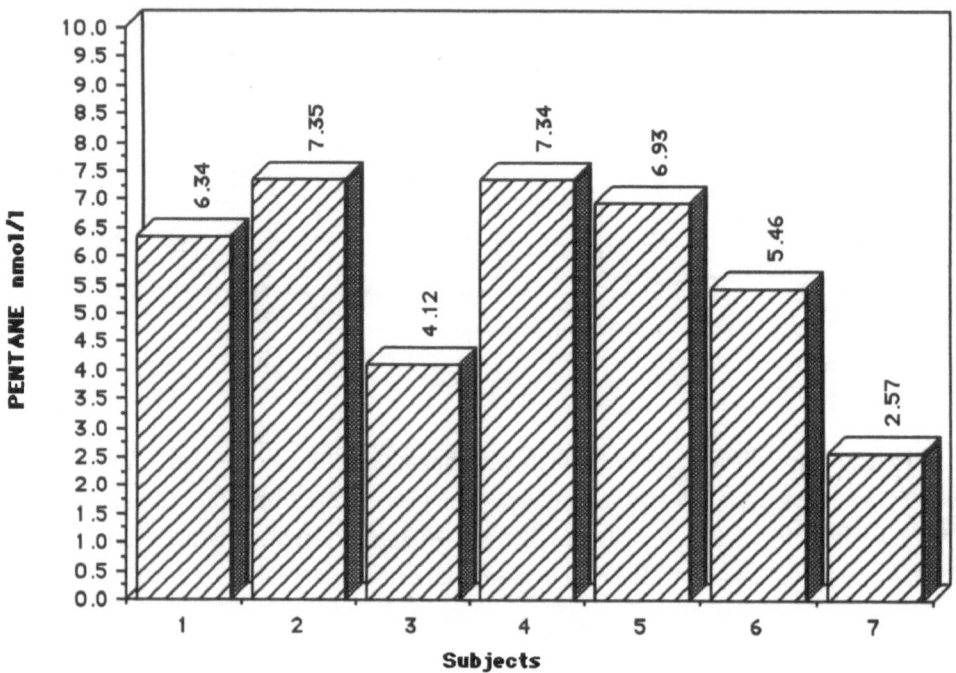

Fig. 4. Expired pentane in human immuno deficiency virus infection.

The average concentration of pentane found in room air from different sites and days is 0.43 +/- 0.05 nmol/l of air, and in alveolar breath samples 4.33 +/- 1.67 nmol/l of air .

Pentane determinations in human immunodeficiency virus infection group show a mean value of 5.73 nmol/l with a standard deviation of 1.81 nmol/l.

Discussion

Mean values of pentane concentration found from ten volunteers can be related with pentane concentrations of Zarling and Clapper (1987) (3.7 +/- 1.2 nmol/l of air in alveolar breath) and Lemoyne et al. (1987) (3.8 +/- 0.57nmol/).

Oxidative stress is thought to be involved in the patholphysiology of several diseases ; activated inflammatory cells are capable of stimulating lipid peroxidation, in 27 patients with rheumatoid arthritis, Humad et al. (1988) found highly signifiant correlations between pentane excretion and inflammation (r= 0.8, p < 0.001). In an other study, Weitz et al. (1991) demonstrated that the breath pentane concentration was higher (p < 0.0001) in the acute myocardial infarction group (4.96 +/- 1.15) than in a patient control group (1.96 +/- 1.04). Lipid peroxidation during acute myocardial infarction reflects action of oxygen radicals and their potential for contribution to the pathogenesis of tissue damage. Finally, Toshniwal and Zarling (1992) published recently a study which demontrated that patients with acute exacerbation of multiple sclerosis had significantly higher concentrations of pentane (10.5 +/- 4.2 nmol/l) (p.< 0,01) compared to control subjects (4.3 +/- 1.1 nmol/l).

In our study, there is a significant difference between healthy subjects and HIV patients (p < 0.05) and one patient have an important concentration of pentane (32 nmol/l). This excess of lipid peroxidation is also demonstrated by elevated serum malondialdehyde levels (HIV : 1.625 +/- 0.108 µmol/l ; Control value : 0.90 +/-0.3 µmol/l).

We do not have enough data for conclusion about free radical therapeutics and pentane expiration and our aim is now to include enough patients in a randomized study in order to gain informations

Conclusion

Several studies suggest that the generation of ROS is impaired in mononuclear phagocytes from HIV-infected subjects, and this ROS activate the intracellular transcription factor nuclear factor kB (NF-kB) which stimulates HIV replication, and can produce, by chain reaction, molecular alterations like lipid peroxidation. Determination of breath pentane is a valuable method to assess lipid peroxidation. Breath sample collecting and chromatographic analysis require strict methodology, but this technique is easy and non-invasive.

340

References

Burk, R.F. and Ludden T.M. (1989) Exhaled alkanes as indices of in vivo lipid peroxidation. *Biochem. Pharmacol.* 38, 1029-1032.

Crastes de Paulet, A. (1987) Les lipides membranaires : une cible privilégiée des radicaux libres. *Cah. Nutr. Diet.* 22, 23-33.

Humad, S., Zarling, E., Clapper, M. and Skosey, J.L. (1988) Breath pentane excretion as amarker of oisease activity in rheumatoid arthritis. *Free Rad. Res. Comm.* 5, 101-106.

Lemoyne, M., Van Gossum, A., Kurian, R., Ostro, M., Axler, J. and Jeejeebhoy, K.N. (1987) Breath pentane analysis as an index of lipid peroxidation : a fonctional test of vitamin E status. *Am. J. Clin. Nutr.* 46, 267-272.

Manolis, A. (1983) The diagnostic potential of breath analysis. *Clin. Chem.* 29, 5-15.

Müller, F. (1992) Reactive oxygen intermediates and human immunodeficiency virus (HIV) infection. *Free Rad. Biol. Med.* 13, 651-657.

Pryor, W.A. and Godber, S.S. (1991) Noninvasive measures of oxidative stress status in humans. *Free Rad. Biol. Med.* 10, 177-184.

Riely, C.A., Cohen, G. and Lieberman, M. (1974) Ethane evolution: a new index of lipid peroxidation. *Science.* 183, 208-210.

Schreck, R., Rieber, P. and Baeuerle P.A. (1991) Reactive oxygen intermediates as apparently widely used messengers in the activation of the NF-κB transcription factor and HIV-1. *EMBO J.* 10, 2247-2258.

Segal, A.W. (1989) The respiratory burst in monocytes and macrophages. In: Zembala, M., Asherson, G.L. (eds): *Human monocytes.* Academic Press, London, pp. 89-100.

Toshniwal, P.K. and Zarling, E.J. (1992) Evidence for increased lipid peroxidation in multiple sclerosis, *Neurochem. Res.* 17 (2), 205-207

Van Gossum, A. and Decuyper, J. (1989) Breath alkanes as an index of lipid peroxidation. *Eur. Respir. J.* 2, 787-791.

Wade, C.R. and Van Rij, A.M. 1988 In vivo lipid peroxidation in man as measured by the respiratory. Excretion of ethane, pentane and other low-molecular-weight hydrocarbons. *Anal. Biochem.* 150, 1-7.

Weitz, W.Z., Birnbaum, A.J., Sobotka, P.A., Zarling, E.J. and Skoley, J.L.(1991) High breath pentane concentrations during acute myocardial infarction. *Lancet.* 337, 933-935.

Wilson, H.K. (1986) Breath analysis. *Scand. J. Work Environ. Health* 12, 174-192.

Zarling, E. and Clapper, M. (1987) Technique for gas-chromatographic measurement of volatile alkanes from single-breath samples. *Clin. Chem.* 33, 140-141.

Oxidative Stress, Cell Activation and Viral Infection
C. Pasquier et al. (eds)
© 1994 Birkhäuser Verlag Basel/Switzerland

Plasma antioxidant status (selenium, retinol and α-tocopherol) in HIV infection.

C. Sergeant[1], M. Simonoff[1], C. Hamon[1], E. Peuchant[2], M.F. Dumon[2], M. Clerc[2], M.J. Thomas[2], J. Constans[3], C. Conri[3], J.L. Pellegrin[4], B. Leng[4]

1- CNRS URA 451, Le Haut-Vigneau, 33175 GRADIGNAN. 2- Biochimie Clinique, Hôpital Saint-André, 33075 BORDEAUX. 3- Médecine Interne, Hôpital Saint-André, 33075 BORDEAUX. 4- Médecine Interne, Hôpital Haut-Lévêque, 33600 PESSAC

Summary

Plasma antioxidant status can be evaluated by selenium, retinol and a-tocopherol. The present study concerns 89 HIV-1 positive patients. The plasma level of selenium was determined by PIXE (proton-induced X-ray emission analysis) and by measurement of levels of vitamins A and E by HPLC. The data were compared to controls; among the HIV-1 patients, significant differences were found for plasma selenium between groups selected on the basis of the CD4 lymphocyte count. Differences were also observed for plasma retinol between the HIV-1 patients and the control subjects. Correlations with other nutritional and immunological parameters were shown.

Introduction

Alterations in nutritional status are widespread among the HIV I population and are known to involve vitamins and trace elements. It has also been shown that deficiencies in selenium, zinc, vitamins A, E and B6 contribute significantly to the pathogenesis accompanying severe immunosuppression (Meydani et al., 1992). Among natural antioxidants occurring in cells under physiological conditions, vitamins A and E and selenium protect membranes and cytoplasm respectively (Packer 1991, Spallholtz 1990).

The aim of the present work was to investigate the antioxidant status with respect to selenium and vitamins A and E in the plasma of these HIV patients and in a control group. Selenium was measured by Proton-Induced X-Ray Emission (PIXE) after chemical preconcentration (Simonoff 1988) at the Centre d'Etudes Nucléaires de Bordeaux-Gradignan. Retinol and α-tocopherol were extracted chemically and quantified by High Performance Liquid Chromatography (HPLC) (Thurnham 1988) on a C18 µBondapack column, using a Waters HPLC system. Triglycerids, Cholesterol (Ch) and HDL-Cholesterol (HDL-C) were measured on a multiparameter automate Paramete (Baxter France). Apoprotein apo A1 and B and Lp(a) were determined by nephelometry using a BNA instrument (Behring France). CD4 and CD8 lymphocytes were counted by flow cytometry. P24 antigenemia was determined in 39 patients (Abbott ELISA).

The Student's t-test, Spearman's correlation test and step by step multivariate analysis were used for the statistical analysis.

Material and Methods

The study was carried out on eighty-nine HIV-infected patients presently hospitalized in two internal medical services in Bordeaux. For each patient, the following clinical features were registered: CDC stage, opportunistic infection, body mass index (BMI) and means of contamination. The risk behaviours were established as use of intravenous drugs (UIVD) in 33 cases, homo- or bi-sexuality in 33 cases, blood transfusion in 10 cases, multiple-partner heterosexuality in 12 cases and unknown in one case. The period of contamination time was usually not precisely known. Several patients were undergoing treatment with Zidovudine. The average age was 36 years (standard deviation, SD = 10).

The patients were divided into 4 groups according to the CD4 lymphocyte count:

I $(CD4 < 50/mm^3)$
II $(50 < CD4 < 200/mm^3)$
III $(200 < CD4 < 400/mm^3)$
IV $(400 < CD4 < 800/mm^3)$

The age, sex and means of contamination of these patients are indicated in Table I.

Table I. Characteristics of the HIV-1 infected patients

	I (N = 31)	II (N = 25)	III (N = 21)	IV (N = 12)
Age (years)	34 ± 6	40 ± 14	36 ± 10	33 ± 8
Sex M/W	22/9	18/7	13/7	9/3
UIVD	15 (48%)	9 (36%)	5 (25%)	4 (33%)
Homosexuality	13 (42%)	6 (24%)	10 (50%)	4 (33%)
Transfusion	3 (10%)	6 (24%)	0	1 (8%)
CDC stage D/B or C	30/1	16/9	10/10	2/10
Zidovudine treatment	25	22	13	2

Fourteen patients had one or more active AIDS-defining diseases (DD) or malignancies at the time of sampling. These involved the Mycobacterium avium complex (MAC) infection in 6 cases, Pneumocystis carinii pneumonia (PcP) in 8 cases, toxoplasmosis in 3 cases, Cytomegalovirus infection in 3 cases, disseminated tuberculosis in one case, progressive multifocal leuko-encephalitis in one case, Kaposi's sarcoma in 2 cases, tuberculosis and non-Hogkin's high- grade lymphoma in 1 case each. Results for these cases were compared to those for the 75 patients not showing AIDS defining disease or malignancy.

Results

Selenium

The plasma selenium values are presented in Table II and Figure 1

Groups	N	Se (ng/ml)
Controls (HIV-)	25	83(17)
I	31	55(14)
II	25	66(17)
III	21	69(16)
IV	12	76(12)

Table II. Plasma selenium values
Mean value (Standard deviation)

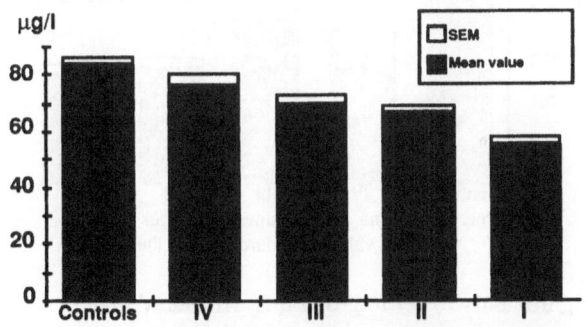

Figure 1. Plasma selenium values in the four groups
Mean value (Standard error of the mean)

Plasma selenium was significantly lower in group I than in group II (p = 0,007), III (p = 0,001), IV (p = 0,008) or in controls (p < 0,0001). Selenium was significantly lower in group II than in controls (p < 0,001) and in group III than in controls (p = 0,002).

To our knowledge, this is the first time that such a regular plasma selenium decrease with CD4 lymphocyte count has been observed.

No significant difference was found between the 14 patients with DD (56 ± 11 ng/ml) and the others (65 ± 17 ng/ml).

Plasma selenium was positively correlated to CD4, CD8 and total lymphocytes (p = 0,0001; 0,007; 0,0001) and also to BMI (body mass index; p = 0,0001), protidemia (p = 0,0001), cholesterol (p = 0,002), HDL-Cholesterol (p = 0,002) and apo A1 (p = 0,0001).

In a multivariate analysis, plasma selenium was correlated only to protidemia (p = 0,006).

Retinol

The plasma retinol values are presented in Table III and Figure 2.

Groups	N	Retinol (mg/l)
Controls (HIV-)	57	0,510(0,16)
I	31	0,258(0,15)
II	25	0,275(0,16)
III	21	0,387(0,25)
IV	12	0,314(0,18)

Table III. Plasma retinol values
Mean value (Standard deviation)

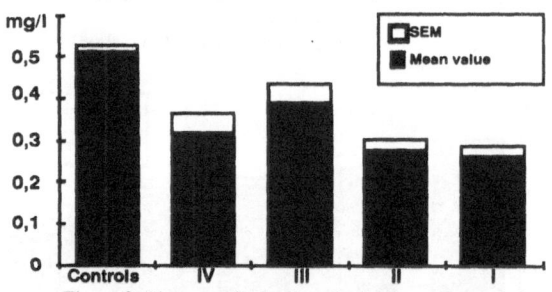

Figure 2 Plasma retinol values in the four groups
Mean value (Standard error of the mean)

Plasma retinol was significantly lower in group I, II, III and IV than in controls (p<0,0001 in the four cases). It was significantly lower in group I than in group III (p= 0,019).

These results point to a very low plasma retinol status for patients with less than 200 CD4/mm^3.

A significant difference in retinol value was found between the 14 patients with DD(0,20 ± 0,10 mg/l) and the others (0,32 ± 0,19 mg/l with p = 0,05).

Plasma retinol was positively correlated to plasma α-tocopherol (p < 0,0001), Se (p = 0,04), CD4 (p = 0,02), cholesterol (p = 0,03), HDL-Cholesterol (p = 0,04), apo A1 (p = 0,02), apoB (p < 0,02). From multivariate analysis, plasma retinol was found to be correlated only to plasma α-tocopherol (p = 0,0001) and cholesterol (p < 0,05).

α - tocopherol

The plasma a-tocopherol values are given in Table IV and Figure 3.

Groups	N	α-tocopherol (mg/l)
Controls (HIV-)	57	8,75 (3,00)
Patients (HIV+)	89	8,25 (4,66)

Table IV. Plasma α-tocopherol values
Mean value (Standard deviation)

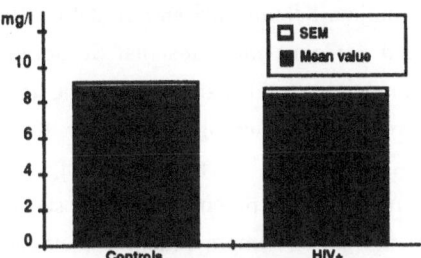

Figure 3.Plama α-tocopherol values
Mean value (Standard error of the mean)

No significant difference in plasma α-tocopherol values was found between the four HIV groups according to the CD4 lymphocyte count, and between the HIV- infected patients and the controls. There was no significant difference in α-tocopherol values between the patients with DD and the others (6,33 ± 2,33 mg/l vs 8,55 ± 4,90 mg/l).

Plasma α-tocopherol was negatively correlated to P24 antigenemia (p = 0,02), and positively correlated to plasma retinol (p < 0,0001), Se (p = 0,001), BMI (p = 0,04), cholesterol (p = 0,003), apoB (p = 0,005) and protidemia (p = 0,03).

In the multivariate analysis, plasma α-tocopherol was correlated to plasma retinol (p = 0,0001), selenium (p = 0,002), cholesterol (p = 0,003) and apoB (p = 0,02).

Plasma lipids

Results for plasma lipids have been described elsewhere (Constans et al., 1993). Briefly, triglycerides were found to be significantly higher; cholesterol and HDL-cholesterol were significantly lower in HIV-infected patients with CD4 lymphocytes below 50/mm³ than in the others. The same was observed for patients with CD4 cells below 200/mm³ with respect to the controls (HIV- negative individuals).

Discussion

Our results clearly show that selenium deficiency is present in HIV-1 infected patients and this is correlated with a decrease in the CD4 lymphocyte count. These low values are corroborated by the results of other studies. Plasma and erythrocyte selenium were found to be significantly lower in 21 AIDS or ARC patients than in controls; the erythrocyte Se-dependent glutathione peroxidase was also found to be reduced (Dworkin et al., 1988). In a multielemental study, Se was low in 59 HIV-infected men (Beck et al., 1990). The effective role of selenium in immunity is now well established (Kiremidjian-Schumacher and Stratzky, 1987, Kiremidjian-Schumacher et al., 1992) and accumulated evidence indicates that Se deficiency is associated with an impairment of immune responsiveness and that Se supplementation abrogates the depressed response and results in an increase in immunologic efficiency. Results for trials on the effect of selenium supplementation in AIDS patients appear to be lacking. A study with 10 AIDS subjects (with nonobstructive cardiomyopathy) supplemented with sodium selenite during 3 weeks revealed that eight of the patients showed improvement of their cardiac function during therapy (Zazzo et al., 1988). In order to establish whether intestinal absorption of dietary selenium is impaired in AIDS or ARC, a supplementary trial was performed during 70 days with 19 symptomatic HIV patients. The whole blood selenium level of the treated subjects was found to increase to that of the controls, and the selenium supplements were well tolerated (Olmsted et al., 1989).

Selenium is related to antioxidant activity via the glutathione peroxidase enzyme that scavenges free radicals, and several experiments have provided evidence for the role of oxidative stress in viral infection (Halliwell and Cross 1991).

HIV infection is characterized by a long latent period between the time of infection and the onset of AIDS. In experimental systems the activation of HIV-1 transcription in T-cells may result from viruses, cytokines or physical or chemical agents following stress mediated by various activated oxygen species. Identification of factors that induce viral expression could be useful for controlling the progress of the disease. Observations of the effects of selenium on retroviruses such as Murine Mammary Tumor Virus (MMTV), Leukemia Virus Rausher (LVR) and viruses (Coxsackie B and hepatisis B) in animals and humans might well be useful in enhancing the effects of selenium in the therapy of HIV infections.

It has been shown experimentally that selenium is an essential nutritional factor for protection against the development of neoplasms (Medina and Morrison, 1988). Se-containing compounds may be used as chemoprotective agents against cancer (Morrison et al., 1988). The participation of the immune system in cancer prevention and treatment appears to be related in part to the ability of a host to generate natural killer cells, CTL (cytotoxic-T-lymphocytes) and LAK (lymphokine-activated killer cells), which emphazises the potential of Se as a cancer-chemopreventive agent though enhancement of the clonal expansion of immunocompetent cells (Kiremidjian-Schumacher et al., 1992). A long-term selenium deficiency as observed in HIV-1 infected patients, may contribute to the development of associated cancers and this factor should be taken into account when new aspects of cellular immune responses based on selenium are considered.

Selenium deficiency impairs thyroid hormone metabolism by inhibiting the synthesis and activity of the iodothyronine deiodinases, which convert T4 (Thyroxine) to the more metabolically active T3 (3,3'-5 triiodothyronine). The type II iodothyronine 5'-deiodinase is a selenoenzyme. Selenium deficiency causes a decrease in total iodine, T4 and T3 in the thyroid gland in the rat, despite adequate quantities of iodine in the diet (Arthur et al., 1991). Selenium deficiency may influence many functions in thyroid hormone metabolism and contribute in this way towards amplification of biochemical disorders in HIV infected patients.

Vitamin E and glutathione peroxidase may modulate arachidonic acid metabolism and the activity of the cyclooxygenase enzymes by affecting peroxide concentration. The balance between arterial wall prostacyclin (PGI_2) production and platelet thromboxane TXA_2 directly influences platelet activity. It has been shown that selenium deficiency does not alter platelet TXB_2 synthesis but significantly decreases aortic PGI_2 synthesis (Meydani 1992). Reports on humans and animals indicate an inverse relationship between plasma selenium and platelet aggregation. Arterial thrombosis in HIV infected patients having no cardiovascular risk factor has been reported (Tabib et al., 1992 , Capron et al., 1992). In our prospective study, lipoprotein (a) in HIV-positive patients (Constans et al., 1993) was significantly increased. Patients with opportunistic infections had significantly higher TNF (Tumor Necrosis Factor) and lower HDL-C and ApoA1. Low plasma selenium and imbalance in circulating lipids may favor thrombosis observed in HIV-infected patients.

Selenium has biochemical effects that cannot be explained solely on the basis of the selenoenzyme glutathione peroxidase. Selenoprotein P is a newly characterized selenoprotein (Burk etm al., 1991). It contains multiple selenocysteines. Secreted by the liver into the plasma, its concentration is very sensitive to the selenium status. However, at levels of dietary selenium below the nutritional requirement, selenoprotein P concentrations are higher than normal with respect to the gluthatione peroxidase activities. This indicates that selenoprotein P has a greater demand on limiting quantities of selenium than does glutathione peroxidase. At present, the function of this protein is unknown.

Bio-availability of selenium from selenite is enhanced by the presence of various thiols (L and D cysteine, cysteamine, thioglycolate) in the gastrointestinal tract, whereas others (D-penicillamine, mercaptosuccinate, thiosulfate) have no stimulatory effect. For selenium, the last

Recommended Dietary Allowances (Recommended Dietary Allowances 1989) is 50-70 µg/day. In France, the plasma selenium level for normal healthy subjects is about 80 ng/ml for a marginally deficient intake between 45 and 50 µg/day (Simonoff and Simonoff, 1991); HIV-1 infected patients have a plasma selenium decrease corresponding to a daily intake in the range 30-35 µg/day for groups I and II. A moderate supplementation (at least 30-40 µg/day) seems justified for these patients in order to prevent selenium deficiency and its consequences.

It is clear from this work that there is a significant plasma retinol deficiency in the HIV-1 infected groups, especially in the subjects with less than 200 CD4/mm^3. The patients with AIDS defining diseases have a significantly lower plasma retinol value than the others. In the United States, two authors did not find a significant difference in plasma retinol of HIV-1 patients. In one case, the latter were asymptomatic and consumed dietary amounts greater than the recommended allowances for vitamin A (Baum et al., 1992). In the other case, the CD4 lymphocyte count was unknown and the authors report a large dose in self-supplementation (Bogden et al., 1990). Other studies suggest that retinoids may play a role in host defense against viral and bacterial pathogens and that low intakes of retinol by mice have been associated with increased disease susceptibility (Watson et al., 1988).

In a previous study we evaluated antioxidant status in aging (Simonoff et al., 1992). For healthy eighty-years old subjects all the levels were diminished (plasma selenium 25%, vitamin A 25%, α-tocopherol 40%). Our dietary survey revealed a significantly reduced intake with aging. In the case of HIV-infected patients, for groups I and II, only plasma selenium and vitamin A are lowered (respectively 28% and 50%) and associated with a zinc deficiency (Simonoff et al., 1993). Vitamin A is mostly stored in liver (90%) and transported by RBP (retinol binding protein). Low RBP can be due to zinc deficiency. Precursor of vitamin A, β-carotene has been shown in epidemiological studies in humans to aid in cancer prevention, and, in vitro to inhibit neoplastic transformation induced physically or chemically in cells. Carotenoids deactivate $^1O_2$ by quenching and inhibit free radical reactions. Vitamin A has a known toxicity specially when used over a long period of time at doses in excess (>20 000 UI/day). So, beta-carotene is an attractive agent because of its lack of toxicity and ready availability.

Our work does not provide evidence for a significant difference in plasma α-tocopherol between the HIV-1 infected patients and the controls, nor between the four HIV groups. However, 20% of the patients have a plasma α-tocopherol value below the normal range (5-15 mg/l). Only 7 patients had plasma α-tocopherol values above the normal range. These results are substantiated by other studies on the micronutrient and vitamin status (Bogden et al., 1990, Baum et al., 1992) in HIV infection. These normal values of antioxidant vitamin E are important because of the essential role it plays in the defense mechanisms of the body against infectious diseases (Odeleye and Watson, 1991).

The role of vitamin E in immune response and disease resistance has been studied (Tengerdy 1990) and depends on nutritional interactions with others antioxidant nutrients such as polyunsaturated fatty acids, beta-carotene, vitamin C, selenium, copper and zinc. We have recently obtained lower values for plasma copper and zinc in the patients of groups I and II (Simonoff et al., 1993). The nutritional status is implicated in the immune response and supplementation may be necessary in some cases: dose levels of 2 to 10 times higher than the recommended values of vitamin E for human and animal diets significantly increased humoral and cell-mediated immune responses and decreased mortality (Bendich 1988).

Conclusion

Antioxidant parameters (selenium, retinol and α-tocopherol) have been measured in the plasma of 89 HIV-1 infected patients at different stages of the disease in order to investigate the hypothesis of an oxidative imbalance.

No significant difference was observed for α-tocopherol between the four HIV groups according to the CD4 lymphocyte count and between the HIV infected patients and the controls.

Evidence was obtained for a significant decrease of plasma retinol in the full group of 89 patients in comparison to controls, suggesting that HIV infected patients are exposed to a vitamin A subdeficiency. The protective role of vitamin A and β-carotene (highly effective in quenching singlet oxygen) against cancer must be considered but with caution in order to avoid toxic hypervitaminosis A from self medication.

Plasma selenium decreased regularly with the CD4 lymphocyte count. Recent studies suggest that oxidants, and particularly H_2O_2, stimulate HIV transcription and replication. Such a plasma selenium decrease exposes blood cells to an accumulation of H_2O_2 and peroxidation products of lipids and proteins; consequently HIV-1 infected patients are exposed to an oxidative imbalance. In a multivariate analysis, plasma selenium is only correlated to protidemia. Thus, the oxidative stress that accompanies the disease seems to be related to the nutritional status, suggesting a requisite for moderate selenium supplementation.

Acknowledgements

The authors are grateful to Dr MacCordick for assistance in the preparation of the manuscript.

350

References

Arthur, J.R., Nicol, F., Grant, E. and Beckett, G.J. (1991) The effects of selenium deficiency onhepatic type-I iodothyronine deionidase and protein disulphide-isomerase assessed by activity measurements and affinity labelling. *Biochem. J.*, 274 (Pt 1), 297-300.

Baum, M.K., Shor-Posner, G., Bonvehi, P., Cassetti, I., Lu, Y., Mantero-Atienza, E., Beach, R.S., Sauberlich, H.E. (1992) Influence on HIV infection on vitamin status and equirements. *Ann. N.Y. Acad. Sci.*, 669, 165-74.

Beck, K.W., Schramel, P., Held, A., Jaeger, H., Kaboth, W. (1990) Serum trace-element levels in HIV-infected subjects. *Biol. Trace Elem. Res.*, 25(2), 89-96.

Bendich, A. (1988) Vitamin E and immune functions. *Basic Life Sci.*, 49,615-20.

Bogden, J.D., Baker, H., Frank, O., Perez, G. (1990) Micronutrient status and human immunodeficiency virus (HIV) infection. *Ann. N.Y. Acad. Sci.*, 587, 189-95.

Burk, R.F., Hill, K.E., Read, R., Bellew, T. (1991) Response of rat selenoprotein P to selenium administration and fate of its selenium. *Am. J. Physiol.*, 261 (1 Pt 1), E26-30.

Capron, L., Kim, Y.U., Laurin, C., Bruneval, P., Fiessinger, J.N. (1992) Atheroembolism in HIV-positive individuals. *Lancet*, 340, 1039-40.

Constans, J., Pellegrin, J.L., Peuchant, E., Dumon, M.F., Simonoff, M., Clerc, M., Leng, B., Conri, C. (1993) High plasma lipoprotein (a) in HIV-positive patients. *Lancet*, 341, 1099-1100.

Constans, J., Pellegrin, J.L., Peuchant, E., Dumon, M.F., Sergeant, C., Simonoff, M., Brossard, G., Barbeau, P., Clerc, M., Leng, B., Conri, C. (1993) Prospective study of plasma lipids in 95 HIV infected patients: relationship with immunological and nutritional status. (Submitted to AIDS)

Dworkin, B.M., Rosenthal, W.S., Wormser, G.P., Weiss, L., Nunez, M., Joline, C., Herp, A. (1988) Abnormalities of blood selenium and glutathione peroxidase activity in patients with acquired immunodeficiency syndrome and aids-related complex. *Biol. Trace Elem. Res.*, 15, 167-77.

Halliwell, B., Cross, C.E. (1991) Reactive oxygen species, antioxidants, and acquired immunodeficiency syndrome. *Arch. Intern. Med.*, 151, 29-31.

Kiremidjian-Schumacker, L., Statzky, G. (1987) Selenium and immune responses. *Environ. Res.*, 42, 277-303.

Kiremidjian-Schumacker, L., Roy, M., Wishe, H.I., Cohen, M.W., Stozky, G. (1992) Regulation of cellular immune response by selenium. *Biol. Trace Elem. Res.*, 33, 23-35.

Medina, D., Morrison, D.G. (1988) Current ideas on selenium as a chemopreventive agent. *Pathol. Immunopathol. Res.*, 7 (3), 187-99.

Meydani, S.N., Hayek, M., and Coleman, L. (1992) Influence of vitamins E and B6 on immune response. *Ann. N.Y. Acad.Sci.*, 669, 125-40.

Meydani, M. (1992) Modulation of the platelet thromboxane A_2 and aortic prostacyclin synthesis by dietary selenium and vitamin E. *Biol. Trace Elem. Res.*, 33, 76-86.

Morrison, D.G., Dishart, M.K., Medani, D. (1988) Serine and methionine enhancement of selenite inhibition of DNA synthesis in a mouse mammary epithelial cell line. *Carcinogenesis*, 9 (10), 1811-15.

Odeleye, O.E., Watson, R.R. (1991) The potential role of vitamin E in the treatment of immunologic abnormalities during acquired immune deficiency syndrome. *Prog. Food Nutr. Sci.*, 15, 1-19.

Olmsted, L., Schrauzer, G.N., Flores-Arce, M., Dowd, J. (1989) Selenium supplementation of symptomatic human immunodeficiency virus infected patients. *Biol. Trace Elem. Res.*, 20 (1-2), 59-65.

Packer, L. (1991) Protective role of vitamin E in biological systems. *Am. J. Clin. Nutr.*, 53, 1050S-55S.

Recommended Dietary Allowances (1989) 10° edition, National Academy of Sciences, Washington.

Simonoff, M.,Simonoff, G. (1991) Le sélénium et la vie (280 pages) Masson Editor, Paris.

Simonoff, M., Sergeant, C., Garnier, N., Moretto, P., Llabador, Y., Simonoff, G., Conri, C.(1992) Antioxidant status (selenium, vitamins A and E) and aging. *Free Radicals and Aging*, ed. by I. Emerit & B. Chance, Birkhäuser Verlag Basel / Switzerland, 368-97.

Simonoff, M., Sergeant, C., Razafindrabe, L., Hamon, C., Pellegrin, J.L., Brossard, G., Barbeau, P., Leng, B., Constans, J., Conri, C., Peuchant, E., Clerc, M. (1993) Plasmatrace-element levels in 89 HIV infected patients: correlation with nutritional and immunological status. TEMA-8 (Eight International Symposium on trace elements in men and animals), Dresden may 1993, in press.

Spallholtz, J.E. (1990) Selenium and glutathione peroxidase: essential nutrient and antioxidant component of the immune system. *Adv. Exp. Med. Biol.*, 262, 145-58.

Tabib, A., Greenland, T., Mercier, I., Loire, R., Mornex, J.F. (1992) Coronary lesions in young HIV-positive patients at necropsy. *Lancet*, 340, 730.

Tengerdy, R.P. (1990) The role of vitamin E in immune response and disease resistance . *Ann. N.Y. Acad. Sci.*, 587, 24-33.

Watson, R.R., Yahya, M.D., Darban, H.R., Prabhala, R.H. (1988) Enhanced survival by vitamin A supplementation during a retrovirus infection causing murine AIDS. *Life Sci.*, 43(6), xiixviii.

Zazzo, J.F., Chalas, J., Lafont, A., Camus, F., Chappuis, P. (1988) Is nonobstructive cardiomyopathy in AIDS a selenium deficiciency-related disease? *J. Parent. Enteral Nutr.*, 12(5), 537-38.

Author Index

Key Words

Cytotoxic Cells:
Recognition, Effector Function, Generation, and Methods

Edited by
M. Sitkovsky and **P. Henkart**
National Institutes of Health, Bethesda, Maryland, USA

1993. 544 pages. Hardcover. ISBN 3-7643-3608-0

This collection of papers from the most important researches of cytoxic cells will make an excellent introduction to the study of cytotoxic T lymphocytes and other cytotoxic cells, including CTL, NK, LAK, TIL ADCC, macrophages, mast cells, and platelets.

These topics are covered comprehensiviely, including generation, recognition, effector functions, and important methodologies. It will provide a state-of-the-art review of this important field. Special chapters cover the mechanisms of lethal hit delivery and immunopharmacological manipulations of cytotoxic cells which will be of interest to pharmacological researchers as well as cancer specialists. Adoptive immunotherapy is covered by experts in each field. The book is divided into brief, very readable chapters by experts in each field.

Leading the way as the first comprehensive work available in the field of Cytotoxic Cells and Cytotoxicity assays, this outstanding collection of papers by internationally renowned immunologists will serve as the reference source for immunology and cell biology laboratories, as well as allied fields of research.

Cytotoxic Cells provides an essential collection of methodologies, which are invaluable to every immunologist and cell biologist studying cellular regulation. The historical, molecular, cell biological, and clinical aspects of cell-mediated cytotoxicity are thoroughly covered by the leading researchers in their respective fields.

Over 50 chapters cover the following topics: • Introduction and Overview • Target Cell Recognition • Generation of Cytotoxic Cells • Molecular Mechanisms of Cellular Cytotoxicity • Granule Proteases • Alternate Mechanisms of Cytolysis • Biochemical and Immunopharmaceutical Manipulation of Cytotoxic Cells • Functions of Cytotoxic Cells In Vivo • Macrophage-Mediated Cytotoxicity • Methods

This is an invaluable introduction to the field for students, as well as a reference tool for practicing researchers, and a must for every laboratory bookshelf.

Please order through your bookseller or directly from:
Birkhäuser Verlag AG
P.O. Box 133
CH-4010 Basel / Switzerland
Fax ++41 / 61 721 79 50
Orders from the USA or Canada should be sent to:
Birkhäuser Boston
44 Hartz Way, Secaucus, NJ 07096-2491 / USA
Call Toll-Free 1-800-777-4643

For more information on recent and forthcoming books and journals you can order the Birkhäuser Life Sciences Bulletin, published twice a year and free of charge.

Birkhäuser

Birkhäuser Verlag AG • Basel • Boston • Berlin

The Search for Antiviral Drugs
Case Histories from Concept to Clinic

Edited by
J. Adams and **V. Merluzzi**
Boehringer Ingelheim Pharmaceuticals, Inc

1993. 250 pages. Hardcover. ISBN 3-7643-3606-4

There are many ways in which projects and programs get started as well as many nuances and differences in approaches and progression of a research project. This is a personalized account of the antiviral drug discovery process by contributors from the leading research laboratories in the world. The authors present an exciting behind-the -scenes picture that describes how projects are initiated, and reveals the problems, hurdles and successes encountered. Included in these case histories of the drug discovery process are those involving herpes virus, rhinovirus and human immunodeficiency virus. Also provided are the different ways in which problems are solved along with the decision processes, the go / no go decisions and critical issues, as well as the actual data.

Contents:
Introduction: The First Effective Antiviral, by Herbert E. Kaufman • The Discovery and Development of Zidovudine as the Cornerstone of Therapy to Control Immunodeficiency Virus Infection • Discovery and Development of 2-Pyridinone HIV-1 Reverse Transcriptase Inhibitors • Discovery of Pirodavir, A Broad-Spectrum Inhibitor of Rhinoviruses • Discovery and Developmentof the HIV Proteinase Inhibitor, Ro 31-8959 • sICAM-1 as a Receptor Antagonist for Rhinoviruses: A Model System of Adhesion Molecules as Cell Receptors for Viruses • Intracellular Receptors as Targets for Drug Discovery • Inactivation of Herpes Simplex Virus Ribonucleotide Reductase by Subunit Association Inhibitors: A Potential Antiviral Strategy • A New Approach to Antiviral Chemotherapy: Intervention in Viral Gene Expression by HIV-TAT Antagonists • Discovery of Nevirapine, A Nonnucleoside Inhibitor of HIV-1 Reverse Transcriptase

"...I was a resident in ophthalmology at the Massachusetts Eye and Ear Infirmary in Boston. At my disposal, I had a small laboratory in a decaying, condemned building, one technician, and no specific funds for the study of viruses."
~ Herbert E. Kaufman, M.D.
LSU Eye Center, New Orleans, LA

This excerpt from the introductory chapter sets the tone for this outstanding collection of personal accounts by researchers from the leading pharmaceutical laboratories throughout the world. From the earliest days of antiviral research, which produced the nucleoside analogs for the effective treatment of viral infections, to the latest antiviral strategies in preclinical testing, such as HIV-TAT antagonists, this informative book examines the pitfalls and successes of large-scale drug research.
Readers will enjoy the unusual behind-the-scenes look into the ways in which projects and programs get started, including the exciting development of AZT, as well as the strategies and problems faced by biomedical researchers looking for answers to Herpes, AIDS, and the common cold.

Please order through your bookseller or directly from:
Birkhäuser Verlag AG
P.O. Box 133
CH-4010 Basel / Switzerland
Fax ++41 / 61 721 79 50
Orders from the USA or Canada should be sent to:
Birkhäuser Boston
44 Hartz Way, Secaucus, NJ 07096-2491 / USA
Call Toll-Free 1-800-777-4643

For more information on recent and forthcoming books and journals you can order the Birkhäuser Life Sciences Bulletin, published twice a year and free of charge.

Birkhäuser
Birkhäuser Verlag AG • Basel • Boston • Berlin